ASHRAE GreenGuide

Design, Construction,
and Operation
of Sustainable Buildings

This publication was developed under the auspices of TC 2.8, Building Environmental Impacts and Sustainability. TC 2.8 is concerned with the impacts of buildings on the local, regional, and global environment; means for identifying and reducing these impacts; and enhancing ASHRAE member awareness of the impacts.

Tom Lawrence, PhD, PE, LEED-AP, is the chair of the editorial revision committee and coeditor of *ASHRAE GreenGuide*, Fourth Edition. He is a senior public service associate with the University of Georgia and has over 30 years of experience in engineering and related technical fields. He is a past chair of ASHRAE Technical Committee (TC) 2.8 and is a member of the committee that wrote ANSI/ASHRAE/USGBC/IES Standard 189.1, *Standard for the Design of High-Performance Green Buildings*. Dr. Lawrence is also a member of ASHRAE's Technical Activities Committee. As an ASHRAE Distinguished Lecturer, he gives seminars on green building design at venues around the world.

Abdel K. Darwich, PE, CEng, LEED-AP, HFDP, is a coeditor of *ASHRAE GreenGuide*, Fourth Edition. He is an associate principal with Guttmann and Blaevoet Consulting Engineers in their Sacramento office. Darwich has more than 15 years of experience in the design of mechanical systems for healthcare, commercial, industrial, K-12, mixed-use, and high-end residential uses with emphasis on low-energy and zero-energy design. He is a member of ASHRAE TC 2.8 and Standing Standard Project Committee 62.1, an ASHRAE certified Healthcare Facility Design Professional, and a recipient of a 2013 ASHRAE Technology Award.

Janice K. Means, PE, LEED-AP, is the third coeditor of *ASHRAE GreenGuide*, Fourth Edition. She is an associate professor in the College of Architecture and Design at Lawrence Technological University in Southfield, Michigan. Means has over 25 years of experience in the industry. She is a trained presenter for ANSI/ASHRAE/IESNA 90.1 2001 and 2004, a voting member of ASHRAE TC 2.8, Chair of ASHRAE TC 6.7, and was a contributing author to Chapter 36 of the 2008 *ASHRAE Handbook—HVAC Systems and Equipment*.

Sarah Boyle, assistant editor in ASHRAE Special Publications, served as staff editor for *ASHRAE GreenGuide*, Fourth Edition.

ASHRAE Staff		
	Special Publications	**Mark S. Owen, Editor/Group Manager of Handbook and Special Publications** **Cindy Sheffield Michaels, Managing Editor** **James Madison Walker, Associate Editor** **Roberta Hirschbuehler, Assistant Editor** **Sarah Boyle, Assistant Editor** **Michshell Phillips, Editorial Coordinator**
	Publishing Services	**David Soltis, Group Manager of Publishing Services and Electronic Communications** **Jayne Jackson, Publication Traffic Administrator** **Tracy Becker, Graphics Specialist**
	Publisher	**W. Stephen Comstock**

Updates/errata for this publication will be posted on the ASHRAE Web site at www.ashrae.org/publicationupdates.

ASHRAE GreenGuide

Design, Construction, and Operation of Sustainable Buildings

Fourth Edition

Atlanta

ISBN 978-1-936504-55-8 (Hbk)
ISBN 978-1-939200-40-2 (PDF)

© 2013 ASHRAE
1791 Tullie Circle, NE
Atlanta, GA 30329
www.ashrae.org

All rights reserved.
First edition published 2003. Second edition published 2006.
Third edition published 2010. Fourth edition published 2013.
Printed in the United States of America

Cover design by Tracy Becker

ASHRAE is a registered trademark in the U.S. Patent and Trademark Office, owned by the American Society of Heating, Refrigerating and Air-Conditioning Engineers, Inc.

ASHRAE has compiled this publication with care, but ASHRAE has not investigated, and ASHRAE expressly disclaims any duty to investigate, any product, service, process, procedure, design, or the like that may be described herein. The appearance of any technical data or editorial material in this publication does not constitute endorsement, warranty, or guaranty by ASHRAE of any product, service, process, procedure, design, or the like. ASHRAE does not warrant that the information in the publication is free of errors, and ASHRAE does not necessarily agree with any statement or opinion in this publication. The entire risk of the use of any information in this publication is assumed by the user.

No part of this book may be reproduced without permission in writing from ASHRAE, except by a reviewer who may quote brief passages or reproduce illustrations in a review with appropriate credit, nor may any part of this book be reproduced, stored in a retrieval system, or transmitted in any way or by any means—electronic, photocopying, recording, or other—without permission in writing from ASHRAE. Requests for permission should be submitted at www.ashrae.org/permissions.

Library of Congress Cataloging-in-Publication Data

ASHRAE greenguide : design, construction, and operation of sustainable buildings. -- Fourth edition.
　　pages cm
　Summary: "Provides information on green-building design. Concerned with sustainable, high-performance projects"-- Provided by publisher.
　Includes bibliographical references and index.
　ISBN 978-1-936504-55-8 (hardcover)
　1. Sustainable buildings--Design and construction. 2. Sustainable buildings--United States--Design and construction. 3. Sustainable architecture. 4. Buildings--Environmental engineering. 5. Sustainable construction. I. American Society of Heating, Refrigerating and Air-Conditioning Engineers. II. Title: ASHRAE green guide.
　TH880.A83 2013
　720'.47--dc23

2013028021

Tomorrow's Child

Without a name, an unseen face,
And knowing not the time or place,
Tomorrow's Child, though yet unborn,
I saw you first last Tuesday morn.
A wise friend introduced us two,
And through his shining point of view
I saw a day, which you would see,
A day for you, and not for me.
Knowing you has changed my thinking,
Never having had an inkling
That perhaps the things I do
Might someday threaten you.
Tomorrow's Child, my daughter-son,
I'm afraid I've just begun
To think of you and of your good,
Though always having known I should.
Begin I will to weigh the cost
Of what I squander, what is lost,
If ever I forget that you
Will someday come to live here too.

by Glenn Thomas, ©1996

Reprinted from
*Mid-Course Correction:
Toward a
Sustainable Enterprise:
The Interface Model*
by Ray Anderson.
Chelsea Green
Publishing Company, 1999.

Contents

Foreword — xv
Preface to the Fourth Edition — xvii
    Who Should Use *ASHRAE GreenGuide* — xviii
    How to Use *ASHRAE GreenGuide* — xviii
    Background on *ASHRAE GreenGuide* — xix
Acknowledgments — xxi

SECTION 1: BASICS

Chapter One: Introduction and Background — 3
    Introduction — 3
    Relationship to Sustainability — 5
    Commitment to Green/Sustainable
        High-Performance Projects — 5
    What Drives Green Projects — 6
    The Impact of Carbon Considerations — 7
    The Engineering/Energy Conservation Ethic — 8
    Sustainability in Architecture — 14
    References and Resources — 20

Chapter Two: Sustainable Sites — 23
    Location of the Building Project — 23
    Urban Heat Island Effect — 24
    Stormwater Management — 27
    References and Resources29

SECTION 2: THE DESIGN PROCESS

Chapter Three: Project Strategies — 39
- Ingredients of a Successful Green Project Endeavor — 39
- Incentives for Green Design — 40
- Successful Approaches to Design — 45
- References and Resources — 51

Chapter Four: The Design Process—Early Stages — 53
- Overview — 53
- The Owner's Role — 56
- The Design Team — 57
- The Engineer's Role — 61
- Project Delivery Methods and Contractor Selection — 62
- Concept Development — 67
- Expressing and Testing Concepts — 70
- Building Information Modeling (BIM) — 72
- References and Resources — 78

Chapter Five: Architectural Design and Planning Impacts — 79
- Overview — 79
- Design Process for Sustainable Architecture — 79
- Prioritization and Studies of Existing Buildings' Energy Resource Utilization — 86
- Intentions in Architecture and Building System Development — 87
- Building-Type GreenTips — 89

Chapter Six: Commissioning — 123
- Commissioning Phases — 125
- Selection of Systems to Commission — 135
- Commissioning Models — 136
- The Growth of Commissioning — 137
- References and Resources — 138

Chapter Seven: Green Rating Systems, Standards, and Other Guidance — 139
- Green-Building Rating Systems — 139
- The LEED Rating System — 141
- Guidelines and Other Resources — 144
- Building Energy Quotient (bEQ) — 145
- Implementation in the Form of Standards and Building Codes — 147

ASHRAE/USGBC/IES Standard 189.1	147
International Green Construction Code (IgCC)	153
References and Resources	153

Chapter Eight: Conceptual Engineering Design— Load Determination   157

The Role of Energy Modeling During Conceptual Design	158
Determining the Load Drivers with Parametric Simulations	160
Energy Impacts of Architectural Features	161
Thermal/Mass Transfer of Envelope	162
Engineering Internal Load-Determining Factors	163
System/Equipment Efficiencies	164
Life-Cycle Cost Analysis (LCCA)	165
References and Resources	179

Chapter Nine: Indoor Environmental Quality   181

Introduction	181
Green-Building Design and Indoor Environmental Quality	182
Integrated Design Approaches and Solutions	183
Thermal Comfort and Control	223
Light and Illumination	225
Acoustics	230
References and Resources	232

Chapter Ten: Energy Distribution Systems   235

Energy Exchange	235
Energy Delivery Methods	236
Steam	240
Hydronics	243
Air	247
Electric	249
References and Resources	250

Chapter Eleven: Energy Conversion Systems   273

Heat Generators (Heating Plants)	273
Cooling Generators (Chilled-Water Plants)	273
Cooling System Heat Sinks	276
Cooling Tower Systems	276
District Energy Systems	281
Water Consumption for Cooling System Operation	283

Distributed Electricity Generation	284
References and Resources	290

Chapter Twelve: Energy/Water Sources — 337

Renewable/Nonrenewable Energy Sources	337
Solar	339
Wind	349
Hydro	350
Biomass	350
References and Resources	351

Chapter Thirteen: Lighting Systems — 369

Overview	369
Electric Lighting	369
Daylight Harvesting	384
Light Conveyors (Tubular Daylighting Devices)	388
Lighting Controls	388
Cost Considerations	392
References and Resources	393

Chapter Fourteen: Water Efficiency — 399

The Energy-Water Balance	400
Water Supply	400
Cooling Tower Systems	402
Domestic Water Heating	407
Sanitary Waste	408
Rainwater Harvesting	408
Fire Suppression Systems	408
Water Recovery and Reuse	408
References and Resources	409

Chapter Fifteen: Building Automation Systems — 439

Control System Role in Delivering Energy Efficiency	440
The Interaction of a Smart Building with the Coming Smart Grid	441
Control System Role in Delivering Water Efficiency	443
Control System Role in Delivering IEQ	444
Control System Commissioning Process	446
Designing for Sustained Efficiency	449
References and Resources	450

Chapter Sixteen: Completing Design and Documentation for Construction — 453

Drawings/Documentation Stage — 453
Specifying Materials/Equipment — 453
Cost Estimating and Budget Reconciliation — 454
Bidding — 454
Managing Risk — 455
References and Resources — 458

SECTION 3: POSTDESIGN—CONSTRUCTION AND BEYOND

Chapter Seventeen: Construction — 463

Site Planning and Development — 463
The Engineer's Role in Construction Quality — 463
Construction Practices and Methods — 466
Commissioning During Construction — 468
Moving into Occupancy and Operation — 468
References and Resources — 468

Chapter Eighteen: Operation, Maintenance, and Performance Evaluation — 471

Plans for Operation — 471
Commissioning — 471
Energy Efficiency in Existing Buildings — 473
Retrofit Strategies for Existing Buildings — 474
Measurement and Verification (M&V) — 475
International Performance Measurement and Verification Protocol — 475
Federal Energy Management Program (FEMP) Guidelines Version 2.2 — 476
FEMP M&V Option A Detailed Guidelines — 476
ASHRAE Guidelines and Standards — 476
Building Labeling — 477
Occupant Surveys — 478
References and Resources — 480

References and Resources — 483

Terms, Definitions, and Acronyms — 505

Index — 509

GREENTIPS

ASHRAE GreenTip #2-1: Rain Gardens	30
ASHRAE GreenTip #2-2: Green-Roof Systems	34
ASHRAE GreenTip #8-1: Night Precooling	168
ASHRAE GreenTip #8-2: Night-Sky Cooling	173
ASHRAE GreenTip #8-3: Plug Loads	177
ASHRAE GreenTip #10-1: Variable-Flow/Variable-Speed Pumping Systems	251
ASHRAE GreenTip #10-2: Variable-Refrigerant Flow (VRF) Systems	255
ASHRAE GreenTip #10-3: Displacement Ventilation	260
ASHRAE GreenTip #10-4: Dedicated Outdoor Air Systems	263
ASHRAE GreenTip #10-5: Ventilation Demand Control Using CO_2	266
ASHRAE GreenTip #10-6: Hybrid Ventilation	269
ASHRAE GreenTip #11-1: Pulse-Powered, Chemical-Free Water Treatment	291
ASHRAE GreenTip #11-2: CHP Systems	295
ASHRAE GreenTip #11-3: Low-NO_x Burners	300
ASHRAE GreenTip #11-4: Combustion Air Preheating	302
ASHRAE GreenTip #11-5: Combination Space/Water Heaters	304
ASHRAE GreenTip #11-6: Ground-Source Heat Pumps (GSHPs)	307
ASHRAE GreenTip #11-7: Water-Loop Heat Pump Systems	311
ASHRAE GreenTip #11-8: TES for Cooling	314
ASHRAE GreenTip #11-9: Double-Effect Absorption Chillers	320
ASHRAE GreenTip #11-10: Gas Engine-Driven Chillers	323
ASHRAE GreenTip #11-11: Desiccant Cooling and Dehumidification	326
ASHRAE GreenTip #11-12: Indirect Evaporative Cooling	330
ASHRAE GreenTip #11-13: Condensing Boilers	333
ASHRAE GreenTip #12-1: Passive Solar Thermal Energy Systems	355
ASHRAE GreenTip #12-2: Active Solar Thermal Energy Systems	358
ASHRAE GreenTip #12-3: Solar Energy System—PV	361
ASHRAE GreenTip #12-4: Solar Protection	365
ASHRAE GreenTip #13-1: Light Conveyors (Tubular Daylighting Devices)	395
ASHRAE GreenTip #14-1: Water-Conserving Plumbing Fixtures	411
ASHRAE GreenTip #14-2: Graywater Systems	419
ASHRAE GreenTip #14-3: Point-of-Use	

Domestic Hot-Water Heaters	422
ASHRAE GreenTip #14-4: Direct-Contact Water Heaters	425
ASHRAE GreenTip #14-5: Rainwater Harvesting	428
ASHRAE GreenTip #14-6: Air-Handling Unit (AHU) Condensate Capture and Reuse	432

Building-Type GreenTips

ASHRAE Building-Type GreenTip #5-1: Performing Arts Spaces	90
ASHRAE Building-Type GreenTip #5-2: Health Care Facilities	93
ASHRAE Building-Type GreenTip #5-3: Laboratory Facilities	101
ASHRAE Building-Type GreenTip #5-4: Student Residence Halls	104
ASHRAE Building-Type GreenTip #5-5: Athletic and Recreation Facilities	107
ASHRAE Building-Type GreenTip #5-6: Commercial Office Buildings	109
ASHRAE Building-Type GreenTip #5-7: K–12 School Buildings	111
ASHRAE Building-Type GreenTip #5-8: Existing Buildings	113
ASHRAE Building-Type GreenTip #5-9: Data Centers	117

DIGGING DEEPER SIDEBARS

International Perspective: Regulations and Commentary	8
Some Definitions and Views of Sustainability from Other Sources	13
One Design Firm's Ten Steps to a Net Zero Energy Building	48
Justifications for Green Design	49
National Renewable Energy Laboratory's Nine-Step Process for Low-Energy Building Design	76
One Firm's Green-Building Design Process Checklist	77
One Firm's Commissioning Checklist	137
CALGreen Code: America's First Statewide Green-Building Code	148
Key Considerations in the HVAC Design Process	166
Example Calculation to Compute a Baseline Predicted Water Consumption for a Building	416
How Much Water Will Collect at Design Conditions?	436
One Design Firm's Materials Specification Checklist	457
Construction Factors to Consider in a Green Design	469
One Design Firm's Operations, Maintenance, and Performance Evaluation Checklist	479

FOREWORD

by William Coad

Mechanical engineering has been defined as "the applied science of energy conversion." ASHRAE is the preeminent technical society that represents engineers practicing in the fields of heating, refrigeration, and air conditioning—the technology that utilizes approximately one-third of the global nonrenewable energy consumed annually.

ASHRAE membership has actively pursued more effective means of utilizing these precious nonrenewable resources for many decades from the standpoints of source availability, efficiency of utilization, and technology of substituting with renewable sources. One significant publication in *ASHRAE Transactions* is a paper authored in 1951 by G.W. Gleason, Dean of Engineering at Oregon State University, titled "Energy—Choose it Wisely Today for Safety Tomorrow." The flip side of the energy coin is the environment and, again, ASHRAE has historically dealt with the impact that the practice of the HVAC&R sciences have had upon both the indoor and the global environment.

However, the engineering community, to a great extent, serves the needs and desires of accepted economic norms and the consuming public, a large majority of whom have not embraced an energy/environmental ethic. As a result, much of the technology in energy effectiveness and environmental sensitivity that ASHRAE members have developed over this past century has had limited impact upon society.

In 1975, when ASHRAE published ASHRAE Standard 90-75, *Energy Conservation in New Building Design* (ASHRAE 1975), that standard served as our initial outreach effort to develop an awareness of the energy ethic and to extend our capabilities throughout society as a whole. Since that time, updated revisions of Standard 90 have moved the science ahead. In 1993, the chapter on "Energy Resources" was added to the 1993 *ASHRAE Handbook—Fundamentals*. In 2002, ASHRAE entered into a partnering agreement with the US Green Building Council, and it is intended that this and future editions of this design guide will continue to assist ASHRAE in its efforts at promoting sustainable design, as well as the many other organizations that have advocated for high-performance building design.

The consuming public and other representative groups of building professionals continue to become more and more aware of the societal need to provide buildings that are more energy resource effective and environmentally compatible. This publication, authored and edited by ASHRAE volunteers, is intended to complement those efforts.

The reader is cautioned that a successful green design, like any other successful design, must achieve a high level of environmental comfort and air quality. In addition, the building must be designed so that it can be operated and maintained in such a way as to keep the high level of performance expected.

PREFACE TO THE FOURTH EDITION

by Tom Lawrence

This new and fourth edition of *ASHRAE GreenGuide* represents another update and revision to what has become one of ASHRAE's primary products and contributions toward sustainable design of the built environment. In early 2012, ASHRAE introduced a major rebranding effort for the society. The Society President at this time was Ron Jarnagin, and in his letter to the members he gave the following summary of this effort: "For more than 100 years, ASHRAE has provided guidance for HVAC&R. As time and technology have changed, so has ASHRAE, moving from focusing solely on HVAC&R to providing guidance for total building design, reconstruction, construction and operation" (www.ashrae.org/news/ashrae-announces-rebranding).

This new fourth edition is being released at the ten year anniversary of the release of the very first edition in January 2004. It has been ASHRAE's general plan to update and maintain this document on a regular basis, because the technology and entire concept of how green building practices are done and considered within the industry is rapidly changing.

Since the release of the previous third edition of the *GreenGuide*, a number of developments have occurred in the green building arena. ASHRAE has continued to refine and modify Standard 189.1, with a new release of that standard occurring in late 2011. ASHRAE also partnered with the International Code Council for the release of the *International Green Construction Code*™(IgCC) in March of 2012. Adoption and use of these codes and standards is beginning to pick up pace, and other organizations and jurisdictions are using these as the basis for their own codes and design standards. Thus, the industry is witnessing the continued evolution of green building programs from strictly voluntary to being both more in the industry mainstream as well as being mandatory in jurisdictions that have adopted these for their building codes.

The fourth edition features new information and GreenTips. This follows the example set by ASHRAE's rebranding in 2012. For example, there is a new chapter that outlines the key components for designing for sustainable sites. A nonessential chapter that reviewed engineering fundamentals was removed to make room for additional material. New GreenTips were created for topics such as condensing boilers, rain gardens, green roofs, and a Building-Type GreenTip for data centers. Many of the chapters were extensively rewritten, in particular the chapter on indoor environmental quality and the chapter on architectural design and planning impacts.

One of the goals for the editorial committee for this revision was to bring in a large number of additional outside reviewers. In this process, it is likely that we have missed giving credit to the individuals that contributed their time and effort in reviewing and editing as well as the creation of new material for this edition; for this I do apologize. The revision process spread out through a wide number of channels and distributions as ASHRAE members and the dedicated volunteers all pitched in to help create the best product possible. Finally, the current version could not have been possible without all the hard work and dedication put into it by others who created the previous three editions. This book truly represents the collaborative nature of the work done by dedicated volunteers within ASHRAE. All work performed—by the authors, editors, developing subcommittees, other reviewers, and TC participants—was voluntary.

WHO SHOULD USE *ASHRAE GREENGUIDE*

ASHRAE GreenGuide is primarily for HVAC&R designers, but it is also a useful reference for architects, owners, building managers, operators, contractors, and others in the building industry who want to understand some of the technical issues regarding high-performance design from an integrated building systems perspective. Considerable emphasis is placed on teamwork and close coordination between parties.

The *GreenGuide* was originally intended for use by younger engineers or architects or more experienced professionals about to enter into their first green design projects. However, a survey taken of those who purchased one of the earlier editions of this publication revealed that it was being used by more experienced individuals primarily. The survey also indicted a higher percentage of the readership from countries outside of North America, perhaps reflecting the growing internationalization of ASHRAE.

HOW TO USE *ASHRAE GREENGUIDE*

This document is intended to be used more as a reference than as something one would read in sequence from beginning to end. The table of contents is the best place for any reader to get an overall view of what is covered in this publication. Throughout the *GreenGuide,* numerous techniques, processes, measures, or special systems are described succinctly in a modified outline or bullet form. These are called ASHRAE GreenTips. Each GreenTip concludes with a listing of other sources that may be referenced for greater detail. (A list of GreenTips and Digging Deeper sidebars can be found in the Table of Contents.)

All readers should take the time to read Chapter 1, "Introduction and Background," which provides some essential definitions and meanings of key terms. Chapter 2, "Sustainable Sites" provides a brief overview of the relationship of the building project to the site and surroundings. Some may question the need for this, but since a successful green building project is the collaboration of many disciplines, it is felt that this topic should not be overlooked.

Chapter 3 provides an overview of project strategies. Chapter 4 covers the early stages of the design process, and Chapter 5 highlights architectural design and planning impacts. These chapters are essential reading for all who are interested in how the green design process works. Building-Type GreenTips are included at the end of Chapter 5. Chapter 6 provides an overview of the commissioning process, a critical component that needs to be addressed from the beginning on all truly successful high-performance building projects. Chapter 7 describes green rating systems and the relevant standards and paths to compliance as they relate to the work of the mechanical engineers.

The next nine chapters deal with virtually all of the practical suggestions for possible strategies and concepts to be appropriately incorporated into a green building design. Chapters 17 and 18 cover what happens after the design documents for the project have been completed—that is, during construction, final commissioning, and the postoccupancy phases of a building project.

At the end of the guide is a comprehensive "References and Resources" section, which compiles all the sources mentioned throughout the guide, and an index for rapid location of a particular subject of interest.

BACKGROUND ON *ASHRAE GREENGUIDE*

The idea for the publication was initiated by 1999–2000 ASHRAE President Jim Wolf and carried forward by then President Elect (and subsequently President) William J. Coad. Members of that first subcommittee were David L. Grumman, Fellow ASHRAE, chair and editor; Jordan L. Heiman, Fellow ASHRAE; and Sheila Hayter, chair of TC 1.10 (the precursor to TC 2.8 of today).

The *GreenGuide* subcommittee responsible for the second and third editions consisted of John Swift, Tom Lawrence, and the people noted in the Acknowledgments section. Work on the fourth edition was overseen by a subcommittee of TC 2.8 chaired by Tom Lawrence and also including Janice Means and Abdel Darwiche.

ACKNOWLEDGMENTS

The following individuals served as coeditors on this edition of *ASHRAE GreenGuide,* provided written materials and editorial content, and formed the Senior Editorial Group of the ASHRAE TC 2.8 *GreenGuide* Subcommittee for the second and third editions:

Thomas Lawrence
University of Georgia
Athens, GA

Abdel K Darwich
Guttmann and Blaevoet Consulting Engineers, Sacramento, CA

Janice K. Means
Lawrence Technological University
Southfield, MI

The committee is deeply thankful for all the individuals who helped with or contributed in the first three editions of the *GreenGuide*. The fourth edition would not be where it is without their help.

The following individuals contributed new written materials on various topics for the fourth edition of *ASHRAE GreenGuide*. All or portions of these contributions were incorporated, with minor editing.

Constantinos A. Balaras
Institute for Environmental Research & Sustainable Development,
National Observatory of Athens (NOA)
Athens, Greece

Jason Bedgood
Student, University of Georgia
Athens, GA

Daniel Faoro
Lawrence Technological University
Southfield, MI

Steven Guttmann
Guttmann and Blaevoet Consulting Engineers
San Francisco, CA

R. Ryan Hammond
Formerly with Guttmann and Blaevoet Consulting Engineers
Sacramento, CA

Michael Meteyer
Erdman Company
Madison, WI

Ashish Rakheja
AECOM
New Delhi, India

Nadia Sabeh
Guttmann and Blaevoet Consulting Engineers
Sacramento, CA

Sara Schonour
Cannon Design
Boston, MA

Jennifer Wehling
Lionakis
Sacramento, CA

A number of people from the ASHRAE Environmental Health Committee participated in the complete rewriting of Chapter 9 on Indoor Environmental Quality. These are:

Robert Baker
BBJ Environmental Solutions
Riverview, FL

Terry Brennan
Camroden Associates
Westmoreland, NY

David Grimsrud
University of Minnesota
Minneapolis, MN

Roger Hedrick
Architectural Energy Corporation
Boulder, Colorado

Martha Hewett
Center for Energy and Environment
Minneapolis, MN

Josephine Lau
University of Nebraska-Lincoln
Lincoln, Nebraska

Hal Levin
Building Ecology Research Group
Santa Cruz, CA

Dennis Lovejoy
Loughborough University
Loughborough, UK

Chandra Sekhar
National University of Singapore
Singapore, Singapore

Zuraimi Sultan
National Research Council of Canada
Ottawa, Canada

Lily Wang
University of Nebraska-Lincoln
Lincoln, Nebraska

Pawel Wargocki
Technical University of Denmark
Lyngby, Denmark

Jianshun Zhang
Syracuse University
Syracuse, NY

In addition, the following individuals have contributed material to prior editions of *ASHRAE GreenGuide* or served on the Senior Editorial Group for the second and third editions:

Ainul Abedin
Jerry Ackerman
John Andrepont

Kimberly Barker
Steven Baumgartner
David Bearg

Bill Becker
James Benya
James Bones
Dean Borges
Gail S. Brager
Stu Brodsky
Amy Butterfield
Stephen Carpenter
Daryn Cline
Dimitri Contoyannis
Kevin Cross
Len Damiano
Michael Deru
Kevin Dickens
Rand Ekman
H. Jay Enck
Michael Forth
Guy S. Frankenfield
Glenn Friedman
Michael Gallivan
Krishnan Gowri
David L. Grumman
Michael Haggans
Jordan L. Heiman
Mark Hertel
Bion Howard
Mark Hydemann
Brad Jones
James Keller
John Kokko
Wladyslaw Jan Kowalski
John Lane
Nils L. Larsson
Eddie Leonardi
Malcolm Lewis
Mark Loeffler
Dunstan Macauley
Garrick Maine
Blair McCarry
Paul McGregor
Mark Mendell
Neil Moiseev
Vikas Patnaik
Ron Perkins
Jason Perry
B. Andrew Price
Douglas T. Reindl
Wayne Robertson
Brian A. Rock
Steven Rosen
Marc Rosenbaum
Sara Schonour
Mick Schwedler
Eugene Stamper
Karl Stum
E. Mitchell Swann
John M. Swift, Jr.
Paul Torcellini
Stephen Turner
Charles Wilkin

Section 1: Basics

CHAPTER ONE

Introduction and Background

INTRODUCTION

There continues to be a growing awareness about the impact of the built environment on the natural environment. The use of green engineering concepts has evolved quite rapidly in recent years and is now a legitimate and spreading movement in the HVAC&R and related engineering professions. Much of this recent work has been driven by the emergence of green architecture, also commonly referred to as sustainable or environmentally conscious architecture. This, in turn, is being encouraged by increased client demand for more sustainable buildings.

Interest in green buildings has been particularly evident in the concern about energy and water resource consumption, but also includes broader concerns such as material use, "smart" development and planning, etc. Many countries in the world now have green-building rating systems (voluntary) and/or codes (mandatory in some form or other). Organizations devoted specifically to this issue are now in existence in most countries. Not only have the messages contained in this outpouring of information attempted simply to explain what this issue is, they have promoted the concept of green design, exhorted to action, strived to motivate, warned of consequences from ignoring it, and instructed how to do it.

While this vast amount of promotion has been helpful, much has not been directly useful to the practicing designer for buildings (i.e., to the ASHRAE member involved on a day-to-day basis in the mechanical/electrical building system design process). ASHRAE identified a need for guidance on the green-building concept specifically directed toward such practitioners. One key motivation for the development of this guide is that it contains information that has direct practical use. This guide is ASHRAE's way to provide information and guidance to the industry and practicing professionals.

Green is one of those words that can have many meanings, depending on the circumstances. One of these is the greenery of nature (e.g., grass, trees, and leaves). This symbolic reference to nature is the meaning this term relates to in this publication. The difference between a green and sustainable design is the degree to which the design helps to minimize the building impact on the environment while simultaneously providing a healthy, comfortable indoor environment. This guide is not intended to cover the full breadth of sustainability, but it is a good overview of many of the main topics. For key characteristics and more detailed discussion of sustainability, refer to the "Sustainability" chapter in the 2013 *ASHRAE Handbook—Fundamentals* (ASHRAE 2013).

The definition of *green buildings* inevitably extends beyond the normal daily concerns of HVAC&R designers alone, since the very concept places an emphasis on integrated design of mechanical, electrical, architectural, and other systems.

Specifically, the viewpoint held by many is that a green/sustainable building design is one that achieves high performance, over the full life cycle, in the following areas:

- Minimizing natural resource consumption through more efficient utilization of nonrenewable energy and other natural resources, land, water, and construction materials, including utilization of renewable energy resources to strive to achieve net zero energy consumption.
- Minimizing emissions that negatively impact our global atmosphere and ultimately the indoor environment, especially those related to indoor air quality (IAQ), greenhouse gases, global warming, particulates, or acid rain.
- Minimizing discharge of solid waste and liquid effluents, including demolition and occupant waste, sewer, and stormwater, and the associated infrastructure required to accommodate removal.
- Minimizing negative impacts on the building site.
- Optimizing the quality of the indoor environment, including air quality, thermal regime, illumination, acoustics/noise, and visual aspects to provide comfortable human physiological and psychological perceptions.
- Optimizing the integration of the new building project within the overall built and urban environment. A truly green/sustainable building should not be thought of or considered in a vacuum, but rather in how it integrates within the overall societal context.

Ultimately, even if a project does not have overtly stated green/sustainable goals, the overall approaches, processes, and concepts presented in this guide provide a design philosophy useful for any project. Using the principles of this guide, an owner or a member of his or her team can document the objectives and criteria to include in a project, forming the foundation for a collaborative integrated project delivery approach. This can lower design, construction, and operational costs, resulting in a lower total cost for the life of the project.

RELATIONSHIP TO SUSTAINABILITY

The related term *sustainable design* is very commonly used, almost to the point of losing any consistent meaning. While there have been some rather varied and complex definitions put forth (see the sidebar titled "Some Definitions and Views of Sustainability from Other Sources"), we prefer a simple one (very similar to the third one in the sidebar). Sustainability is providing for the needs of the present without detracting from the ability to fulfill the needs of the future.

The preceding discussion suggests that the concepts of green design and sustainable design have no absolutes—that is, they cannot be defined in black-and-white terms. These terms are more useful when thought of as a mindset: a goal to be sought and a process to follow. This guide is a means of (1) encouraging designers of the built environment to employ strategies that can be used in developing a green/sustainable design, and (2) setting forth some practical techniques to help practitioners achieve the goal of green design, thus making a significant contribution to the sustainability of the planet.

Another method for assessing sustainability is through the concept of The Triple Bottom Line (Savitz and Weber 2006). This concept advances the idea that monetary cost is not the only way to value project design options. The Triple Bottom Line concept advocates for the criteria to include economic, social, and environmental impacts of building design and operations decisions.

COMMITMENT TO GREEN/SUSTAINABLE HIGH-PERFORMANCE PROJECTS

Green projects require more than a project team with good intentions; they require commitment from the owner and the rest of the project team, early documentation of sustainable/green goals documented by the Owner's Project Requirement document, and the designer's documented basis of design. The most successful projects incorporating green design are ones with dedicated, proactive owners who are willing to examine (or give the design team the freedom to examine) the entire spectrum of ownership—from design to construction to long-term operation of their facilities. These owners understand that green buildings require more planning, better execution, and better operational procedures, requiring a firm commitment to changing how building projects are designed, constructed, operated, and maintained to achieve a lower total cost of ownership and lower long-term environmental impacts.

Implementing green/sustainable practices could indeed raise the initial design soft costs associated with a project, particularly compared to the code minimum building design. First cost is an important issue and often is a stumbling block in moving building design from the code minimum ("good design") to one that is more truly sustainable. Implementing the commissioning process early in the predesign phase of a project adds an initial budget line item but can often actually reduce overall total design/construction costs and the ultimate cost of ownership.

In addition, significant savings and improved productivity of the building occupants can be realized for the life of the building, lowering the total cost of ownership and/or providing better value for tenants. To achieve lifelong benefits also requires operating procedures for monitoring performance, making adjustments (continuing commissioning) when needed, and appropriate maintenance.

WHAT DRIVES GREEN PROJECTS

Green-building advocates can cite plenty of reasons why buildings should be designed utilizing integrated green concepts. The fact that these reasons exist does not make it happen, nor does the existence of designers—or design firms—with green design experience. The main driver of green-building design is the motivation of the owner—the one who initiates the creation of a project, the one who pays for it (or who carries the burden of its financing), and the one who has (or has identified) the need to be met by the project in question. If the owner does not believe that green design is needed, thinks it is unimportant, or thinks it is of secondary importance to other needs, then it will not happen. In addition, recent trends in the industry are moving toward green-building practices being made mandatory, either through local adoption of new codes and standards or through an organizational policy. These trends are discussed in more detail in Chapter 7.

In the very early stages of a building's development—perhaps during the designer interview process or before a designer has even been engaged—an owner may become informed on the latest trends in building design. This may occur through an owner doing research, conferring with others in the field, or discussing the merits of green design with the designer/design firm the owner intends to hire.

This initial interaction between the owner and the design professional is where the design firm with green design experience can be very effective in turning a project not initially so destined into one that's a candidate for green design. When an owner engages a designer, it is because the owner has faith in the professional ability of that designer and is inclined to listen to that designer's ideas on what the building's design direction and themes should be. As the designer works with the owner to meet his or her defined objectives and criteria for the project, the designer has the opportunity to identify approaches that can meet those objectives and criteria in a green/sustainable way. Thus, designers should regard the very early contacts with a potential owner as a golden opportunity to steer the project in a green direction.

THE IMPACT OF CARBON CONSIDERATIONS

The attention paid to concerns about greenhouse gas emissions has certainly increased in much of the world. During the first decade of the twenty-first century, two organizations issued challenges to the industry to design and implement buildings that had a significantly lower energy consumption compared to current typical designs. The Architecture 2030 Challenge (see the "References and Resources" section at the end of the chapter for more information) is one of these. Architecture 2030 was initiated by Edward Mazria in 2002, and has a chief goal of net zero energy and net zero carbon buildings by the year 2030. This goal is realized by achieving substantially better energy results on a sliding scale from 2010 through 2030. The near-term focus of the challenge has been adopted by the American Institute of Architects (AIA). The Architecture 2010 Imperative set a goal of having buildings built by the year 2010 that would show a 50% improvement in energy efficiency compared to those built using the 1999 version of ANSI/ASHRAE/IES Standard 90.1-1999, *Energy Standard for Buildings Except Low-Rise Residential Buildings* (ASHRAE 1999).

ASHRAE took the lead in meeting these challenges in several ways. To address the Architecture 2010 Imperative, significant effort was put into modifying Standard 90.1 (ASHRAE 1999) to drastically improve energy efficiency. The 2010 version of Standard 90.1, in essence, met the AIA challenge for 2010 by introducing requirement changes that were developed and introduced during that decade. Although the specific requirements may differ in some cases, ANSI/ASHRAE/USGBC/IES Standard 189.1, *Standard for the Design of High-Performance Green Buildings* (ASHRAE 2011) has energy efficiency levels even better than Standard 90.1.

To meet the Architecture 2030 Challenge, these ASHRAE standards will be continually updated to raise the bar for building energy performance. One way this is being accomplished is through the production of the ASHRAE Advanced Energy Design Guide series. The series covers prescriptive measures that result in significant energy efficiency improvements, with the first series dealing with measures that should achieve a 30% savings over Standard 90.1 (ASHRAE 1999). A total of six Advanced Energy Design Guides for 30% savings are available for free download from the ASHRAE website. A continuation of that series has already published four additional guides for achieving a 50% energy savings over Standard 90.1-2004, with even more strenuous improvements in the planning process. Guidance on achieving net zero energy performance is planned to be completed by the end of this decade.

THE ENGINEERING/ENERGY CONSERVATION ETHIC

Since the 1973 oil embargo, the HVAC&R industry has continued to improve the efficiency of air-conditioning systems and equipment, promulgated energy conservation standards, developed energy-efficient designs, experimented with a wide variety of design approaches, strived for good IAQ, and shared the lessons learned with industry colleagues. As in the past, efforts must continue to find new and better solutions to improve energy efficiency, further reduce dependence on nonrenewable energy sources, and increase the comfort of people in the buildings they occupy. In addition, concerns about the human impact on climate makes the need to work toward greater energy efficient even more imperative.

Most designers have guided owners through life-cycle analyses of various options, identified approaches to improve building efficiency, and developed strategies to meet the stated goals of owners. Owners have often rejected their ideas because payback periods are too long. Despite those setbacks, progress toward green/sustainable design is becoming more prevalent, and is becoming an industry standard practice. It is incumbent on our industry to recognize the impact its work has on the environment, which goes beyond matters of first cost, recurring costs, and even life-cycle cost. The ethics of the industry requires practitioners to strive to identify these environmental costs and assign values to them—values that represent the total cost to society rather than just conventional measurements of capital..

INTERNATIONAL PERSPECTIVE: REGULATIONS AND COMMENTARY

Society has recognized that previous industrial and developmental actions caused long-term damage to our environment, resulting in loss of food sources and plant and animal species, and changes to the Earth's climate. As a result of learning from past mistakes and studying the environment, the international community identified certain actions that threaten the ecosystem's biodiversity, and, consequently, it developed several governmental regulations designed to protect our environment. Thus, in this sense, the green design initiative began with the implementation of building regulations. An example is the regulated phasing out of fully halogenated chlorofluorocarbons (CFCs) and partially halogenated refrigerant hydrochlorofluorocarbons (HCFCs).

In Europe, the main regulatory instrument for tackling the energy consumption of buildings is the Directive on the Energy Performance of Buildings (EPBD) recast (European Commission 2010), which took effect in 2012 and replaced the original EPBD Directive (European Commission 2002). All EU member states introduced national laws, regulations, and administrative provisions for setting minimum requirements on the energy performance of new and existing buildings that are subject to major renovations and for energy performance certification of buildings. Additional requirements include regular inspection of boilers and air conditioning systems in buildings, an assessment of the existing facilities, and provision of advice on possible improvements and alternative solutions. Moreover, the EPBD recast strengthens the energy performance requirements and clarifies and streamlines some of the original EPBD provisions to reduce the large differences between EU Member States' practices. In particular, it requires that EU Member States lay down the requirements so that new buildings are nearly zero energy by 2020 (2018 for public buildings) and the application of cost-optimal levels for setting minimum energy performance requirements for both the building's thermal envelope and technical systems.

Energy performance certificates (EPC) are issued when buildings are constructed, sold, or rented out (see Figures 1-1a through 1-1d for examples). The EPC documents the energy performance of the building and is expressed as a numeric indicator or a letter grade that allows benchmarking of primary energy consumption. The certificate also includes recommendations for cost-effective improvement of the energy performance, and is valid for up to ten years.

The Concerted Action EPBD that was launched by the European Commission provides updated information on the implementation status in the various European countries (www.epbd-ca.org).

It is not just in the developed countries that green-building design and energy efficiency concerns are taking hold. The later part of the past decade has seen an explosion of adopting building energy efficiency standards and green-building design programs. For example, India was the first expansion of the Leadership in Energy and Environmental Design programs outside of the United States, with the establishment of the India Green Building Council in 2003. India also created a nationwide energy efficiency standard, the Energy Conservation Building Code, in 2008. This code was based on ANSI/ASHRAE/IES Standard 90.1 but modified for the local climates and situations.

Energy efficiency standards throughout the world generally adopt two approaches. One is to have a set of mandatory requirements and then offer a prescriptive or a performance-based path for compliance (examples include the approaches taken for energy codes in the U.S., Canada, India, and Australia). Another approach is to have a set of mandatory items, then build on this with a point system for other features, with a minimum number of points required. This approach is the one taken by Japan and South Korea, for example.

Digging Deeper

ΠΙΣΤΟΠΟΙΗΤΙΚΟ ΕΝΕΡΓΕΙΑΚΗΣ ΑΠΟΔΟΣΗΣ	

Α.Π.: 1/2011 Α.Α.: E005F-KXMNT-FMMNJ-Z

ΧΡΗΣΗ: Πολυκατοικία
Κτίριο ☐   Τμήμα κτιρίου ☑
Αριθμός ιδιοκτησίας: Z1H1
Κλιματική Ζώνη: Β
Διεύθυνση: ΣΕΧΟΥ 8
Τ.Κ.: 11524
Πόλη: ΝΕΑ ΦΙΛΟΘΕΗ ΑΜΠΕΛΟΚΗΠΩΝ ΑΘΗΝΑ
Έτος κατασκευής: 2009
Συνολική επιφάνεια [m²]: 183.82
Θερμανόμενη επιφάνεια [m²]: 152.44
Όνομα ιδιοκτήτη: Λ & Χ ΜΠΕΤΣΗΣ & ΣΙΑ Ο.Ε

ΒΑΘΜΟΛΟΓΗΣΗ ΕΝΕΡΓΕΙΑΚΗΣ ΑΠΟΔΟΣΗΣ

	ΕΝΕΡΓΕΙΑΚΗ ΚΑΤΗΓΟΡΙΑ
ΜΗΔΕΝΙΚΗΣ ΕΝΕΡΓΕΙΑΚΗΣ ΚΑΤΑΝΑΛΩΣΗΣ	
EP ≤ 0.33·R_R **A+**	
0.33·R_R < EP ≤ 0.5·R_R **A**	
0.5·R_R < EP ≤ 0.75·R_R **B+**	**B+**
0.75·R_R < EP ≤ 1.0·R_R **B**	
1.0·R_R < EP ≤ 1.41·R_R **Γ**	
1.41·R_R < EP ≤ 1.82·R_R **Δ**	
1.82·R_R < EP ≤ 2.27·R_R **E**	
2.27·R_R < EP ≤ 2.73·R_R **Z**	
2.73·R_R < EP **H**	
ΕΝΕΡΓΕΙΑΚΑ ΜΗ ΑΠΟΔΟΤΙΚΟ	
Υπολογιζόμενη ετήσια κατανάλωση πρωτογενούς ενέργειας κτιρίου αναφοράς [kWh/m²]:	104.7
Υπολογιζόμενη ετήσια κατανάλωση πρωτογενούς ενέργειας [kWh/m²]:	75.5
Υπολογιζόμενες ετήσιες εκπομπές CO_2 [$kgCO_2/m^2$]:	19.45

Πραγματική ετήσια κατανάλωση ενέργειας & Εκπομπές CO_2		
Ηλεκτρική ενέργεια [kWh/m²]: ----	Καύσιμα [kWh/m²]: ----	Θερμική άνεση ☑
		Οπτική άνεση ☑
Συνολική ετήσια κατανάλωση πρωτογενούς ενέργειας [kWh/m²]: ----		Ακουστική άνεση ☐
Συνολικές ετήσιες εκπομπές CO_2 [kg/m²]: ----		Ποιότητα αέρα ☐

Image courtesy of Dr. Gerd Hauser.

Figure 1-1a Example of Greece's EPC.

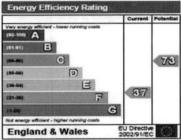
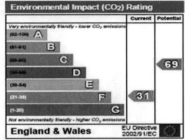

Image courtesy of Communities and Local Government (2010).

Figure 1-1b Example of England and Wales' EPC.

Image courtesy of NINN Studio MILANO - Arch. Nicola Ingenuo, Arch. Antonella Puopolo.

Figure 1-1c Example of Italy's EPC.

Thus, green-buildings programs exist or are in development in many countries across the world, establishing that the green-building movement is not just a fad, but truly is transforming the marketplace worldwide.

SOME DEFINITIONS AND VIEWS OF SUSTAINABILITY FROM OTHER SOURCES

- "The best chance we have of addressing the combined challenges of energy supply and demand, climate change and energy security is to accelerate the introduction of new technologies for energy supply and use and deploy them on a very large scale." *(Thomas Friedman, Hot, Flat and Crowded.)*
- "Humanity must rediscover its ancient ability to recognize and live within the cycles of the natural world." *(The Natural Step for Business)*
- Development is sustainable "if it meets the needs of the present without compromising the ability of future generations to meet their own needs." *(Brundtland Commission of the United Nations)*
- To be sustainable, "a society needs to meet three conditions: Its rates of use of renewable resources should not exceed their rates of regeneration; its rates of use of nonrenewable resources should not exceed the rate at which sustainable renewable substitutes are developed; and its rates of pollution emissions should not exceed the assimilative capacity of the environment." *(Herman Daly)*
- "Sustainability is a state or process that can be maintained indefinitely. The principles of sustainability integrate three closely intertwined elements – the environment, the economy, and the social system – into a system that can be maintained in a healthy state indefinitely." *(Design Ecology Project)*
- "In this disorganized, fast-paced world, we have reached a critical point. Now is the time to rethink the way we work, to balance our most important assets." *(Paola Antonelli, Curator, Department of Architecture and Design, New York City Museum of Modern Art)*

SUSTAINABILITY IN ARCHITECTURE

The emergence of green-building engineering is best understood in the context of the movement in architecture toward sustainable buildings and communities. Detailed reviews of this movement appear elsewhere and fall outside the scope of this document. A brief review of the history and background of the green design movement is provided, followed by a discussion of its applicability. Several leading methodologies for performing and evaluating green-building design efforts are reviewed.

Prior to the industrial revolution, building efforts were often directed throughout design and construction by a single architect—the so-called Master Builder Model. The master builder alone bore full responsibility for the design and construction of the building, including any engineering required. This model lent itself to a building designed as one system, with the means of providing heat, light, water, and other building services often closely integrated into the architectural elements. Sustainability, semantically if not conceptually, predates these eras, and some modern unsustainable practices had yet to arise. Sustainability in itself was not the goal of yesteryear's master builders. Yet some of the resulting structures appear to have achieved an admirable combination of great longevity and sustainability in construction, operation, and maintenance. It is interesting to compare the ecological footprint (a concept discussed later in this book) of Roman structures from two millennia ago heated by radiant floors to a twentieth century structure of comparable size, site, and use.

In the nineteenth century, as ever more complicated technologies and the scientific method developed, the discipline of engineering building systems and design emerged separate from architecture. This change was not arbitrary or willful but rather was due to the increasing complexity of design tools and construction technologies and a burgeoning range of available materials and techniques. This complexity continued to grow throughout the twentieth century and continues today. With the architect transformed from master builder to lead design consultant, most HVAC&R engineering practices performed work predominantly as a subcontract to the architect, whose firm, in turn, was retained by the client. Hand-in-hand with these trends emerged the twentieth century doctrine of Buildings Over Nature, an approach still widely demanded by clients and supplied by architectural and engineering firms.

Under this approach (buildings are designed under the architect, who is prime consultant, following the Buildings Over Nature paradigm) the architect conceives the shell and interior design concepts first. Only then does the architect turn to structural engineers, then HVAC&R engineers, then electrical engineers, etc. (Not coincidentally, this hierarchy and sequence of engineering involvement mirrors the relative expense of the subsystems being designed.)

With notable exceptions, this sequence has reinforced the trend toward Buildings Over Nature: relying on the brute force of sizable HVAC systems that are resource-intensive—and energy-intensive to operate—to build and maintain conditions acceptable for human occupancy. In this approach to the design process, many opportunities to integrate architectural elements with engineered systems are missed—often because it's too late. Even with an integrated design team to bridge back over the gaps in the traditional design process, a sustainable building with optimally engineered subsystems will not result if not done by professionals with appropriate knowledge and insight.

When Green Design is Applicable

Perhaps the obvious answer is "When is green design not applicable?" However, practicalities do exist in the design process, funding, and expectations of stakeholders in the process that may, in some people's opinion, preclude consideration of green design. This book is intended to help overcome these impediments.

One leading trend in architecture, especially in the design of smaller buildings, is to invite nature in as an alternative to walling it off with a shell and then providing sufficiently powerful mechanical/electrical systems to perpetuate this isolation. This situation presents a significant opportunity for engineers today. Architects and clients who take this approach require fresh and complementary engineering approaches, not tradition-bound engineering that incorporates extra capacity to overcome the natural forces a design team may have invited into a building. Natural ventilation and hybrid mechanical/natural ventilation, radiant heating, and radiant cooling are examples of the tools with which today's engineers are increasingly required to acquire fluency. One example of a building designed with this in mind is the GAP, Inc. building in San Bruno, California. (See Chapter 10 for GreenTips relating to alternative ventilation techniques and see Watson and Chapman [2002] for radiant heating/cooling design guidance.)

Fortunately, there is a great deal of information available about green-building design. Further, new tools for understanding and defending engineering decisions in such projects are emerging, for example, a revised ASHRAE thermal comfort standard, ANSI/ASHRAE Standard 55-2010, *Thermal Environmental Conditions for Human Occupancy* (ASHRAE 2010), that includes an adaptive design method that is more applicable to buildings that interact more freely with the outdoor environment. ASHRAE Standard 55 also accommodates an increasing variety of design solutions intended both to provide comfort and to respect the imperative for sustainable buildings.

Another more widely demanded approach to green HVAC engineering presents a significant opportunity for engineers. This approach applies to projects ranging from flagship green-building projects to more conventional ones where the client has only a limited appetite for green. The demand for environmentally conscious engineering is evidenced by the expansion of engineering groups, either within or outside architectural practices that have built a reputation for a green approach to building design. In addition, many younger engineers and architects just entering the profession are more committed to the concept of sustainable design than their more established predecessors.

Work done by others on such projects can inform engineers' building designs. In addition, many informative resources are available to help engineers better understand the principles, techniques, and details of green-building design, and to raise environmental consciousness in their engineering practices. By learning and acquiring appropriate resources, obtaining project-based experience, and finding like-minded professionals, engineers can reorient their thinking to deliver better services to their clients.

Green HVAC engineering can be provided, for its own sake, independent of any client or architect demand. Ideally, the end result is an energy-efficient system that is more robust and provides for better thermal control and indoor environment than the cookie-cutter conventional design. The appetite for environmentally conscious engineering must be carefully gauged, and opportunities to educate the design team carefully seized. In this way, engineers can bring greater value to their projects and distinguish themselves from competing individuals and firms.

Embodied Energy and Life-Cycle Assessment

Building materials used in the construction and operation of buildings have energy embodied in them due to the manufacturing, transportation, and installation processes of converting raw materials to final products. The material selection process should consider the environmental impact of demolition and disposal after the service life of the products. Another new type of building life-cycle assessment (beyond life cycle costing) focuses on the environmental impact of products and processes. This is termed *life-cycle environmental assessment* or simply *life-cycle assessment* (LCA). This is a cradle-to-grave approach that evaluates all stages of a product's life to determine its cumulative environmental impact. In the case of a building, the structure is a product itself, but it also is comprised of a large number of other individual products. Thus, the combined impact of the entire building as a system is very difficult to quantify; however, many studies are underway to develop a standard approach to measuring LCA. A detailed description of the LCA approach has been developed by the U.S. Environmental Protection Agency (EPA), and more detail can be found on their website on LCA. Life-cycle assessment (LCA) databases and tools are used to calculate and compare the embodied energy of common building materials and products. Designers should give preference to resource-efficient materials and reduce waste by recycling and reusing whenever possible.

The building design team has a variety of options to consider if conducting an LCA analysis. The International Organization for Standardization 14000 series of standards on environmental management serves as a method to govern the development of these tools. LCA tools are available from private commercial as well as governmental or public domain sources. The Building for Environmental and Economic Sustainability (BEES) tool was developed by the National Institute for Standards and Technology in the United States, with support from the U.S. Environmental Protection Agency. The Tools for the Reduction and Assessment of Chemical and Other Environmental Impacts (TRACI) from the EPA focuses on chemical releases and raw materials usage in products. Some commercial firms also offer tools.

Climate Issues and the Carbon Economy

Depending on whom you talk with, the controversy either continues or has been settled regarding the extent of concern about anthropogenic (human-caused) emissions of greenhouse gases and the consequent impact these emissions have on the environment and society. Some think that the issue needs further study before significant investments and changes are mandated, while others, such as the Intergovernmental Panel on Climate Change (IPCC), think a consensus has been reached (reports supporting this stance can be found at www.ipcc.ch.) One thing that cannot be argued is the documented fact that CO_2 levels in the atmosphere have been steadily rising (definitive records have been

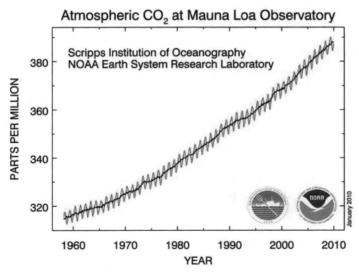

Image courtesy of C.D. Keeling of the Scripps Institute of Oceanography, La Jolla, California, and Pieter Tans of NOAA's Earth System Research Laboratory in Boulder, CO.

Figure 1-2 Historical trends in atmospheric CO_2 levels.

kept since at least early in the twentieth century), as seen in Figure 1-2. In 2013, we saw the first time when concentrations that exceeded 400 ppm. While CO_2 is not the only greenhouse gas that human activity causes, it is one where we do have the ability to control it more readily.

There is bad news and good news when we look at how buildings are involved with greenhouse gas emissions. First the bad news: buildings (commercial and residential) are responsible for approximately 30% of the greenhouse gas emissions in the United States and most developed countries, and the trend is also holding up in key developing nations. The good news is that buildings have also been identified as the economic sector with the best potential for cost-effective mitigation of greenhouse gas emissions, as highlighted in Figure 1-3. Therefore, the building industry can and should take responsibility for reducing greenhouse gas emissions, primarily through a reduction in energy consumption.

Past experience shows that is it unrealistic to expect voluntary reduction in greenhouse gas emissions to occur in large enough quantities to stem the documented rise in CO_2 levels. The reason for moving beyond voluntary control can be attributed to the following public policy measures: (1) a tax on emissions of carbon (directly for a source or indirectly through the purchase of fuels, etc.) or (2) a cap and trade program.

The concept of a tax on carbon is relatively simple to grasp, although implementation would require a lot of concern for details. Many economists and business leaders prefer a carbon tax over a cap and trade system, because it is a known measure that can be planned for within a business model. However, politicians tend to prefer a cap and trade system, since that provides a more politically safe approach.

In the past, a cap and trade system for other air pollutants has been implemented in the United States. It addressed air pollution concerns such as SO_x emissions and involved major polluters such as electric utilities. However, one factor that helped this succeed was that there were fewer emissions sources. Initially, the cap and trade programs in the European Union have stumbled, as implementation problems were worked out. A cap and trade system has been proposed in the United States for CO_2, and this too would focus on major emissions sources such as electrical utilities, although politics is most certainly involved as well.

The HVAC&R engineer can provide a significant benefit to society (as well as to the building project's owners) via CO_2 emissions reduction associated with energy use. All new building projects that would want to be considered green should at least estimate the CO_2 equivalent emissions footprint of the building (of which a large part is through energy consumption). Using simply emissions factors, these calculations are not complicated and can provide insight. For existing buildings, compute the reduction in emissions associated with energy conservation measures being proposed. In both cases, the greenhouse gas emissions factor used should be based on source energy and not on energy consumed on site alone. A good reference source for emissions factors is contained in a National Renewable Energy Laboratory (NREL) report released in 2007 (Deru and Torcellini).

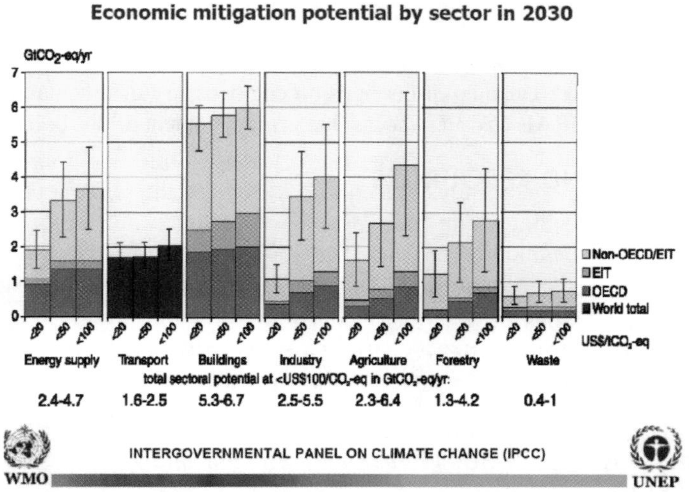

Economic mitigation potential by sector in 2030

Source: Climate Change 2007: Synthesis Report. Contribution of Working Groups I, II and III to the Fourth Assessment Report of the IPCC, Figure 4.2. IPCC, Geneva, Switzerland.

Figure 1-3 The building economic sector has the greatest potential for economical mitigation of greenhouse gases.

Green-Building Rating Systems and Standards

The introduction of green-building practices has been significantly helped by the emergence of rating systems and, more lately, through the development of green-building standards intended for adoption as part of building codes. The creation and implementation of green-building programs such as LEED have helped provide incentives for green-building concepts beyond that which would have happened from just the early adopter or sustainability friendly owners. More recently, the release of ANSI/ASHRAE/USGBC/IES Standard 189.1 and the *International Green Construction Code* (IgCC) and the adoption of Standard 189.1 as the reference point for design by organizations such as branches of the U.S. military have all begun the transition of green design from one-off voluntary adoption to being part of the mainstream design. Chapter 7 in the *GreenGuide* provides a much more extensive summary of this area.

Applications of Fundamental Concepts of Engineering

Green design requires the design team focus on achieving optimum efficiency levels, particularly in the consumption of energy and water resources. New technologies in energy efficiency and other environmental concerns often are based on advanced engineering fundamentals. For example, many now advocate the consideration of exergy concepts in the design of HVAC and related systems. It is assumed that the reader has a good foundation in engineering fundamentals or at least a beginning understanding of the concepts involved. Detailed discussions on engineering fundamental concepts are not included in ASHRAE *GreenGuide*, as that is not the intent of this book.

REFERENCES AND RESOURCES

Published

AIA. 1996. *Environmental Resource Guide*. Edited by Joseph Demkin. New York: John Wiley & Sons.

ASHRAE. 2010. ANSI/ASHRAE Standard 55-2010, *Thermal Environmental Conditions for Human Occupancy*. Atlanta: ASHRAE.

ASHRAE. 1999. ANSI/ASHRAE/IES Standard 90.1-1999, *Energy Standard for Buildings Except Low-Rise Residential Buildings*. Atlanta: ASHRAE.

ASHRAE. 2013. *ASHRAE Handbook—Fundamentals*, Ch. 35. Atlanta: ASHRAE.

ASHRAE. 2011. ANSI/ASHRAE/USGBC/IES Standard 189.1-2011, *Standard for the Design of High-Performance Green Buildings*. Atlanta: ASHRAE.

ASHRAE. 2010. ANSI/ASHRAE/IES Standard 90.1-2010, *Energy Standard for Buildings Except Low-Rise Residential Buildings*. Atlanta: ASHRAE.

Deru, M., and P. Torcellini. 2007. *Source Energy and Emissions Factors for Energy Use in Buidlings*. Technical Report NREL/TP_550-38617, National Renewable Energy Laboratory, Golden, CO.

European Commission. 2002. Energy Performance of Buildings, Directive 2002/91/EC of the European Parliament and of the Council. Official Journal of the European Communities, Brussels.

European Commission. 2010. Energy Performance of Buildings, Directive 2010/31/EU of the European Parliament and of the Council. *Official Journal of the European Communities,* Brussels.

Grondzik, W.T. 2001. The (mechanical) engineer's role in sustainable design: Indoor environmental quality issues in sustainable design. HTML presentation available at www.polaris.net/~gzik/ieq/ieq.htm.

Savitz, A.W., and K. Weber. 2006. *The Triple Bottom Line*. San Francisco: John Wiley & Sons, Inc.

Online

The American Institute of Architects
  www.aia.org.
Architecture 2030 Challenge
  www.architecture2030.org.
Advanced Energy Design Guides
  www.ashrae.org/aedg.

Building for Environmental and Economic Sustainability (BEES)
  www.nist.gov/el/economics/BEESSoftware.cfm.
BuildingGreen (for purchase)
  www.greenbuildingadvisor.com.
Building Research Establishment Environmental Assessment Method
  (BREEAM®) rating program
  www.breeam.org.
Center of Excellence for Sustainable Development, Smart Communities Network
  www.smartcommunities.ncat.org.
European Commission, Concerted Action Energy Performance of Buildings Directive
  www.epbd-ca.org.
Green Globes
  www.greenglobes.com.
GreenSpec® Product Guide
  www.buildinggreen.com/menus.
The Hannover Principles
  www.mindfully.org/Sustainability/Hannover-Principles.htm.
International Living Building Institute, The Living Building Challenge
  https://ilbi.org/lbc.
International Organization for Standardization family of 14000 standards,
  www.iso.org/iso/home/standards/management-standards/iso14000.htm.
Intergovernmental Panel on Climate Change
  www.ipcc.ch.
Lawrence Berkeley National Laboratories, Environmental Energy Technologies Division
  http://eetd.lbl.gov/.
Minnesota Sustainable Design Guide
  www.sustainabledesignguide.umn.edu.
National Renewable Energy Laboratory, Buildings Research
  www.nrel.gov/buildings/.
Natural Resources Canada, EE4 Commercial Buildings Incentive Program
  http://canmetenergy-canmetenergie.nrcan-rncan.gc.ca/eng/software_tools/ee4.html.
The Natural Step
  www.naturalstep.org.
New Buildings Institute
  www.newbuildings.org/.

Oikos: Green Building Source
   www.oikos.com.
Rocky Mountain Institute
   www.rmi.org.
Sustainable Building Challenge
   www.iisbe.org/.
Sustainable Buildings Industry Council
   www.sbicouncil.org.
Tools for the Reduction and Assessment of Chemical and Other Environmental Impacts (TRACI)
   www.epa.gov/nrmrl/std/traci/traci.html.
U.S. Green Building Council, Leadership in Energy and Environmental Design, Green Building Certification System
   www.usgbc.org/leed.
U.S. Environmental Protection Agency, Life Cycle Assessment
   www.epa.gov/nrmrl/std/lca/lca.html.
The Whole Building Design Guide
   www.wbdg.org.

CHAPTER TWO

SUSTAINABLE SITES

One interesting phenomenon that has come out of the encompassing of the sustainable design and integrated design processes is that design professionals are becoming more aware of the work done by disciplines outside their own. ASHRAE *GreenGuide* is primarily focused for the needs of its members and related disciplines, but for those involved with green-building design projects it is worthwhile to have at least a basic knowledge of all aspects of green design. In 2012, ASHRAE also announced a rebranding of the society to becoming a sustainability resource for the industry, including changing the Society's tagline to "Shaping Tomorrow's Built Environment Today."

Sustainability will be part of the design in all features of the future built environment, and this includes the site development as well. For example, in addition to the LEED program, ANSI/ASHRAE/USGBC/IES Standard 189.1 includes a whole section on "Sustainable Sites," and much of the material in this chapter relates to how those issues are treated in Standard 189.1.

This chapter provides a summary of the key issues in the following topical areas:

- Where to locate the building project
- Landscaping
- Urban heat island effect
- Exterior lighting/light as a pollution source
- Stormwater management

Each of these is briefly summarized in the following sections.

LOCATION OF THE BUILDING PROJECT

For most locations, at least in most of the U.S., Canada and Europe, local or regional regulations and agencies control many decisions as to exactly where development would be allowed and where it would be not allowed. These rules and regulations are designed to, as a minimum, prevent building development from taking place in environmentally sensitive areas (such as near a wetland). Other local regulations govern where land-use planning is a primary method for determining where new development is allowed. Depending on the region, these regulations can be fairly strict or more lenient. However, it is safe to say that this is primarily a local or regionally controlled decision.

The building's location may be allowable under minimum codes or local regulations, but not desirable for a building project that was intended to being "green." For example, consider a highly efficient office building with many sustainability features incorporated but where the project developer has decided to pass up a more urban location for a rural setting. Would that building be considered truly a green building if all the occupants had to drive private vehicles to the site each day? There is no right or wrong answer to that question as all projects are unique, but certainly transportation to the site is a consideration. Thus, urban sprawl is another topic for consideration to moving toward a more sustainable built environment. Urban sprawl affects more than just environmental impact considerations. It effects issues such as quality of life, but these issues are beyond the scope of this guide.

The decisions and issues associated with this topic go well beyond just the design of the physical building structure and systems and thus are somewhat outside the scope of this book. However, the design professional of the future should be at least aware of these issues and, when possible, contribute to the overall societal discussion on these topics.

URBAN HEAT ISLAND EFFECT

The urban heat island effect is the tendency of urban areas to have ambient air temperatures higher than the surrounding rural areas in the same region. This is a complex topic, and the contributors to the urban heat are many. Those who have participated in a building project that was working for LEED certification, ANSI/ASHRAE/USGBC/IES Standard 189.1, or one of the other green-building programs are most likely aware of the credit points offered that are intended to address the urban heat island effect. Those credit points address one of the main contributors to the urban heat island effect, that is, the tendency of buildings and the site hardscape areas to absorb and retain heat from the sun.

There are many other contributors, or potential contributors, to the urban heat island effect that are beyond the control of the building designers. These include items like industrial activity and motor vehicle transportation. And of course, outdoor condensing units play a big role in the development of the urban heat island effect as all the thermal energy from inside the building (plus the energy used to power the refrigeration system that provided that cooling) is transported to the ambient air.

Why be Concerned About the Urban Heat Island Effect?

In some circumstances, and from the thinking of some people, the urban heat island effect may be considered a good thing. For example, if the urban heat island effect in a major metropolitan area means that precipitation falls as rain rather than snow, many motorists in that area might be thankful. However, in summer that same urban heat island effect, and the resulting increase in ambient air temperature beyond what it normally would be, means that every air-cooled condensing unit is working that much harder to provide the required cooling. The additional load placed on the compressor means additional thermal heat load to the ambient air, thus increasing the heat island effect even more. For example, a simple, air-cooled air-conditioning unit operating with a 100°F (38°C) ambient air temperature will have approximately 10% lower compressor power input than for a comparable unit providing the same tons of cooling and operating in an ambient air temperature of 105°F (40°C).

In this regard, the heat island effect should therefore be looked at as a societal issue and not just evaluated from the cost effectiveness of reductions in building heat gain from the installation of roof with a lower solar absorption rate on the building. In fact, much of the benefit from measures that reduce the urban heat island effect, such as a cool roof or lower solar reflective index (SRI) hardscape, is experienced by all in the local area and not by just that particular building's owner. Therefore, this is a subject that can be considered to be a benefit to society as a whole.

Heat Island Mitigation Methods Associated with a Building Project

Users of this guide are already likely familiar with many methods that the building designer has available to help minimize the contribution of their building project to the heat island in the area where this building is to be built. One of the primary methods includes the use of materials that have a higher SRI. The SRI is a metric that indicates the ability of a material to absorb the sun's heat (in terms of the absorptivity of the material) as well as the ability to lose heat by thermal radiation, which is determined by the thermal emissivity of the material. Other methods include the use of shading to prevent a building wall or the hardscape from absorbing the sun's heat, green-roof systems, or porous paving materials.

The Interrelationship of Urban Heat Island and Building Heat Gain

There is a complicated relationship between the use of materials that have a higher SRI value and their impact on the building heat gain (and hence cooling load) and the urban heat island effect. The use of higher-SRI materials will have varying impact on the building heat gain, depending on how thermally coupled the building interior is with the roof and walls. In some structures, the reduction in heat gain is practically insignificant, while in other cases it will help reduce the overall cooling load to a noticeable degree. However, green-building standards and programs are concerned not only with the building heat gain but also with the contribution of the building structure and associated hardscape with the heat island effect. The heat island effect is a sites issue, rather than strictly a building energy issue, although criteria such as a high SRI roof can affect both. That is why criteria are included in the sites section in green-building rating systems, programs, and standards such as ANSI/ASHRAE/USGBC/IES Standard 189.1. This also helps explain the differences in why the heat island requirements, specifically those for the roof materials, differ between ANSI/ASHRAE/IESNA Standards 90.1 and ANSI/ASHRAE/USGBC/IES Standard 189.1. Standard 90.1 is solely focused on the impact on the building heat gain and cooling load, while Standard 189.1 is intended to provide a balanced approach to treating environmental issues. Thus, Standard 189.1 contains more stringent requirements in the heat island section because it is intended to consider both building energy as well as the heat island effect.

EXTERIOR LIGHTING

Similar to the heat island issue, the design of exterior lighting levels and lighting fixture selection for a green building involves both consideration of lighting power (for total energy consumption) and the impact of the building and its systems on the surrounding locality. In the case of exterior lighting and the building site, it is also a matter of how the lighting is directed to the area intended to be lit.

To prevent light from escaping the building project property boundary to areas that would be adversely affected by that light, the lighting design should specify fixtures that meet certain standards. Exterior lighting fixtures are rated on how well they perform in terms of backlight, uplight, and glare (the BUG"ratings). It also is important to consider the locality where the building is proposed to be built. In some areas, there already is a light pollution problem that exists from all the other surrounding developments, and, thus, one more building contributing to that problem will have minor impact on the total problem. In other localities away from city lights and near areas where additional light in the surrounding environment would cause unwanted consequences, special concern should be given to the lighting fixture design to prevent unwanted light from escaping the property boundary.

A number of references are available from organizations that are concerned with exterior lighting. For sustainable site considerations (above and beyond just the energy consumption aspects), the International Dark-Sky Association provides a series of resources. See the references listed at the end of this chapter for a link to this valuable resource.

STORMWATER MANAGEMENT

Development of properties for buildings and the associated infrastructure has led to the creation of impervious surfaces that contribute to increased surface runoff into surface waterways. That increased runoff is the cause, or at least a significant contributor, to flooding problems downstream. In most localities, limited requirements are in place for addressing this issue. This is one area where the design and features of the building project can be included that provide a measurable positive contribution to the environment. While many of these measures are outside the normal responsibility of the building systems design engineers and are more the responsibility of the landscape architect, some of the measures can be used to also provide an alternative source of water that can be used to reduce the overall demand for potable water.

Prescriptive Versus Performance Focus

Existing green-building rating systems, codes, and standards approach the stormwater management issue from various perspectives. One approach is to specify or require the building project achieve a particular performance basis. This approach will seek to address the amount and quality of stormwater runoff based on comparison to a set criterion. For example, there could be a target for the project design team to end up with the postdevelopment runoff amount to not exceed that which would have occurred from the same site if left in an undisturbed, predevelopment state. A performance approach is what is used by the U.S. Green Building Council LEED rating systems. Performance is what ultimately counts but this is achieved by the actual design and selection of specific measures.

A different approach could be taken, one prescriptive in nature. Instead of specifying specific performance criteria to be achieved, this alternative approach would require the inclusion of best management practices that have been proven to address the problems caused by unmitigated stormwater from the built environment. This approach is described as being prescriptive in nature, as it prescribes specific technologies or practices to be implemented. Some of these practices are briefly outlined in the section below.

Stormwater Mitigation Techniques

The best designs for minimizing stormwater impact will result in a smaller amount of surface runoff from that property than would have occurred from a natural, undisturbed landscape at that same location. This can be achieved by a combination of one or more of the following design features.

Maintaining a Larger Vegetated Percentage of the Property. Obviously, maintaining as much as possible of the property with some form of vegetated surface is a prime consideration. However, some properties offer limited, if any, potential for this, so additional measures are needed.

Rain Gardens. Rain gardens are low-lying regions on the property located to collect surface runoff from impervious surfaces (such as parking areas) slowing down that runoff enough to allow for a large portion of it to naturally infiltrate into the soil or leave by evaporation or transpiration from plants in the rain garden. In this regard, rain gardens provide a similar function as a traditional retention pond on site, but with the additional benefits of providing a more natural landscape look and avoiding some of the negative aspects of retention ponds such as liability concerns and the need to fence that area off. Rain gardens are the subject of GreenTip # 2-1 at the end of this section.

Rainwater Harvesting. Rainwater collection for later use in the building or on the site as irrigation water has two sustainability-related benefits. First, this water would typically be collected from impervious surfaces, such as a rooftop, and thus all water collected is that much water that is prevented from quickly running off the property and contributing to stormwater problems downstream. (This is thus a sustainable site issue.) If the collected water is used to displace potable water, for example, as makeup water to a cooling tower, then rainwater harvesting also contributes toward water efficiency. Rainwater harvesting is a subject of GreenTip #14-5 in this guide.

Green Roofs. Green roofs offer the ability to absorb at least a portion of the precipitation that falls on the roof. That amount is a function of soil depth, condition of the soil before the storm event (whether dry or already containing some water), and the type of soil in the green-roof system. Green roofs provide additional environmental benefits, but can also be a source of additional water consumption if not properly designed and irrigation is used.

Figure 2-1 Example of rain garden used to manage stormwater runoff from a building roof.

REFERENCES AND RESOURCES

Online

Cool Roof Rating Council
    www.coolroofs.org.

Greenroof Industry Information Clearinghouse and Database
    www.greenroofs.com.

International Dark-Sky Association
    www.darksky.org/index.php.

U.S. Environmental Protection Agency, Heat Island Effect
    www.epa.gov/hiri.

U.S. Environmental Protection Agency, Stormwater Management
    www.epa.gov/oaintrnt/stormwater/.

ASHRAE GreenTip #2-1

Rain Gardens

GENERAL DESCRIPTION

In 1990, in response to a demand for better stormwater management in Prince George's County, Maryland, bioretention systems, or rain gardens, were used as a financially and aesthetically pleasing solution (EPA 2013a). Since their utilization in Maryland, case studies and research have taken place to determine the potential impacts that rain gardens can have on decreasing the amount of stormwater runoff in local areas, as well as the amount of pollutants that can be absorbed by the vegetation incorporated in rain garden designs. The results of these studies have been positive.

The concepts of a rain gardens are basic. Where there are impervious structures that rain water encounters, there will be runoff. This runoff is often directed to infrastructure designed to carry runoff (and the associated contaminants) to nearby waterways. Rain gardens intercept runoff before reaching those systems and filter a portion of the runoff using soil, mulch, and appropriate vegetation. In order for runoff to be captured, rain gardens must be on a shallow slope away from the aforementioned impervious structures in the direction runoff flows toward stormwater management systems (EPA 2013b).

Rain gardens most often include a variety of native plants which are drought tolerant and wet tolerant that are found in low areas (Cleanwater Campaign 2012). While rain garden designs will vary from owner to owner, there are some specifics. Most are dug approximately four to six inches deep with mulch and specific vegetation used to absorb water and filter pollutants. During a typical rain fall, the first inch of rain washes away the majority of pollutants from roof tops, roads, parking lots and other structures. Rain gardens are generally designed to hold water from a one inch rainstorm event. During that time, pollutants are absorbed and filtered through the mulch and vegetation instead of entering the stormwater drains and ultimately the local waterways (University of Rhode Island 2013). The design is shallow enough that rain will not collect and stay for more than

24 hours so that mosquitoes and other insects are unable to breed, nest, and lay eggs (Cleanwater Campaign 2012). These designs will provide more absorption and filtration than common grass lawns while providing a more diverse landscape (University of Rhode Island 2013).

WHEN AND WHERE THEY ARE APPLICABLE

While some alterations may need to be made in colder climates (snow and salt content must be considered), rain gardens are applicable almost anywhere in the United States (EPA 2013b). In most cases, the existing soil type will not prohibit designs, since the areas are dug out first and a specific soil mixture is added onto the cleared ground. An ideal soil mixture suitable for rain gardens is a mix of 50%–60% sand, 20%–30% topsoil, and 20%–30% compost (Cleanwater Campaign 2012).

PRO

- Development and design is similar to that of common landscape projects which require no major technology or infrastructure.
- Due to the specific plant type requirements, rain gardens require little maintenance over time other than observation for erosion and clogging.
- With adequate planning and implementation, they can be built to serve most regions and soil types.
- While providing water management and pollutant filtration, rain gardens are most often aesthetically pleasing to the owner.
- Rain gardens can be designed on new developments or retrofitted to existing developments (Oklahoma Farm to School 2013).

CONS

- If not designed properly, standing water or increased erosion can occur (Oklahoma Farm to School 2013). These issues should be avoidable by following guidelines provided by local, state, or national entities such as the Environmental Protection Agency.

KEY ELEMENTS OF COST

First Cost

- Mulch
- Desired vegetation
- Initial watering (Daily for approximately two weeks) (Engineering Technologies Associates and Biohabitats 1993)

Recurring Cost

- Remulching void areas
- Treating diseased vegetation
- Replacement of tree stakes and wiring (Engineering Technologies Associates and Biohabitats 1993)
- Weeding (potentially, depending on the circumstance and location)

REFERENCES USED IN THIS GREENTIP

Environmental Protection Agency—Stormwater Case Study. 2013a. http://cfpub.epa.gov/npdes/stormwater/casestudies_specific.cfm?case_id=14

Environmental Protection Agency—Bioretention (Rain Gardens). 2013b. http://cfpub.epa.gov/npdes/stormwater/menuofbmps/index.cfm?action=browse&rbutton=detail&bmp=72

Cleanwater Campaign: Solutions to Stormwater Pollution. 2012 www.cleanwatercampaign.com/html/636.htm

University of Rhode Island, Healthy Landscapes. 2013 www.uri.edu/ce/healthylandscapes/raingarden.htm

Oklahoma Farm to School. 2013. www.okfarmtoschool.com/

Engineering Technologies Associates and Biohabitats. 1993. *Design Manual for Use of Bioretention in Stormwater Management.* Prepared for Prince George's County Government, Watershed Protection Branch, Landover, MD.

SOURCES OF FURTHER INFORMATION

Environmental Protection Agency—Experimental Rain Gardens
www.epa.gov/oaintrnt/stormwater/edison_rain_garden.htm State University of New York College of Environmental Science and Forestry
www.esf.edu/sustainability/action/raingarden.htm

University of Rhode Island Cooperative Extension

Rain Gardens: A Design Guide for Homeowners in Rhode Island

Helping to Improve Water Quality in Your Community
www.uri.edu/ce/wq/nemo/publications/pdfs/sw.rgbrochure_text.pdf

ASHRAE GreenTip #2-2

Green-Roof Systems

GENERAL DESCRIPTION

Living vegetative surfaces on a building rooftop, green roofs are classified in two categories based upon soil depth:

Intensive:
- Minimum soil depth: 12 in. (30 cm)
- Capable of housing large trees, shrubs and recreational facilities
- Heavy; requiring structural reinforcement, 80–150 lb/ft^2 (400–750 kg/m^2)

Extensive:
- Average soil depth: 1–6 in. (3–15 cm)
- Often installed using modular plots
- Lighter; less of everything, 10–15 lb/ft^2 (50–75 kg/m^2)

Green-roof systems are composed of a series of layers used to house the vegetation, filter and collect water, protect the building from water damage, and insulate the building. The uppermost layer is the vegetation layer. This layer varies the most in application due to climatic factors and depth of soil. Maintenance required is determined by the types of plant life used; an intensive system will require more attention than an extensive system. The second layer is the growth media, the soil mix layer in which the vegetation grows. This layer will also vary with location as different soil types will require a variation in the soil mix ratio.

In extensive systems, under the growth media is typically a layer of filter fabric. This filter allows for water penetration, but must be tough to withstand pressure. Typical materials for filter fabric systems are geosynthetic fabrics or geotextiles.

In intensive systems, the filtration system is also a part of the drainage system. A layer of gravel beneath the fabric-protected growth media allows tree roots to grow down and increase the stability of the plant layer above. A root barrier is layered beneath the gravel. Water can be stored in the gravel, drained to a cistern, or, in the case of extensive systems, be contained in a molded modular storage unit system. This is also the layer in which irrigation can take place to access and encourage deep roots.

Beneath the drainage layers of green-roof systems is a waterproofing layer. Watertight materials can be applied as a liquid-applied membrane, a single-ply specialty sheet, or a three-layer laminated roof system. One or more layers of insulation provide reduced heat transfer through the roof as well as lower noise transmission.

WHEN AND WHERE THEY ARE APPLICABLE

Green-roof systems are particularly well-suited for dense urban application. Unused rooftops can become productive spaces. Structural constraints make retrofit projects more difficult for the implementation of a green-roof system, but a new building can handle the structural details in the design phase.

Climate plays a crucial role in the actual return on investment figures. For example, stormwater retention may be less of a benefit for a green roof located in low rainfall areas or regions where rainfall tends to be concentrated in the winter months.

KEY ELEMENTS OF COST

What climate region a green-roof system is implemented in will determine actual cost savings as the same system will not perform identically in different locations. Using L to represent lower cost, H to represent higher cost, and S to represent the same cost; here is a brief comparison of several expenditure types:

First Cost
- Installation   H/S (H in retrofits)
- Design fees   S

Recurring Cost
- Overall energy costs   L
  - Water costs   L (if rainwater harvesting is done)
- Training for operations   H
- Overall maintenance costs   H

Table 2-1: Potential Benefits and Drawbacks of Green Roofs

Advantages	Disadvantages
Stormwater runoff reduction	Additional structural load
Reduced heat gains (in summer) and heat loss (in winter) to building structure	Cost
Longer life for the base roofing system (may not apply to an intensive green roof)	Additional maintenance, ranging from limited for an extensive green roof with low-maintenance plants to high for a manicured landscape intensive roof
Reduced noise transmission from outside	Optimal roof type, plant materials, and soil depths will vary depending on climate
Aesthetic benefits to people in or around the building with the additional green space	Documentation of benefits such as reduction in heat island effect has not been proven
Other general environmental benefits, such as reduced nitrogen runoff (source: bird droppings), air pollutant absorption, potential carbon sink, bird habitat	

SOURCES OF FURTHER INFORMATION

The City of London, Greater London Authority, Design for London, London Climate Change Partnership. *Living Roofs and Walls – Technical Report: Supporting London Plan Policy.* London: Greater London Authority, 2008.

"Green-Roof System Components" Greenroofs.com, 2011. Web. 15 Sep 2012.
www.greenroofs.com/Greenroofs101/.

Section 2:
The Design Process

CHAPTER THREE

PROJECT STRATEGIES

INGREDIENTS OF A SUCCESSFUL GREEN PROJECT ENDEAVOR

The following ingredients are essential in delivering a successful green design:

- Commitment from the entire project team, starting with the owner.
- Establishing Owner's Project Requirements (OPR), including green design goals, early in the design process.
- Integration of team ideas.
- Effective execution throughout the project's phases—from predesign through the end of its useful service life.

Establishing Green Design Goals Early

Establishing goals early in the project planning stages is a key to developing a successful green design and minimizing costs. It is easy to say that goals need to be established, but many designers and owners struggle with what green design is and what green/sustainable goals should be established. The following are typical questions to ask:

- What does it cost to design and construct a green project?
- Where do you get the best return for the investment?
- How far should the team go to accomplish a green design?

Today there are many guides a team can use with ideas on which green/sustainable principles should be considered. Chapter 7 of this guide presents several rating systems and references on environmental performance improvement. The essence of these documents is to provide guidance on how to reduce the impact the building will have on the environment. While the approaches and goals contained in each differ, all suggest common principles that designers may find helpful to apply to their projects.

Integration of Team Ideas

No green project will be successful if the various project stakeholders are not included in the process. These stakeholders include the owner, the owner's operations staff, the commissioning authority, design disciplines, contractors, and users. These stakeholders, if known, should work in close coordination, beginning in the earliest stages. Use the commissioning process, as discussed in Chapter 6, to obtain input from the various stakeholders and develop the OPR document that defines the owner's objectives and criteria (which includes the green/sustainable project goals that set the foundation for what the team is tasked with delivering and benchmarks for success). Based on the owner's stated objectives and criteria, the integrated team works together with a clear direction and focus to deliver the project. Gone are the days when the mechanical, electrical, and plumbing engineers and landscape architects become involved only after the building's form and space arrangements are set (i.e., end of schematic design). That is far too late for the necessary cross-pollenization of ideas between engineering and architectural disciplines that must occur for green design to be effective. Ultimately, the owner is the most important member of the team. On the other hand, within the owner's organization there are often different departments or staff that focus on user activities, design and construction, or operation and maintenance and do not necessarily communicate (or are not invited to communicate) their objectives and criteria to the design and construction team. The success of the project and the long-term benefits of established green/sustainable goals can be easily lost if all of the owner's staff is not included. For example, operators with insufficient training and understanding of the project's intended operation can inadvertently negate the benefits through improper operation.

Effective Execution

Still, all the good intentions an owner and project team may have during the early stages are meaningless without follow-through both during the construction process and over the entire life of the facility. Ideally, success requires a committed owner, use of the commissioning process, the owner's defined objectives and criteria, including green/sustainable project goals established in predesign, knowledgeable design practitioners working in concert to develop a green/sustainable design, competent contractors who buy into the green concept, and, finally, operators armed with the necessary tools and who are properly trained and dedicated to keeping the facility operating at peak performance for its life.

INCENTIVES FOR GREEN DESIGN

For both individuals and firms in the design and construction profession, many incentives exist to develop green/sustainable projects. As with any aspect of business practice that adds value to a project, fees and client expectations must be carefully managed. While some clients may balk at added fees charged for the commissioning, additional coordination, and studies necessary to meet green/sustainable goals for systems that are part of a traditional project scope, others welcome these services. First, the appetite of the project client and full project team for a green/sustainable project must be gaged. Then, the commensurate level of commissioning, design, construction, and operational services can be provided. The OPR, defined early and documented in writing, facilitate such understandings and provide a more defined initial fee proposal and acceptance.

Individuals and firms are finding that green project capabilities can positively impact building professionals' careers and the firms that employ them. Firms can enhance these capabilities by providing leadership on green issues, building individual competencies, providing ongoing support for professional development in relevant areas, rewarding accomplishments, marketing or promoting green success stories, and building their clientele's interest in green/sustainable design.

When properly delivered, green design/construction/operation capabilities can lead to enhanced service to clients, repeat business, increased public relations and marketing value, and increased demand for these green design services, especially among the architects or owners that represent a substantial proportion of many firms' billings. In addition, green capabilities can also improve employee retention and employee satisfaction. Finally, green/sustainable project competency can reduce risks in practice, meaning that knowledge of green issues is necessary to manage risk when participating in aggressively green projects.

Human Productivity. While difficult to measure, the benefits of improving the learning, living, and workplace environment (all aspects of indoor environmental quality [IEQ]) and feelings of well-being can yield big gains in human productivity. Figure 3-1 shows the typical relationship among the various categories of the costs of operating a business with human costs outstripping all other costs several times over.

Below are some examples of some data generated by private commercial organizations that support the concept that green design increases productivity:

Lockheed
- 15% rise in production
- 15% drop in absenteeism

West Bend Mutual Insurance
- 16% increase in claims processes

NG Bank
- 15% drop in absenteeism

Verifone
- 5% increase in productivity
- 40% drop in absenteeism

Similar results have been shown in learning and living environments. Some projects may lend themselves to the incorporation of green concepts without incurring additional costs beyond the business's own investment in the knowledge of, and experience with, green projects. Other sustainable design projects may require additional services for a greater number of design considerations not customarily included in ordinary design fees, such as extended energy analysis, daylighting penetration, and solar load analysis, or extended review (of materials and components, environmental impacts such as embodied energy, transportation, construction, composition, and indoor air quality [IAQ] performance). Other categories of additional services may include project-specific research or training of the engineering team (where required) and client education and communication.

Financial Incentives. The economic benefits of green design strategies as discussed earlier in this chapter are enhanced by identifying funding sources for the high-performance design concepts that are being considered for a building design project. Having the knowledge of where the opportunities can be found and assisting the team in obtaining the funding enhances the engineer's role and importance in the green design process and significantly improves the owner's return on investment (ROI). Funding sources include utility company rebates, state and federal grant programs for energy efficiency and renewable energy measures, and private foundation grants for design enhancements and green rating system implementation.

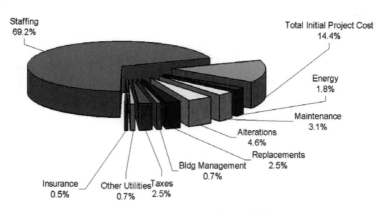

Image courtesy of Michael Dell'Isola, Faithful + Gould

Figure 3-1 People cost vs. other costs in business operation.

Two broad categories of incentives, external and internal, can each encompass both direct and indirect financial benefits and nonfinancial advantages.

The familiar range of energy incentives offered by many utility-administered, state-mandated, demand-side management programs represents one obvious form of external financial incentive—the rebate check available to owners or utility ratepayers who pursue energy conservation measures in buildings and mechanical/electrical systems. Of course, the availability of these and other financial incentives varies from place to place and project to project. But many other external financial incentives can reduce project costs where applicable, such as the following:

- Sustainable design tax credits
- Marketable emissions credits
- Tax rebates
- Brownfield funds
- Historic preservation funds
- Community redevelopment funds
- Economic development funds
- Charitable foundation funds

In addition, future public policy initiatives associated with climate concerns may direct attention toward low-carbon or net zero carbon designs, as discussed later in this guide.

U.S. For example, a 26-story, 960-bed residence hall project completed in 2009 at Boston University (BU) used the NSTAR Comprehensive Design Rebate Program (NSTAR is a Massachusetts-based electric and gas utility). The program provides incentives for purchasing and installing high-efficiency equipment for use in commercial and industrial operations. NSTAR offers incentives of 90% of the incremental cost differential for comprehensive design or up to 75% of the incremental cost differential between standard base line and high-efficiency equipment. In addition, customers are eligible to receive cost sharing for engineering services and design and commissioning services.

On the BU Housing project, NSTAR and the design team held an energy conservation measure charrette during the schematic design phase. Measures incorporated into the design included energy recovery for the 100% outdoor air ventilation units, chilled-water system energy enhancements, energy conservation measure motors and variable-speed control logic for the fan-coil units, upgraded glazing performance, and variable-speed drives on all pump motors (with integrated system control logic) on the project. The net result was that for an investment of just over $500,000 on a $130 million project, BU received a rebate from NSTAR for $240,000. In addition, the estimated energy savings from incorporation of these measures is estimated to be $210,000 annually. The result is an annual ROI of greater than 70%.

The following link provides information on incentive programs for all 50 states: www.dsireusa.org/.

International. The European Union (EU) is aiming for a 20% cut in Europe's annual primary energy consumption by 2020 (http://ec.europa.eu/energy). The European Commission has proposed several measures to increase efficiency at all stages of the energy chain, targeting final consumption and the building sector, where the potential for savings is greatest. The European Directive 2002/91/EC on the energy performance of buildings (EPBD) and the EPBD recast (Directive 2010/31/EC) constitute the main legislative instruments for improving the energy efficiency of the European building stock. EU member states are strengthening the energy performance requirements and setting more stringent goals for reducing the energy performance of buildings. Accordingly, by 2021 all new buildings must be "nearly zero-energy buildings" while new buildings occupied/owned by public authorities should comply by 2019. National implementation efforts are underway to meet these ambitious goals while determining the optimal cost level for the life cycle of the buildings. All existing buildings that undergo major refurbishment (25% of building surface or value) should meet minimum energy performance standards, while national policies and specific measures should stimulate the transformation of refurbished buildings into "nearly zero energy buildings." Austria, Denmark, Germany, and Sweden have emerged as successful leaders in adopting and implementing the EPBD into national initiatives to promote green buildings. This is well aligned with the EU's goals set in Energy 2020, "A strategy for competitive, sustainable, and secure energy" (http://ec.europa.eu/energy/energy2020), known as the 20-20-20 targets: a reduction in EU greenhouse gas emissions of at least 20% below 1990 levels; an increase to 20% of RES contribution to EU's gross final energy consumption; and a 20% reduction in primary energy use by improving energy efficiency, by 2020.

Over the years, the Intelligent Energy Europe program has provided additional funding sources for projects that support the promotion of energy efficiency and the use of alternative energy sources in buildings. For example, the ecobuildings projects use a new approach when it comes to the design, construction, and operation of new and/or refurbished buildings. This approach is based on the best combination of the double approach: to reduce substantially, and, if possible, to avoid, the demand for heating, cooling, and lighting and to supply the necessary heating, cooling, and lighting in the most efficient way based as much as possible on renewable energy sources, polygeneration (combined heat and power), and trigeneration (combined heat, power, and cooling). Information on existing ecobuilding demonstration projects throughout Europe is available at www.sara-project.net.

In 2009, the European Commission launched its BUILD UP initiative to increase the awareness of all parties in the building chain of the savings potential, which remains untapped. BUILD UP promotes better and smarter buildings across Europe by connecting building professionals, local authorities, and citizens. Its interactive web portal supports Europe's collective intelligence for an effective implementation of energy-saving measures in buildings. The BUILD UP web portal (www.buildup.eu/) shares and promotes existing knowledge, guidelines, tools, and best practices for energy-saving measures, while it also informs and updates the market about the legislative framework in terms of goals, practical implications, and future revisions.

Mesures d'Utilisation Rationnelle de l'Energie, MURE, (www.muredatabase.org) provides information on energy efficiency policies and measures that have been carried out in EU member states along with information on national financial instruments like grants, subsidies, loans, and other related fiscal issues. Dena, the German Energy Agency's (www.dena.de/en/) Subsidy Overview EU-27 offers additional information on grant programs and regulatory frameworks in the EU member states. At this site, information on technologies that generate heat from renewable energies is presented in concise table form.

Incentives that are internal to a project can be created through contractual arrangements. The simplest example is an added fee for added scope in doing the work needed to evaluate green design alternatives. At the other end of the spectrum, some proactive firms offer design review and commissioning services on a performance basis. Construction project insurance offerings have included rebates if no IAQ claims are made a predetermined number of years after construction is completed. If both client and firm are willing, there are many ways to create internal incentives for green engineering design on a project-specific basis. Many individuals and groups have done valuable work in exploring various incentives for green design. *ASHRAE Journal, High Performing Buildings* magazine, and many other industry periodicals carry case studies with green engineering elements. Assembling and familiarizing oneself with the many successfully completed green projects is useful when you must sell and deliver green engineering services.

SUCCESSFUL APPROACHES TO DESIGN

For the design of a sustainable or green project to be successful, it is very important to understand the owner's goals and reasons for him to seek such a level of performance for the project. Once the goal is clear, the design team will have a much better chance to meet it. Possible goals could be certain level of certification of certain rating systems (e.g., an A rating for a high performance building according to ASHRAE's Buidling Energy Quotient [bEQ] energy labeling program, LEED Gold, BREEAM® Excellent, or Green Globes 3 Globes), a certain level of energy savings (e.g., 25% better than ANSI/ASHRAE/IES Standard 90.1), a net zero energy building (e.g., A+ rating according to ASHRAE's bEQ energy labeling program), or a sustainable building with system commissioning without official certification from a rating system. This list is by no means inclusive and is just an illustration of examples encountered on some projects. Below are three examples of client goals and the strategies used to meet those goals.

Sustainable Design with System Commissioning

Institutions and real estate managers throughout the country, recognizing the long-term cost benefits of sustainable design with system commissioning, are successfully implementing changes in how their projects are designed, constructed, and operated. Emory University's Whitehead Biomedical Research Building, for example, uses enthalpy wheels that recover 83% of the energy exhausted by the general exhaust system, captures air-conditioning condensate to displace potable water otherwise used for cooling tower makeup, and captures all the rainwater from the roof for site irrigation, displacing more than 3 million gallons of potable water usage per year. Refer to Chapter 6 for a more extensive discussion on the benefits of the commissioning process and how it can be successfully implemented on a project.

The following sections describe other approaches to the green-building design process that have been successful.

Low-Energy Design Process:
National Renewable Energy Laboratory's Experience

This process was used to design and construct the visitor's center at Zion National Park. The visitor's center was targeted to use 70% less energy than a comparable visitor's center built to Federal Energy Code 10 CFR 435 (CFR 1995). The authors used an integrated design process, including extensive simulations, to minimize the building's energy consumption. The result was a passive solar commercial building that makes good use of the thermal envelope, daylighting, and natural ventilation. Passive, downdraft cool towers provide all the cooling. Two Trombe walls provide a significant amount of heat for the building. After two years of metering, the results show a net energy use intensity of 24.7 kBtu/ft^2·yr (281 MJ/m^2·yr) and a 67% energy cost saving. Low energy use and aggressive demand management result in an energy cost intensity of \$0.43/ft^2 (\$4.63/m^2). For more information, Torcellini et al. (2006) discusses lessons learned related to the design process, the daylighting, the PV system, and the HVAC system.

Zero Energy Design

A net zero building can be a net zero energy building (NZEB), a net zero water building, and/or a net zero wastewater. Although there are various definitions and classifications for NZE in the literature, the best definition is provided in the NREL "Zero energy buildings: A critical look at the definition," 2006 report which defines four types of NZE buildings:

1. A site NZEB produces at least as much energy as it uses in a year, when accounted for at the site.

   For example, if a building consumes

   - 3,000 kWh of electricity, and
   - 100 therms of natural gas (= 2,930 kWh),

   then the building's renewable energy system has to produce 5,930 kWh for the building to be site NZE.

2. A source NZEB produces at least as much energy as it uses in a year, when accounted for at the source.

   For example, if a building consumes

   - 3,000 kWh of electricity (site) = 10,020 kWh source (3000 × 3.34), and
   - 100 therms of natural gas site (or 2,930 kWh site) = 3,076 kWh source (2,930 × 1.05),

   then the building's renewable energy system has to produce 13,096 kWh at source for the building to be source NZE. Assuming PV (electricity), 13,096 kWh/3.34 = 3,920 kWh will need to be produced at site or 33% less than site NZEB. The site-to-source conversion factors used (3.34 for electricity and 1.05 for gas) are U.S. averages as reported by EPA. These factors vary throughout the U.S. and the world.

3. A net zero energy cost building is one where the amount of money the utility pays the building owner for the energy the building exports to the grid is at least equal to the amount the owner pays the utility for the energy used over the year.

   For example, if the building utility bill is

   - 3,000 kWh of electricity at 0.12$/kWh = $360, and
   - 100 therms of natural gas at $1/therm = $100,

   then the building's renewable energy system has to export energy in the cost of $460. Assuming PV (electricity) and the utility company buying at same rate of 0.12$/kWh, the PV system needs to produce 3833 kWh.

4. A net zero emission building produces at least as much emissions-free renewable energy as it uses from emissions-producing energy sources.

To perform the calculations, the emission factors need to be determined, which maybe be difficult to obtain in some regions (e.g., in the State of California, the emission factors are 0.524 lb CO_2/kWh electricity and 13.446 lb CO_2/therm [0.21 kg/kWh] of gas). In addition, although emissions encompass a variety of pollutants, this definition is still limited primarily to carbon dioxide.

An example of a NZEB is the Aldo Leopold Legacy Center in Baraboo, Wisconsin. The building is a carbon-neutral NZEB that uses 39.6 kW rooftop photovoltaic (PV) array to produce about 10% more electricity than that needed annually to operate the building. Specifically, the Legacy Center qualifies as a site NZEB, source NZEB, and emissions NZEB. Another recent example is the National Renewable Energy Laboratory Research Support Facility (NREL RSF) that has demonstrated the production of a net zero energy building that includes plug loads and their data center while still maintaining a high degree of indoor air and environmental quality.

For additional information on the classification of net zero energy buildings, see the online resource listing for the net zero energy buildings database by the U.S. Department of Energy at the end of this chapter.

The following link provides information on incentive programs for all 50 states: www.dsireusa.org/.

ONE DESIGN FIRM'S TEN STEPS TO A NET ZERO ENERGY BUILDING

1. Define the zero you are seeking (site, source, carbon or cost)
2. Establish the design criteria for system performance.
3. Understand the microclimate the of the project site (local climate surrounding the project site).
4. Design the siting, form, and thermal mass of the building to maximize the use of natural energy flows and reduce external loads to the absolute practical minimum.
5. Minimize lighting energy use through effective daylighting.
6. Reduce plug loads as far as possible
7. Employ extremely efficient HVAC systems to handle the remaining loads (including natural ventilation and mixed mode where feasible).
8. Develop accurate predictions of annual building energy consumption.
9. Provide renewable energy systems to offset resulting energy use
10. Commission your systems and ensure proper hand-over to O&M staff

JUSTIFICATIONS FOR GREEN DESIGN

Doing the Right Thing

The motivation and reasons for implementing green buildings are diverse but can be condensed into essentially wanting to do the right thing to protect the Earth's resources. For some, a wake-up call occurred in 1973 with the oil embargo—with the embargo came a realization that there may be a need to manage our planet's finite resources.

Regulations

Society has recognized that previous industrial and developmental actions caused long-term damage to our environment, resulting in loss of food sources and plant and animal species, and changes to the Earth's climate. As a result of learning from past mistakes and studying the environment, the international community identified certain actions that threaten our ecosystem's biodiversity. Consequently, it developed several governmental regulations designed to protect our environment. Thus, in this sense, the green design initiative began with the implementation of building regulations.

Lowering Ownership Costs

A third driver for green design is lowering the total cost of ownership in terms of construction costs, resource management and energy efficiency, and operational costs, including marketing and public relations. Ways to lower costs include providing a better set of construction documents that reduce or eliminate change orders, controlling site stormwater for use in irrigation, incorporating energy-efficiency measures in HVAC&R design, developing maintenance strategies to ensure continued high-level building performance and higher occupant satisfaction, and reducing marketing and administrative costs.

Case studies on commissioning show that construction and operating costs can be reduced from 1 to 70 times the initial cost of commissioning. A recent study by an international engineering firm indicated that treating stormwater on site cost one-third as much as having the state or local government treat stormwater at a central facility, significantly lowering the burden to the tax base.

A 123,000 ft^2 (11,427 m^2) higher education building constructed in 1997 in Atlanta, Georgia—a building that already had many sustainable principles applied during its design—provides an example of how commissioning, measurement, and verification play a critical role in ensuring that the sustainable attributes designed into the building are actually realized at $1.00/ft^2 ($10.80/m^2) savings when recommissioned, lowering the total cost of ownership. Recommissioning identified several seemingly inconspicuous operational practices that were causing higher-than-needed consumption in the following areas: chilled-water cooling, by 40%; steam, by 59%; and electricity, by 15%. Implementing continuing measurement and verification ensures that the building will continue to perform as designed, again, lowering the total cost of ownership.

More esoteric is the cost of unsatisfied occupants, which includes administrative costs, marketing costs due to more frequent tenant turnover, or increased business cost due to absenteeism or reduced productivity, as well as the impact on marketplace image. Marketplace image is a significant driving force in promoting green/sustainable design. Green/sustainable projects provide owners an opportunity to distinguish themselves in public as well as promote their business or project in order to obtain the desired result. Promotion of green/sustainable attributes of a project can help with public relations and help overcome community resistance to a new project.

Increased Productivity

Another driver for green design is the recognition of increased productivity from a building that is comfortable and enjoyable and provides healthy indoor conditions. Comfortable occupants are less distracted, able to focus better on their tasks/activities, and appreciate the physiological benefits good green design provides.

A case study conducted by Pacific Northwest Laboratory points out many interesting observations about human response to daylight harvesting, outside views, and thermal comfort. The study, which compares worker productivity in two buildings owned by the same manufacturer, illustrates both the positive and negative impact the application of green design principles can have on human productivity. (Heerwagen et al. 1995)

The first building is an older, smaller industrial facility, divided into offices, and a manufacturing area. It has high-ribbon windows around the perimeter walls in both office and manufacturing areas, providing only limited daylight harvesting. It has an employee lounge, a small outdoor seating area with picnic tables, and conference rooms.

The newer facility is 50% larger with energy-efficient features, such as large-scale use of daylight harvesting, energy-efficient fluorescent lamps, daylight harvesting controls, direct digital HVAC controls, environmentally sensitive building materials, and a fitness center at each end of the manufacturing area.

The study focuses on individual quality-of-work issues and the manufacturer's own production performance parameters. The study provides a mixed review of green design and gives insight on what conditions need to be avoided. While occupants perceived satisfaction and comfort stemming from daylighting and outside views, they also expressed complaints regarding glare and lack of thermal comfort. The green building studied did not appear to have controlled the quality of daylight through proper glazing selection, which may be the cause of the complaints about glare and thermal comfort.

Subsequent chapters discuss specific design parameters relative to building envelope design, including daylight harvesting, energy efficiency, and thermal comfort.

Filling A Design Need

There are increasing numbers of building owners and developers asking for green design services. As a result, there is considerable business for design professionals who can master the principles of green design and provide leadership in this arena.

Some publications that demonstrate the drivers of green design include *Economic Renewal Guide* (Kinsley 1997), *Natural Capitalism* (Hawken, Lovins, and Lovins 1999), and the *Earth From Above* books (Arthus-Bertrand 2002), and *Green to Gold: How Smart Companies Use Environmental Strategy to Innovate, Create Value, and Build Competitive Advantage* (Esty, Winston 2006).

REFERENCES AND RESOURCES

Published

Arthus-Bertrand, Y. 2002. *Earth From Above*. New York: Harry N. Abrams.

CFR. 1995. Energy conservation voluntary performance standards for commercial and multi-family high rise residential buildings; mandatory for new federal buildings. Code of Federal Regulations 10 CFR 435, Office of the Federal Register, National Archives and Records Administration, Washington, DC.

Esty, D.C., and A.S. Winston. 2006. *Green to Gold: How Smart Companies Use Environmental Strategy to Innovate, Create Value, and Build Competitive Advantage*. New Haven, CT: Yale University Press.

Hawkin, P., A. Lovins, and L.H. Lovins. 1999. *Natural Capitalism*. Little Brown.

Heerwagen, J.H., J.C. Montgomery, W.C. Weimer, and J.G. Heubach. 1995. Assessing the human and organizational impacts of green buildings. Pacific Northwest Laboratory, Richland, WA.

Kinsley, M. 1997. *Economic Renewal Guide*. Colorado: Rocky Mountain Institute.
Torcellini, P., S. Pless, M. Deru, B. Griffith, N. Long, and R. Judkoff. 2006. Lessons learned from case studies of six high-performance buildings. Technical report, NREL/ TP-550-37542. National Renewable Energy Laboratory, Golden, CO.

Online

Building Performance Institute of Europe
  www.bpie.eu/financial_instruments.html.
Dena
  www.dena.de/en/.
EuporeanEuropean Commission, BUILD UP program
  www.buildup.eu/.
European Commission (DG TREN), SARA Project—Sustainable architecture applied to replicable public access buildings
  www.sara-project.net.
High Performing Buildings magazine,
  www.hpbmagazine.org.
MURE (Mesures d'Utilisation Rationnelle de l'Energie), Database on energy efficiency policies and measures,
  www.muredatabase.org.
North Carolina Solar Center, DSIRE
  www.dsireusa.org/.
U.S. Department of Energy, Energy Efficiency and Renewable Energy, Zero Energy Buildings
  http://zeb.buildinggreen.com.

CHAPTER FOUR

THE DESIGN PROCESS— EARLY STAGES

OVERVIEW

The design process is the first crucial element in producing a green building. For design efficiency, it is necessary to define the owner's objectives and criteria, including sustainable/green goals, before beginning the design in order to minimize the potential of increased design costs. Once designed, the building must be constructed, its performance verified, and it must be operated in a way that supports the green concept. If it is not designed with the intent to make it green, the desired results will never be achieved.

Figure 4-1 conceptually shows the impact of providing design input at succeeding stages of a project, relative to the cost and effort required. The solid

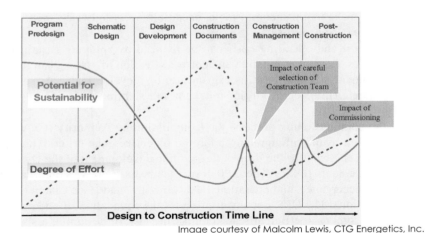

Image courtesy of Malcolm Lewis, CTG Energetics, Inc.

Figure 4-1 The potential for sustainability and the impact of commissioning.

curve shows that it is much easier to have a major impact on the performance (potential energy savings, water efficiency, maintenance costs, etc.) of a building if you start at the very earliest stages of the design process. The available impacts diminish thereafter as you proceed through the subsequent design and construction phases. A corollary to this is that the cost of implementing changes to improve building performance rises at each successive stage of the project (cost is shown as the dotted curve of this graph).

Designers are often challenged and sometimes affronted by the idea of green design, for many feel they have been producing good designs for years. The experience of many is that they have been forced to design with low construction costs in mind, and when they offer opportunities to improve a building's design, they are often blocked by the owner due to budget constraints. The typical experience is that owners will not accept cost increases that do not show a return in potential savings in five years or less, and many demand 18 months or less. Many owners, especially owners who own the project for life, allow longer return on investments and use life-cycle cost parameters to lower the total cost of ownership.

Achieving green or sustainable design goals requires a different approach than has been customarily applied. Engineers and other designers are asked to become advocates, not just objective designers. Some have expressed the view that significant reductions in energy usage and greenhouse gas emissions will never occur by simply tweaking current practice. In other words, simply installing high-efficiency systems or equipment will not reduce energy usage sufficiently. Sustainable design requires designers to take a holistic approach and go beyond designing for just the owner and building occupants. They need to look at the long-term environmental impacts the development of a building will create. This may make many uncomfortable, because it seemingly asks them to go beyond their area of expertise.

Both first cost and operating costs can be reduced by applying sustainable/green principles. Correct orientation and correct selection of glazing can reduce HVAC equipment size and cost, as can the use of recycled materials such as crushed concrete (in place of virgin stone) for soil stabilization and structural fill.

Using the commissioning process, the Commissioning Authority (CxA) can assist the design team with agreeing upon the objectives, the criteria, and the sustainable/green goals for the project through the development of the Owner's Project Requirements (OPR). The OPR forms the basis from which all design, construction, acceptance, and operational decisions are made (see Chapter 6 for more information). The OPR document provides the foundation of understanding for the designers to efficiently accomplish the task of designing a sustainable/green project within the owner's business model and the constraints and limitations of the project. This process allows improved design efficiency and better team integration, because there are clear objectives and criteria established before design begins. While sustainable/green principles can lower first cost as well as operational costs, the soft costs of design do increase slightly, due to the additional design and coordination efforts required of the team.

Starting in predesign and carrying through to postoccupancy is essential for the success of a sustainable/green design. It starts with examining every aspect of the process (e.g., the owner's site selection, building configuration, architectural elements, and efficient construction and operation) and can only occur with an integrated approach. Defining the OPR document containing the project goals, even before site selection if possible, is the suggested starting point.

At the very start of the project, green design goals have to be discussed, correlated to the owner's objectives and criteria, agreed to, and in fact, embraced by the extended project team. This is often done in a charrette format or simply a session spent discussing the issues. As these goals are defined, they are included in the OPR document.

Goals for a project traditionally include the functional program, leasable or usable area, capital cost, schedule, project image, and similar issues. The charrette simply puts environmental goals on a plane with the capital cost and other traditional goals.

One of the goals may be to achieve the environmental goals at the same or similar capital cost. (As with any goals, the environmental goals should be measurable and verifiable.) Another one of the goals may be a specific green-building rating system target and an energy target, or perhaps an energy target alone.

A typical set of goals for a green design project might be as follows:

1. Achieve a level of energy use at least 50% lower than the U.S. Department of Energy (DOE)-compiled average levels for the same building type and region, both projected and in actual operation. (Actual energy numbers may be adjusted for actual versus assumed climatic conditions and hours of usage.)
2. Achieve an actual peak aggregate electrical demand level not exceeding 4.5 kW/ft^2 (50 kW/m^2) of building gross area.
3. Provide at least 15% of the building's annual energy use (in operation) from renewable energy sources. (Such energy usage may be discounted from the aggregate energy use determined under the above-mentioned Goal 1.)
4. Taking into account the determinations of Goals 1, 2, and 3, assess the impact of the lesser net energy use on raw energy resource use (including off-site energy use) compared to that of a comparable but conventional building, including the changed environmental impacts from that resource use, and verify that the aggregate energy and environmental impacts are no greater.
5. Achieve a per capita (city) water usage that is 40% lower than the documented average for this building type and region.

6. Achieve an aggregate up-front capital cost for the project that does not exceed x dollars/ft^2 (m^2) of building gross area, which has been deemed by the project team to be no higher than 102% of what a conventional building would cost.
7. Recycle (or arrange for the recycling of) at least 60% of the aggregate waste materials generated by the building.
8. By means of postoccupancy surveys of building users conducted periodically over a five-year period, achieve an aggregate satisfaction level of 85% or better. Surveys solicit occupant satisfaction with the indoor environment as to the following dimensions: thermal comfort, air quality, acoustical quality, and visual/general comfort.
9. Obtain a gold-level U.S. Green Building Council (USGBC) LEED certification for the building.

(For an example of what one major firm has done, see the Digging Deeper sidebar, "One Firm's Green-Building Design Process Checklist," featured later in this chapter.)

THE OWNER'S ROLE

Of all of the participants, it is the owner who is the most crucial when it comes to making a green building happen. With the owner's commitment, the design, construction, and operating teams will receive the motivation and empowerment needed to create a green design.

Key design team members can—and should—attempt to educate the owner on the long-term benefits of a sustainable/green design, particularly if the owner is unfamiliar with the concept. After all, experienced design team members are in the best position to sell the merits of green design. However, to be effective, such a commitment on the part of the owner must be made early in the design process.

Specific roles that an owner can fill in making a sustainable/green design effort successful include the following:

- Expressing commitment and enthusiasm for the green endeavor
- Establishing a basic value system (i.e., what is important, what is not)
- Selecting a CxA
- Participating in selection of design team members
- Setting schedules and budgets
- Participating in the design process, especially the early stages
- Maintaining interest, commitment, and enthusiasm throughout the project

The owner on a project could be a corporation or small business, hospital, university or college, office building developer, nonprofit organization, or an individual. In any case, that owner will have a designated representative on the building project team, presumably one who is very familiar with the owner's views and philosophy and can speak for that owner with authority.

THE DESIGN TEAM

Setting It Up

One of the first tasks in a sustainable/green design project is forming the design team and the commissioning team. This team should include the design team leader (often the architect), the owner, the CxA, the design engineers, and operations staff. Much of the design team's successful functioning depends not just on having ideas about what should go into the project but on being able to analyze the impact of the ideas quickly and accurately. It's likely that a large part of this analysis will be completed by one of the engineers on the project.

A traditional project team includes the following members:

- Owner
- Project manager
- Architect
- HVAC&R engineer
- Plumbing/fire protection engineer
- Electrical engineer
- Lighting designer
- Structural engineer
- Landscaping/site specialist
- Civil engineer
- Code consultant

An expanded project team for a sustainable/green design with commissioning would also include the following members:

- Energy analyst
- Daylighting consultant
- Environmental design consultant
- CxA
- Construction manager/contractor
- Cost estimator
- Building operator
- Building users/occupants
- Acoustical consultant

The preceding lists the possible roles that might need to be filled on a reasonably large design project. Some roles may not be applicable or even needed on certain types of projects (e.g., civil engineer or landscaping/site specialist), and other roles may not be feasible in the early stages of project (e.g., building operator, building users, or code enforcement official). Further, the variety of roles does not mean that there needs to be an equal number of distinct individuals to fill them; one individual may fill several roles (e.g., the architect often serves as project manager, the HVAC&R engineer as plumbing engineer or energy analyst, the electrical engineer as lighting designer, and a contractor on the team as cost estimator). Likewise, depending on the type of project, there could be other specialists as well.

There are a few roles that are particularly important in green design including energy analyst, environmental design consultant, and CxA, which are discussed below.

Energy Analyst. Although this role has existed for some time, it assumes a much more intense and timely function in sustainable/green design, as there is a need to quickly evaluate various ideas (and interactions between them) in terms of impact on energy. These can range from different building forms and architectural features to different mechanical and electrical systems. The person in this role must be intimately familiar with energy and daylight analysis modeling tools and able to provide feedback on ideas expressed reasonably quickly. In short, he or she is a much more integral part of a sustainable/green design team than in a traditional design effort. In this respect, for a sizable project, it might be difficult for a single person to fill this role plus another.

Environmental Design Consultant (EDC). As owners begin to request sustainable/green buildings from the design professions, a new discipline has emerged: the EDC. The role of this person is to help teams recognize design synergies and opportunities to implement sustainable and green features without increasing construction costs. When the CxA is also the EDC, the owner and designers benefit from an improved integration of the design process across disciplines, with the intent of creating an outcome with much lower environmental impact and higher user satisfaction. Leading projects show that this can often be accomplished without adding cost. The EDC has input in areas such as site, water, waste, materials, Indoor Environmental Quality [IEQ], energy, durability, envelope design, renewable energy, and transportation. The CxA documents these objectives and criteria into the OPR document. Although this guide focuses primarily on those areas pertinent to the HVAC&R design professional, it is becoming evident that this profession must broaden its sphere of concern in order to contribute meaningfully to the creation of sustainable/green buildings.

The CxA already has a collaborative relationship with the designers and, as an EDC, works with the HVAC&R team and others throughout the process to meet the owner's objectives and criteria. For example, a design reviewer may raise questions for the team to consider, such as the following:

- Is the building orientation optimized for minimum energy use?
- Is the combined system of building envelope, including glazing choices and the HVAC system, optimized for minimum energy use and lowest life-cycle cost?

- Have passive, active and renewable strategies for optimizing building energy usage been optimized?
- Are the loads, occupancy, and design conditions properly described?
- Are the proposed analytical tools adequate for the task of computing life-cycle costs and guiding design decisions?
- Is the proposed integrated design approach going to deliver excellent air quality to occupants under all conditions?
- Is the proposed integrated design approach going to deliver thermal comfort to occupants under all conditions?
- Is the proposed integrated design approach going to be easy to maintain? Is there enough space for mechanical equipment and adequate access to service and perhaps to eventually replace it?
- Is the proposed mechanical approach going to give appropriate control of the system to users?
- Is the proposed integrated design approach going to consume a minimum amount of parasitic energy to run pumps and fans?
- Are there site or other conditions likely to impact the mechanical system in unusual ways?
- Has the impact of the building on the site and surroundings been identified and taken into account?
- Are the proposed systems properly sized for the loads?

While it may seem that the role of the EDC is very similar to that of the energy analyst, the roles differ in that the EDC is more of a question-asker or issue-raiser, similar to the CxA asking questions and suggesting strategies or solutions for the team to consider. The EDC's brief is broader and more comprehensive in scope; his or her role is to take a step back from the project and ask the broader questions regarding the environmental impact of the project.

CxA. (Please refer also to Chapter 6.)

The Team's Role

Green design requires owners to make decisions sooner, design documents to be more complete and comprehensive, the construction process to be better coordinated, and operators to be better trained in maintaining facilities. All of this will impact the viability and success of a green project endeavor. Contractors not familiar with this project model may sound the cost and schedule alarm due to their inexperience in the new procedure. First-time application of sustainable development principles can result in slightly higher first cost, but this phenomenon will reverse itself as teams gain experience and improve their learning curve. As the building industry becomes more familiar with applying these principles, lower costs of ownership will result.

In addition to the standard tasks associated with a design project, the design team is responsible for developing and implementing new concepts that will create a green project. For most, this will require learning on their own time, becoming familiar with new advances in software tools, green materials, and alternative systems. There is an abundance of information; advances are occurring daily in the development of green products and materials as well as processes. The speed of these changes requires designers to continuously add to their knowledge base.

The greatest challenge to accomplishing green design is creating a team organizational structure that provides the following:

- Criteria for assessing how green the project should be
- Strong leadership through the green design process to integrate team members
- Clearly defined objectives that result from careful examination of design alternatives, costs, and schedule impacts
- Documentation of success
- Strong leadership by experienced green-building practitioners leading the team through the decision process
- Definition of what tasks are required to accomplish green design
- Identification of who is responsible for each of the tasks
- Identification of when tasks must be completed, so as not to impede the design process or the project schedule
- Establishment of criteria for the selection of green design features considered for incorporation into a project
- Assistance with integrating selected green design goals into the construction documents
- Definition of the level of effort required for each of the green project goals
- Help to enable contractors overcome psychological and physical constraints
- Establishment of how to track, measure, and document the success of accomplished project goals

The designers must also help inform their clients that there are costs for the depletion of resources to be consumed beyond the cost of extraction. The practice of looking only at simple payback when analyzing alternatives based on extraction cost has never been realistic because there is no way to replace many resources at any cost.

Currently, most design teams are eager to develop green designs but lack the experience of actually integrating green design into their projects. The addition of an experienced EDC will shorten the learning curve by helping them integrate their extensive knowledge, which will result in a cost-effective practical green design that meets the owner's requirements. In addition, most teams struggle with what makes a design green, how to incorporate green design principles, and the logistics of incorporating these principles into the design. Green design creates a need for a broader involvement of disciplines and a wider range of experience to ensure that a wider range of input and participation gets factored into the decision-making process.

The project team—from the initial concept to the construction documents, construction, and building operations—must work as an integrated unit to achieve the goals set by the owner's objectives and criteria, creating better project performance, which is a basic principle of green design. This model will require the project team to investigate new approaches and process more information than ever before, as they strive to increase performance and lower the total cost of ownership. The decision-making process must change from the traditional hierarchical method, with an emphasis on lowest first cost, to an integrated method focused on life-cycle cost. To achieve this requires close collaboration of the project team combined with innovative thinking among all disciplines.

The design team's responsibility as part of the project team is to assist the owner with setting sustainability/green goals that often include

- life-cycle cost optimization of energy-consuming systems, materials, and maintenance;
- systems integration and maintainability;
- minimization of environmental impact;
- documenting basis of design; and
- assisting with training of building operations and management staff during commissioning.

THE ENGINEER'S ROLE

The HVAC&R engineer is a crucial player in the design of a green building. In fact, it is virtually impossible (and certainly not cost-effective) to design a green building without major involvement from that discipline. The HVAC&R engineer must get outside the normal box in which he or she lives and become more involved in the why of a design, as well as the how. This means moving beyond responding to questions asked by others and participating in the decision making regarding how project goals will be achieved.

Engineers help analyze the various options to be considered, create mathematical computer models that are used to judge alternatives, provide creative input, and assist with development of new techniques and solutions. The HVAC&R engineer can be invaluable in helping the architect with building orientation considerations, floor plate form and dimension, and deciding which type of glazing will provide the maximum quantity of natural light, while at the same time analyzing the heat transfer characteristics of the glazing options. The HVAC&R engineer can also help the architect select structural systems and exterior walls to use thermal mass features to reduce equipment needs. Working with the electrical engineer and architect, the

HVAC&R engineer can offer ideas and various options, such as incorporating daylighting and lighting controls to reduce artificial light when natural light is available, which, in turn, can result in lower cooling requirements and lowered HVAC requirements to meet peak load. Smaller equipment size translates into reduced structural and electrical requirements, lower operating and maintenance costs, and lower construction costs, all of which lower the total cost of ownership.

The plumbing engineer, working with the structural and civil engineers, and the landscape architect can reduce the facility's potable water, sewer, and stormwater conveyance requirements. Some examples are waterless urinals or use of stormwater or graywater for irrigation of vegetation or for flushing toilets. Depending on the type of building, water from condensate can be used for graywater applications or for cooling tower makeup. The design engineer must weigh the benefits of water-cooled versus air-cooled condensers and the water versus electrical energy consumed by each. The engineer must examine the site climate, determine what alternatives and strategies can best be applied, and develop life-cycle analyses to guide the owner through the decision process posed by the maze of complex issues surrounding green design.

PROJECT DELIVERY METHODS AND CONTRACTOR SELECTION

Successful projects depend upon the entire team of players involved: architect, engineers, program managers, construction managers, owner's representatives, facilities personnel, building users, and contractors. It is assumed that all parties will be ethical, reliable, diligent, and experienced. There are a number of project delivery methods that could be used to deliver the design and construction of a project. The major methods, to be briefly discussed here, are the construction manager approach, the design/bid/build approach (D/B/B), the design/build (D/B) approach, and the public-private partnership (P3) approach. It is important that the project delivery method be chosen early in the project for the same reasons that it is important when considering green design options. The discussion of the three methods below is intended only to relate to the effect this decision will have on the success of the optimized design and operation of the building.

Construction Manager Method

The construction manager method is the process undertaken by public and private owners in which a firm with extensive experience in construction management and general contracting is hired during the design phase of the project to assess project capital costs and constructability issues. This is especially important when considering design alternatives that are being considered in an effort to deliver a high-performance building to the client. The initial design process often includes a project definition stage, or programming, in which the owner works with the design professionals, the CxA, and the construction manager to define the specific scope of the project. The design professional uses this information to prepare a set of bidding documents that the construction manager uses to obtain bids from qualified subcontractors. The lowest responsible price is usually selected and the contractor then constructs the project.

- *Advantages:* Budget control, buy-in of green concepts by contractors.
- *Disadvantages:* Perceived lack of competitive general contractor and subcontractor pricing, innovative systems could be shelved due to overly conservative first-cost estimates.

D/B/B Method

D/B/B construction is the traditional process undertaken by public and private owners. The initial design process often includes a project definition stage, or programming, in which the owner works with the design professionals to define the specific scope of the project. The design professional uses this information to prepare a set of bidding documents. The bid documents are then available for qualified contractors and subcontractors to prepare pricing. The lowest responsible price is usually selected and the contractor then constructs the project.

- *Advantages:* Usually results in the lowest first cost at the outset of the construction of a project.
- *Disadvantages:* No contractor buy-in to green process and concepts, prequalification of contractors is difficult to do well.

D/B Method

D/B construction is typically a response to a request-for-proposal (RFP) developed by an owner. The RFP is usually a document that defines the general scope of the project and then solicits price proposals to accomplish this work. The work effort to prepare the specific design of the project is to be included in the D/B offering. The D/B team usually consists of an architect, engineers for the various disciplines involved, a general contractor, and the trade subcontractors. This entire team should be in place until the project is turned over to the owner.

As the D/B team develops the design, it must respond to the premises defined in the original scope of work.

- *Advantages:* Can result in lowest first cost, agreement by design/construction team with regard to design and building operation concepts.
- *Disadvantages:* Can result in uneven distribution of risk among team members, can result in loss of design team members as owner advisors.

Public-Private Partnerships (P3)

The use of P3 is well established in Canada, and is beginning to gain favor in the United States and other countries. P3 are playing a bigger role in building and capital projects across all areas of government, including power generation, energy delivery, water and wastewater facilities, waste disposal, transportation, communications, education and health facilities, and public service buildings. Recently, its popularity has led to P3 being used more frequently in smaller-scale developments across Canada, such as schools, courthouses, and hospitals.

Bridging documents are provided to guide an integrated and facilitated P3 process by

- providing narratives that describe, in conceptual terms, the requirements for the new facility;
- outlining general performance and prescriptive requirements, operations requirements, programmatic space requirements and room data sheets, space standards, adjacency requirements, conceptual blocking and stacking;
- defining business terms; and
- proposing criteria for evaluation of submitted proposals.

The P3 process allows market forces to play out, with the opportunity for the government entity to decide later on funding options. Projects are delivered quickly. Competition on a wide variety of services is balanced by an integrated approach, while risk transfer from the public to private sector in the areas of pricing, compliance, and operations benefits the stakeholders of the government entity.

Information from this section was taken from website resources listed in the references section of this chapter, specifically Mondaq (2003) and the California Debt and Investment Advisory Commission (2008).

Factors in Choosing an Approach

In all of the above scenarios, the team or contractors should be prequalified to perform the work prior to a request for pricing. It's important to confirm that the team or contractors fulfill the following requirements:

- Experience in similar work.
- Record of past performance by responsible references.
- Financial capability to complete the project.
- Experienced staff available to work on the project.

A record of exemplary performance and fiscal capability may be more important than experience in similar work. In any case, it is valuable to preselect the teams or contractors from whom you request pricing. Successful projects occur due to careful planning and implementation.

What Does *Integrated Design* Mean?

One of the key attributes of a well-designed, cost-effective green building is that it is designed in an integrated fashion, wherein all systems and components work together to produce overall functionality and environmental performance. This has a major impact on the design process for HVAC-related systems, as conceptual development must begin with HVAC system integration into the building form and into the approaches being taken to meet other green-building aspects. For example, consider the following:

- HVAC systems that use natural ventilation and underfloor air distribution, often used in green buildings, can have major impacts on building form.
- Other building energy innovations, such as daylighting, passive solar, exterior shading devices, and active double wall systems, often have significant impacts on the design of the HVAC system.
- On-site energy systems that produce waste heat, such as fuel cells, engine-driven generators, or microturbines, affect the design of HVAC systems so that waste heat can be used most effectively.

Beyond these form-giving elements, there are many other specific features of a green building that affect (or are affected by) HVAC systems to achieve the best overall performance. Some of these strongly impact HVAC system conceptual design, and some require only minor adjustments to HVAC specifications. Such features might include the following:

- Using the commissioning process to document the owner's defined sustainable/green objectives and criteria and assist the project team to deliver
- Effective use of ventilation (and indoor air quality [IAQ] sensors linked to the ventilation system) to improve IAQ
- Provision of user controls for temperature and humidity control
- Reduced system capacities to reflect lower internal loads and building envelope loads
- Use of thermal energy storage systems to reduce the overall size of the chiller plant equipment, such as chillers, cooling towers, and pumps (to save capital costs, ongoing energy, and operational costs, and reduce outdoor noise levels during the daytime)
- Selection of non-ozone-depleting refrigerants
- Reduction and optimization of building energy usage below the levels of ANSI/ASHRAE/IES Standard 90.1-2013, *Energy Standard for Buildings Except Low-Rise Residential Buildings*-based (ASHRAE 2013) codes or other applicable state and local energy codes (reduction of as much as 40% to 50% below Standard 90.1 becoming more common and encouraged)

- Use of reclaimed water for cooling tower makeup, and minimization of cooling tower blowdown discharge to the sanitary sewer system
- Testing of key systems, especially HVAC systems

Key Steps

The integrated design process (IDP) includes the following elements:

- Ensuring that as many of the interested parties as possible are represented on the design team as early as possible—including not only architects, engineers, and the owner (client), but also CxA, construction specialist (contractor), cost estimator, operations/maintenance person, and other specialists (outlined below)
- Interdisciplinary work among architects, engineers, costing specialists, operations people, and other relevant persons right from the beginning of the design process
- Discussion and documentation by the owners and the design team of the relative importance of various performance and cost issues, and the establishment of a consensus on these matters between client and designers, and among the designers themselves
- Provision of a design facilitator (or EDC) to suggest strategies for the team to consider, as well as a CxA to raise performance issues throughout the process and to bring specialized knowledge to the table
- Addition of an energy specialist to test various design assumptions through the use of energy and daylight simulations throughout the process and to provide relatively objective information on a key aspect of performance
- Addition of subject specialists (e.g., for daylighting or thermal storage) for short consultations with the design team
- Clear articulation of performance targets and strategies to be updated throughout the process by the owner and the design team

Iterative Design Refinement

The design process requires the development of design alternatives. To come up with the most effective combination, these alternatives must be evaluated, refined, evolved, and finally optimized. This is the concept of iterative design, wherein the design is progressively refined over time, as shown in Figure 4-2.

Often fee and schedule pressures lead the designer to want to lock in a single design concept at the beginning of the project and stick with it throughout. But this precludes the opportunity to come up with a better system that reflects the unique combination of loads and design integration opportunities for this specific building. This better design usually evolves during schematics and early design development in the iterative process.

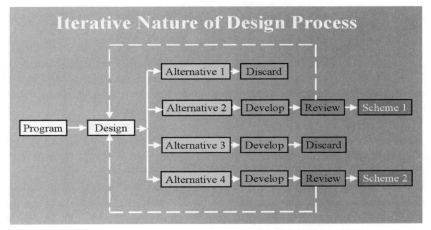

Image courtesy of Malcolm Lewis, CTG Energetics, Inc.

Figure 4-2 The iterative design process.

CONCEPT DEVELOPMENT

The Big Picture

Designers should always keep in mind the three major steps for achieving a sustainable/green design:

- Reduce loads.
- Apply the most efficient systems.
- Look for synergies.

Reduce Loads. If you have a building with a normal 500 ton cooling load, is it possible to provide a more comfortable environment with, for example, only 300 tons? The team would have to really work to achieve this. Solar loads on the building would have to be reduced; lighting loads to the space would also have to be lowered; maybe the building could use daylight rather than electric light during the day, so the building shape would be influenced. The site for the building, its shape, its thermal mass, and its orientation could all work together to reduce the cooling load. Early and quick modeling can provide interesting information to assist decisions.

Similar things could be done with the heating load. Does the winter sun provide much of the building's heating needs during the day? With design changes, could the sun do more?

These considerations in a hierarchal design process are not typically brought to the attention of the HVAC&R engineer, who could significantly help improve the energy efficiency instead of just calculating the loads, and could apply energy-efficient systems. Significant reduction of utility consumption and environmental impact cannot occur by simply doing the same old job just a little bit better.

The HVAC&R or energy engineer can make a positive contribution to the success of a sustainable/green design. The value of the engineer to the project is significantly increased, and the results are reduced heating and cooling loads, as well as overall utility consumption.

One attitudinal approach to building design that the designer should strive toward is making the building inherently work by itself. Building systems are there to fine tune the operation and pick up the extreme design conditions. In contrast, buildings are often traditionally designed like advanced fighter aircraft: if the flight computers are lost, the pilot cannot fly the plane.

Apply the Most Efficient Systems. This is the world of the energy-efficient engineer. This is the area where ASHRAE generally operates. While it is very important, it is not enough by itself. Many times, the most efficient system is not cost effective or the owner O&M personnel are not yet ready to operate and maintain such a system.

Look for Synergies. The preceding two major steps have the potential of increasing capital costs. Therefore, a wonderful, energy-efficient building might not be built due to high first cost, or the value-engineering knives might come out and cut the project back to a traditional, affordable project—proof perhaps that green cannot work. Part of the solution to get around this syndrome is to look for synergies of how building elements can work together. This also relates to the cost transfer mentioned earlier.

If a building has a large amount of southern exposure, exterior-shading devices might significantly reduce the summer solar load, while still admitting lower angle winter sun. Daylight (but not direct sun) would allow for shutting off the electric lights on sunny days. The HVAC system for the south perimeter zones could be significantly reduced in size and cost as the simultaneous solar and electric lighting loads are reduced. Indeed, the very nature of the HVAC system might well be simplified due to the significant load reduction. Resulting cost savings can be used to pay for some or all of the additional treatments.

A major benefit of an integrated design that is on budget is avoiding wasting a lot of time on elemental payback exercises and value engineering (and cost cutting), because you are on budget. Many of the integrated solutions work so that if you save by cutting out an element, such as the exterior shades, there is an additional cost in another area such as the size of the HVAC system.

The Nitty-Gritty

The success of green design starts with establishing the project's goals and objectives, defining roles and responsibilities of each team member, fostering communication among design team members, developing a decision-making process, and clarifying the level of effort that will be required by each member of the team. It is recommended that a workshop be conducted to introduce the team to the process of integrating sustainable development principles and determining how decisions establish objectives and criteria that are documented in the OPR and provide guidance to the team in the delivery of the project. Objectives and criteria are tracked and used to form concise benchmarks to gage and document the team's success in meeting the established goals. Commercially available software can assist with cost-benefit analysis, design coordination, energy and daylight analysis, and organizing process design.

The creation of documentation supporting both the decision-making process and the results of decisions is important in guiding and determining the success of green design efforts, as well as in establishing what was or was not successful and the reasons why. Like a business plan or construction plan, it is important to measure milestones so that adjustments can be made to correct course deviations in reaching the goals. Sustainable/green design objectives and criteria defined in the OPR should also identify the criteria on which life-cycle cost analyses are based. Also, any assumptions made must be recorded in the basis of design document so comparison against actual performance can be measured. Learning from the deviations that occur will allow teams, as well as individuals, to grow from the experience.

The owner's sustainable/green design objectives and criteria documented in the OPR provide the organizational structure required for successful projects. Software tools available today also increase communication within a team, help stimulate innovative thinking, and help teams optimize design trade-offs by grouping related issues.

The team must develop consensus criteria that might include the following:

- Selecting a site that minimizes environmental impacts
- Using existing infrastructure to the maximum extent possible to avoid building additional infrastructure to support the project
- Minimizing the impact of automobiles and the infrastructure required to support them, such as parking, roads, and highways
- Developing high-performance buildings that enhance occupant productivity and comfort, minimize energy and water consumption, and are durable and recyclable at the end of their useful service life

Based on the consensus criteria selected, identify potential goals. Once goals are identified, develop tasks necessary to obtain these goals, including studying the impacts these goals will have upon:

- project cost, schedule, and energy and water usage;
- IEQ, operational and maintenance costs, life of the building, and occupant productivity; and
- environmental impacts at the end of the building's or facility's useful service life.

Next, assign roles and responsibilities by identifying who is responsible for each task, when the task must be completed, and in what chronological order tasks must be completed, so as to facilitate the tracking and management of the sustainable/green design process. The design and construction commissioning plans and checklists can significantly assist a team in accomplishing their goals and minimizing wasted effort.

A good OPR provides the team the information needed to guide the team's decisions, measure the team's success, and document changes and the reasons for change in the project.

EXPRESSING AND TESTING CONCEPTS

Expressing concepts is very important in green design, because that is the way ideas and intentions are communicated to the owner and others on the design team. This is especially true because green design requires the close and active participation of many different parties.

There are three ways of expressing concepts in the design of buildings: the traditional verbal means and the diagrammatic or pictorial means, and the third (which has come of age more recently along with computers) is the modeling means.

Verbal

Both the written and the spoken word play an especially important part in green design. Because there are many meetings or charrettes where ideas are explored and intentions voiced, getting across what is expressed accurately assumes significance. Then, succinctly and clearly putting down on paper what has been expressed (memorializing it) is also critical to the various team members, as they each go about filling their respective roles. To illustrate, read Chapter 6, "Commissioning," to learn how important a written record of what happened during design (i.e., design intent, assumptions, etc.) is to the successful follow-through of a well-executed green design.

Diagrammatic/Pictorial

The use of diagrams, sketches, photos, renderings, etc., a tried and true method of communicating a lot of information, continues to be an essential part of green design (Figure 4-3). The old adage "one picture is worth a thousand words" most certainly applies here. But there is now a relatively new way of creating a picture (Figure 4-4) of a building, an energy system, or a year of operations—modeling.

Modeling

This computer-age technique plays such an important part in green design because of its speed, accuracy, and comprehensiveness.

Everyone is familiar with how speedy computers are, once the input data are entered. The slow part in this process is the human analyst, the one who converts intentions and ideas into computer-modeling program input, which is why it is so important for that analyst to be very conversant with the modeling process. This is especially true for load and energy calculations that impact HVAC&R systems. The team has an idea, and they want an answer fast as to how well that idea would work.

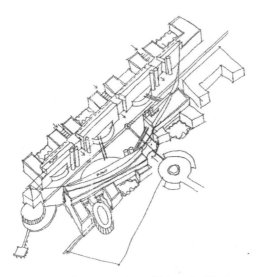

Image courtesy of Integrated Environmental Solutions (IES).

Figure 4-3 Example of diagrammatic building sketch.

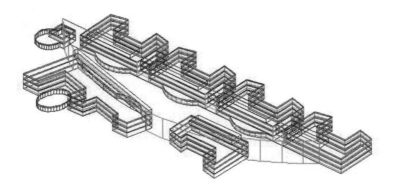

Image courtesy of Integrated Environmental Solutions (IES).

Figure 4-4 Example of computer simulation model derived from Figure 4-3.

Most would also acknowledge that computers are accurate: they do not make careless mistakes. Again, if there are inaccuracies, they usually come from the human side, which is why the analyst must be an expert at avoiding garbage in.

Modeling programs have another advantage, especially the more sophisticated ones: they are comprehensive in what they can analyze simultaneously. The human mind can only accommodate so many ideas or concepts at once without getting confused and bogged down; a properly conceived model will not get confused and can provide answers that may be counterintuitive. As an example, a good modeling program can track heat gain from lights, plug loads, and solar energy, along with heat loss from the building envelope and infiltration, do it for every hour or every day of the year in whatever weather conditions are assumed, take into account mass effects of the structure, and still yield an accurate answer. It would be impossible for the human mind to do this in a reasonable time, unless it was very good at guessing.

See Figure 4-5 for an example of daylighting analyses output.

BUILDING INFORMATION MODELING (BIM)

The approach to building design is moving away from conventional computer-aided design software to follow the way design software has evolved in the manufacturing sector. Your building can now be a working digital prototype. Sustainable design is driving this solution to ensure that the buildings we are erecting are

Image courtesy of Integrated Environmental Solutions (IES).

Figure 4-5 Example of daylight analysis software output—conceptual design.

designed, constructed, and operated in a manner that minimizes their environmental impact and are as close to self-sufficient as possible. This technology is known as *building information modeling* (BIM).

BIM is a software tool that uses a relational database together with a behavioral model that captures and presents building information dynamically. In the same way that a spreadsheet is automatically updated, a change in the parametric building modeler is immediately reflected everywhere. This means, for example, revising windows from one type to another not only produces a visually different graphic representation, in all views of your building, but the insulation value of the glazing (R-values) is also revised. Due to the integration of BIM with existing tools of analysis, running energy calculations is greatly simplified. Visualization tools are sophisticated and allow three-dimensional views and walk-throughs of the building.

Multiple design options can therefore be developed and studied within a single model early in the design process to not only see the building and provide conventional documentation for construction, but also to interact with other software to perform energy analysis and lighting studies. (See Figure 4-6 for an example.)

Using BIM helps with the demanding aspects of sustainable design, such as solar applications and daylight harvesting, and also automates routine tasks such as documentation. Schedules are generated directly from the model. If the model changes, so do the schedules. Architects are able to filter and sort material quantities automatically, bypassing the manual extraction/calculation process required. Determining the percentages of material reuse, recycling, or salvage can be tracked and studied for various sustainability design options.

Engineers can perform year-round sun studies to understand when the building is provided natural shading and, thus, optimize the orientation of the building to maximize afternoon shading from the hot summer sun and properly size roof overhangs to minimize solar heat gain. Engineers can then reduce the capacity of cooling systems, demonstrating the building is exceeding the baseline building energy requirements.

An energy analysis using a two-dimensional computer-aided design file requires manually removing the building values/areas from the floor plans and entering said data into an energy simulation application. The data for supporting green design are captured during the design process in BIM and are extracted as necessary.

The energy performance for the baseline model averages the results of four simulations during one year of operation. One simulation is based on the actual orientation of the building on the site; the others rotate the entire building by 90°, 180°, and 270°, which enables the proposed design to receive credit for a well-sited building. This is easily accomplished when using BIM.

The software carries all the data required for building a structure; it understands the data. It quantifies building materials so designers can move walls or insert windows and almost instantly have the building's data on energy performance, daylight harvesting, and, perhaps most importantly, costs shift accordingly in real time. Designers can test the life-cycle performance of brick walls versus concrete walls. The digital representations behave like buildings and not just drawings.

BIM is transforming the way we work and will enable further endeavors in the practice of creating sustainable, cost-effective buildings. A more detailed overview of how to use these tools during the conceptual design process is provided in Chapter 8.

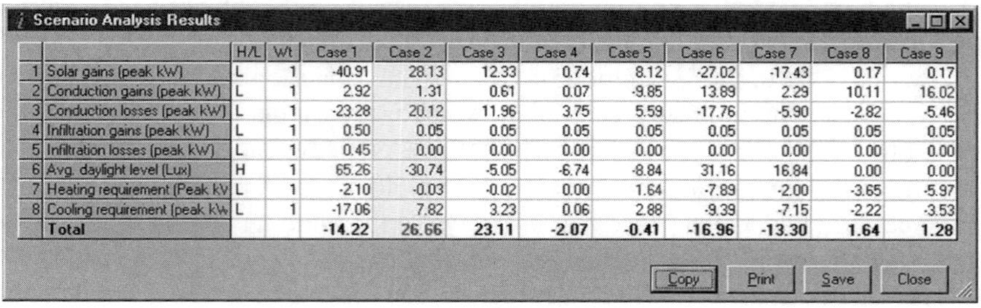

#	Scenario Analysis Results	H/L	Wt	Case 1	Case 2	Case 3	Case 4	Case 5	Case 6	Case 7	Case 8	Case 9
1	Solar gains (peak kW)	L	1	-40.91	28.13	12.33	0.74	8.12	-27.02	-17.43	0.17	0.17
2	Conduction gains (peak kW)	L	1	2.92	1.31	0.61	0.07	-9.85	13.89	2.29	10.11	16.02
3	Conduction losses (peak kW)	L	1	-23.28	20.12	11.96	3.75	5.59	-17.76	-5.90	-2.82	-5.46
4	Infiltration gains (peak kW)	L	1	0.50	0.05	0.05	0.05	0.05	0.05	0.05	0.05	0.05
5	Infiltration losses (peak kW)	L	1	0.45	0.00	0.00	0.00	0.00	0.00	0.00	0.00	0.00
6	Avg. daylight level (Lux)	H	1	65.26	-30.74	-5.05	-6.74	-8.84	31.16	16.84	0.00	0.00
7	Heating requirement (Peak kV)	L	1	-2.10	-0.03	-0.02	0.00	1.64	-7.89	-2.00	-3.65	-5.97
8	Cooling requirement (peak kW)	L	1	-17.06	7.82	3.23	0.06	2.88	-9.39	-7.15	-2.22	-3.53
	Total			-14.22	26.66	23.11	-2.07	-0.41	-16.96	-13.30	1.64	1.28

Image courtesy of Integrated Environmental Solutions (IES).

Figure 4-6 Example of multiple design options analysis.

Evaluating Alternative Designs

Modeling of alternative designs is made easier by the plethora of modeling tools available today. The chosen model should meet specific requirements depending on the level of accuracy needed. Error can be introduced into modeling if users forget the garbage in/garbage out rule. To reduce the chance of inappropriate or misunderstood input, modeling programs can use input and output formats that allow the user quick reality checks. Various stages of design, requirements, and associated tools are shown in Table 4-1.

Table 4-1: Summary of Available Analysis/Modeling Tools

Stage	Requirements	Tools	Reality Checks
Scoping	• Quick analysis • Comparative results • Reduce alternatives to consider • Control strategy modeling	• System Analyzer™ • Modified bin analysis (where load is not entirely dependent on ambient conditions) • eQUEST • IES VE	• Operation cost/ft² (m²) • Payback or other financial measure
System design	• Accurate output • Industry-accepted methods	• HAP • TRACE 700 • Elite Design • IES VE	• cfm/ft² (L/s/m²) • cfm/ton (L/s/kWR)
Energy/cost analysis	• Accurate • Complies with energy cost budget method requirements of Standard 90.1-2013 • Industry-accepted methods • Flexible • Allows modeling of complex control strategies	• EnergyPlus • DOE • HAP • TRACE 700 • SUNREL • IES VE • eQUEST	• Btu/h· ft² (kWh/m²)/yr • Operation cost/ft² (m²) • Payback or other financial measure
Monitoring	• Simplicity • Intuitive interface • Systemwide • Interoperable	• BACnet® compatible automation systems	• Trended operating characteristics • Benchmark comparisons (such as system kW/ton [kW/kWR])

NATIONAL RENEWABLE ENERGY LABORATORY'S NINE-STEP PROCESS FOR LOW-ENERGY BUILDING DESIGN

1. Create a base-case building model to quantify base-case energy use and costs. The base-case building is solar neutral (equal glazing areas on all wall orientations) and meets the requirements of applicable energy efficiency codes such as Standard 90.1 (ASHRAE 2013) and ANSI ASHRAE Standard 90.2-2007, *Energy Efficient Design of Low-Rise Residential Buildings* (ASHRAE 2007).
2. Complete a parametric analysis to determine sensitivities to specific load components. Sequentially eliminate loads from the base-case building, such as conductive losses, lighting loads, solar gains, and plug loads.
3. Develop preliminary design solutions. The design team brainstorms possible solutions that may include strategies to reduce lighting and cooling loads by incorporating daylighting or to meet heating loads with passive solar heating.
4. Incorporate preliminary design solutions into a computer model of the proposed building design. Energy impact and cost effectiveness of each variant are determined by comparing the energy with the original base case building and with the other variants. Those variants having the most favorable results should be incorporated into the building design.
5. Prepare preliminary set of construction drawings. These drawings are based on the decisions made in Step 3.
6. Identify an HVAC system that will meet the predicted loads. The HVAC system should work with the building envelope and exploit the specific climatic characteristics of the site for maximum efficiency. Often, the HVAC system is much smaller than in a typical building.
7. Finalize plans and specifications. Ensure that building plans are properly detailed and that specifications are accurate. The final design simulation should incorporate all cost-effective features. Savings exceeding 50% from a base-case building are often possible with this approach.
8. Rerun simulations before design changes are made during construction. Verify that changes will not adversely affect the building's energy performance.
9. Commission all equipment and controls. Educate building operators. A building that is not properly commissioned will not meet the energy-efficiency design goals. Building operators must understand how to properly operate the building to maximize its performance.

ONE FIRM'S GREEN-BUILDING DESIGN PROCESS CHECKLIST

- Create an integrated, cross-disciplinary design team that is committed to sustainability and is aware of environmental issues, especially those that the building impacts.
- Precharrette and charrette meetings should include the project owner, CxA, architects and landscape architects, engineers, an energy engineer with experience in computer simulation of building energy consumption, environmental consultant, facility occupants and users (including purchasing, human resources, and managers), facility manager, contractors (when hired), interior designer, local utility representatives, cost consultant, and other specialty consultants.
- Review how occupants will use the project and any of the client's unique operational characteristics. Based on the this information, decide which sustainable/green design principles are aligned with the owner's overall objectives, criteria, and activities that will provide the most positive impact, in order to provide focus and priority.
- Define environmental standards, goals, and strategies early in predesign, and clearly state these in the OPR.
- Translate the owner's objectives and criteria into the construction documents.
- Channel development to urban areas with existing infrastructures, protecting greenfields and preserving habitat and natural resources.
- Increase localized density to conform to existing or desired density goals by using sites that are located within an existing minimum development density of 60,000 ft^2/acre (13,779 m^2/ha), or a 0.5 mile (0.8 km) area, with at least ten services, such as banks, restaurants, retail businesses, etc.
- Channel development to areas with existing transportation infrastructure that provides non-automobile-dependent choices.
- Select a location for the project within 0.5 miles (0.80 km) of a rail station (e.g., commuter rail, light rail, or subway), within 0.25 miles (0.40 km) of two or more bus lines, or in a live, work, walk, mixed-use environment.
- Reduce the overall building footprint and development area.
- Make important decisions on the mechanical load, daylighting, solar absorption, response to local climate and environment, and key building elements at the beginning to define subsequent decisions.
- Establish benchmarks/performance targets as a reference point.
- Integrate recycling systems into every aspect, from reusing existing building materials to purchasing new materials.
- Design for disassembly at the end of the building's useful life.

> **Digging Deeper**
> - Educate contractor and subcontractor in sustainable practices.
> - Evaluate and benchmark OPR, including sustainable goals at the same intervals as budgets and schedules.
> - Have a CxA as a team member to help ensure that owners' objectives and criteria, including sustainability goals, are met at each stage; benchmark and document success for future reference; and advocate for environmental choices during the course of the project.
> - Tie compensation for the architect and design team to achieved building performance.

REFERENCES AND RESOURCES

Published

ASHRAE. 2005. ASHRAE Guideline 0-2005, *The Commissioning Process*. Atlanta: ASHRAE.

ASHRAE. 2010. ANSI/ASHRAE/IES Standard 90.1-2013, *Energy Standard for Buildings Except Low-Rise Residential Buildings*. Atlanta: ASHRAE.

Online

California Debt and Investment Advisory Commission, *Public-Private Partnerships: A Guide to Selecting a Private Partner*. March 2008.
www.treasurer.ca.gov/cdiac/publications/p3.pdf.

Mondaq. December 2003.
www.mondaq.com/canada/article.asp?articleid=23737.

CHAPTER FIVE

ARCHITECTURAL DESIGN AND PLANNING IMPACTS

OVERVIEW

The architect and landscape architect play particularly critical roles within any design team whose charge is to meet an owner's sustainability design intent. Careful considerations are important for the site selection, building orientation and form, the structure's envelope, and the arrangement of spaces and zoning. These considerations are covered in this chapter, with the exception of the site selection which is covered in Chapter 2, "Sustainable Sites."

Early site and architectural decisions can positively propel or alternatively severely hinder the ability of the rest of the team to optimize their roles in producing a high-performing building. Architectural design choices directly affect energy-related first costs for construction and the total embodied energy in construction materials. These choices further impact the operational energy use, resulting in carbon emissions as well as the well-being and comfort of the occupants. The importance of working as a team with all of the design stakeholders throughout the entire design/construction process cannot be overemphasized.

This chapter is intended to help designers make design decisions that best effect positive sustainable/green project outcomes. A design process is provided as a suggested procedure.

DESIGN PROCESS FOR SUSTAINABLE ARCHITECTURE

The process in designing energy-conserving architecture involves the following:

- Analyzing climatic conditions at the macro- and micro-site levels
- Identifying passive and active strategies applicable for the climate
- Designing the building envelope, shape, and orientation to reduce energy loads while providing comfort
- Determining room positions and grouping of rooms into HVAC zones to reduce energy loads while, again, providing desired comfort levels

- Researching case studies and building energy load data for similar building types in the same climate zone to identify typical resource-consuming systems and benchmarked energy demands
- Selecting potential optimal strategies
- Applying computer-based simulation studies and related data sources to evaluate energy consumption against accepted metrics and benchmarks to demonstrate the performance-based sustainable outcomes for alternative designs.

The identified steps are not discreet actions but are applied in a more integrative fashion, as described below.

Climate Studies and the Contribution of Passive Strategies and Active Systems

After selecting a site, the next step in sustainable architecture and building systems design is researching the macro- and microclimatic conditions. Desired collected data includes ranges of design dry-bulb and wet-bulb temperatures (or humidity ratios), solar exposure, and wind velocity (predominant direction and intensity). Seasonal, as well as daily ranges should be noted.

The climate will dictate if the building will be either heating dominated or cooling dominated. The dominant need will prescribe the relative importance of heating, cooling, and ventilating conditioning strategies. Once temperature and humidity data is collected, plot their average minimum and maximum ranges on a psychrometric chart or use an available computer program to plot this data with its frequency of occurrence for outdoor conditions. Overall strategies and means to extend the comfort zone for both passive and active systems have historically been identified and plotted on psychrometric charts by several researchers as bioclimatic charts. This was originally done by Victor Olgyay in 1963, modified by Baruch Givoni in 1969, and then further developed in collaboration by Baruch Givoni and Murray Milne in 1979.

The strategies identified on the bio-climatic charts include sun shading, thermal high mass with or without night flushing, direct evaporative cooling, natural ventilation cooling, passive solar direct gain with low or high mass, humidification, wind protection, conventional air conditioning, and conventional heating. Free university-developed software is also available to plot a year's set of points and then evaluate these strategies by the percentages of time they can most effectively be applied. See Figure 5-1 for an example using this software for the city of Santa Rosa, California.

Once identified, key comfort zone strategies can be prioritized based on their percentage contribution to retaining comfort zone conditions. If an active system is the highest-rated strategy for restoring the comfort zone, then reducing the energy load for such equipment is a high priority. Decisions on design strategies should also include occupancy and seasonal use conditions, (e.g., a summer camp project

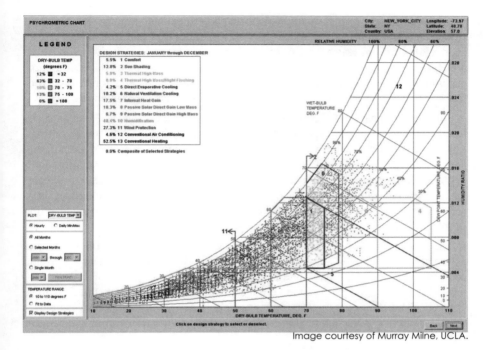

Image courtesy of Murray Milne, UCLA.

Figure 5-1 Climate Consultant Software file report demonstrating the correlation of a location's year-long ambient air state points with applicable HVAC strategies.

in a heating-dominated climate should not be designed for heating loads, as a climate study might initially suggest.)

Strictly speaking, bioclimatic analysis does not take into account the building type, the building's intended use, nor projected internal load conditions; however some software developers are beginning to make efforts in this area, (e.g., the UCLA developed Climate Consultant Software includes internal heat gain as a design strategy.)

A complete climate analysis includes studies of the solar exposure, the seasonal effects of wind, and the availability of rainwater, which are briefly discussed in the next section and in more detail in Chapter 2, "Sustainable Sites."

Building Form, Orientation, and Envelope Considerations

Decisions for designing a building's form, orientation, and envelope construction are informed by climate data, building site's existing or potential topography with its landscape details, and potential optimization of natural on-site resources.

In predominately cold climates, a compact cube shape with a tight, well-insulated envelope is optimal to reduce heat loss from conduction and infiltration. However, elongation in an east-west axis will increase exposure to the winter sun, thus increasing winter heat gain. Reducing overall glazing area, specifying high-performing windows and placing the bulk of windows on the equator facing exposure while minimizing glazing on other orientations is also recommended.

For a predominately hot climate, consider elevating the structure (for under airflow) and otherwise creating large shaded surface areas in contact with the outside. This can be accomplished with the creation of open courtyards, deep wraparound porches and/or multiple building wings with several openings for ventilation paths aligned with prevailing wind patterns. For climates which contribute to overheating, architects also need to reduce direct solar gain on the equator-facing and west-facing surfaces. North- and east-facing surfaces rarely contribute to overheating from solar radiation.

The design of thermal mass and weight of building materials internally and externally is a function of both climate and building use. "Elephant" building designs, those with high mass and heat absorbing materials, are suitable for buildings with high internal gains and in climates with large diurnal temperature swings. "Butterfly" buildings are those with lightweight materials suitable for building uses with low internal gains and moderate climates that would benefit from natural ventilation.

Charting the apparent sun path throughout the year (e.g., on the 21st day of each month) and adjusting the irradiation values using the ASHRAE clear-sky radiation method is useful for determining the potential for solar energy collection (for thermal or photovoltaic collectors) and passive solar heating building design. Sun angle data can be used for designing light shelves and other architectural strategies for daylighting to augment artificial light and also to design shading to reduce unwanted solar heat gain. See Chapter 12 on energy/water sources for more information on using solar energy. The user is advised to use the new ASHRAE clear-sky solar radiation model for estimating potential solar irradiation levels. This model was introduced in the 2009 *ASHRAE Handbook—Fundamentals* and revised in the 2013 *ASHRAE Handbook—Fundamentals*.

Wind rose charts can be consulted to find the seasonal direction and speed of the wind for a particular site. They are used to determine the feasibility for collecting wind energy and to analyze a site for protecting the building from unwanted wind exposure during the heating season or to determine options for channeling wind for increased ventilation for the cooling season. If wind rose charts are not directly available for a location, they can be developed with software or manually drawn using data from a local airport wind velocity-recording station. See Chapter 12 on energy/water sources for more information on using wind energy.

Optionally, earth berms can be created or banks of trees or shrubs can be planted to help block wind to protect a building from wind-induced convective heat losses. Earth berms and shrubbery can also be used to redirect or focus wind for use in cooling or enhancing ventilation. For both cold-dominated and hot-dominated climates, earth sheltering in contact with the building's envelope also diminishes temperature extremes. Sinking the building into the ground or designing the building to extend into a hill can provide cooling when outdoor temperatures are hot. Earth sheltering also acts as a buffer to the cold, when berming is located on the windward side of the building and particularly when the berms or hills are situated such that they allow the equator-facing side of the building to still be exposed (for maximizing solar exposure). Use of a green roof provides multiple benefits for all climates: a tempering effect for cold or hot conditions, an offset for storm water which otherwise would be headed for a storm drain, and reduced heat island effect. See Chapter 2, "Sustainable Sites," for more discussion on further optimizing site conditions.

Rainfall data can be used to determine rainwater collection potential for use in irrigating plants, toilet flushing, or for other uses. This is an opportunity for both saving energy and conserving water. Rainfall data is also used for planning site water retention design schemes such as for rain gardens and bioswales. The collection, use, and maintaining of water on a site reduces or eliminates the need for storm water drain systems infrastructure.

Several computer programs are available to simulate alternatives and to assist in determining the optimal final building shape, alternative footprint, orientations, insulation levels, and glazing distribution for minimizing energy consumption. Computer programs simulating the effects of mass are more difficult to find. For the design of structures over three stories, an additional ASHRAE source that lists parameters and methods for low-energy use is the ANSI/ASHRAE/USGBC/IES Standard 189.1, *Standard for the Design of High Performance Green Buildings*. This standard specifies minimum R-values, maximum U-values, and F-factors for the building-envelope insulation and windows based on climate zone.

A challenge exists for architects to meet constraints in shape, orientation, and form, since these attributes are ones for which they have traditionally had great latitude and design freedom to explore varying options for aesthetic and building-use considerations. For sustainable design, architecture studies must be initiated with the knowledge of preferred shape/envelope/footprint ratios and orientation for optimization of energy use while providing comfort.

Placement and Zoning of Internal Spaces

The placement of rooms and zoning for heating and cooling has a big impact on energy use and comfort of the occupants. However, prior to placing and zoning internal spaces, consideration should be given to minimizing the building size by designing spaces that are flexible or dual purposed, (e.g., sharing an area as both a dining hall and lecture room can potentially reduce the total required built volume for a project).

Arrangement and zoning of internal spaces is ideally based on orientation of exterior walls, space function, and occupancy/equipment diversity. In general, the following guidelines apply:

- Place rooms where occupants are primarily sedentary such that direct sunlight can reach them. Use closets, storage areas and hallways as buffers to other spaces and place them on orientations where solar exposure is scarce.
- Cluster rooms with similar function together.
- Group spaces with similar comfort needs and schedules together. Specify separate zones for infrequently used areas so that they are conditioned fully only when occupied.

Optimizing the placement of internal spaces for energy efficiency is somewhat dependent on their relative conditioning requirements. Consider thermal buffering for cold climates. One way of adding thermal buffering is the placing of unconditioned spaces or those requiring relatively lower comfort levels between those spaces requiring higher comfort conditions and the exterior walls (e.g., placing corridors, storage rooms, datacom rooms, and utility rooms on the non-equator-facing side of a building in a temperate or cold climate). This action reduces thermal losses from the more sensitive conditioned spaces where people occupy the space.

High-energy-consumption areas can also be buffered by locating them as interior spaces. A lab space, one that has high energy use due to air replacement demands, could be located in the interior of a building with offices and other areas bordering it as long as the installation needs and access for ventilation for fume hoods can be met. A recreation pool/natatorium space in a cold climate is another candidate for interior space. It could be bordered by storage and exercise rooms, as shown in Figure 5-2.

A more formal thermal buffer zone (TBZ) can be provided by constructing a double skin façade system (DSF). A DSF, also called a *dual-skin wall system* or *envelope system* is constructed by providing a glazed wall a short distance from a conditioned space-enclosing wall (usually on the equator-facing side of a building but may also be installed on the opposite facing outside wall). Air movement due to natural convection and stack effect between the dual wall skins can contribute to reducing both the cooling and heating load. In summer operation at the Yazaki World Headquarters in Canton, Michigan, cooler air in the below-grade crawlspace is drawn into the interstitial space between the outer and inner wall and rises by natural convection, pulling stale air from conditioned space (through opened windows on the inner wall) up and out of the building. In winter, air is introduced at a low level, is solar heated and rises between the walls (with inside windows closed), providing a thermal blanket. In some buildings, the tempered air is ducted into a conventional air handler or heat pump to eventually supply the conditioned space with heat.

Figure 5-2 Partial plan showing buffering of a high energy use space.

There are many variations in DSF design. The interstitial space varies in dimension and design. Some DSFs have a solar tower at the top to increase the convective action. Movable blinds or photovoltaic panels may be added. Many retrofitted commercial buildings in Europe and elsewhere have successfully used DSF principles.

Glazing and daylighting should be considered in view of total building energy utilization. In general, costly photovoltaic energy sources should not be used to power artificial lighting when daylighting can be provided to reduce electrical lighting loads. Based on studies with whole-building energy-analysis software in more extreme cold climates, providing an overuse of daylighting by increasing windows and skylights will result in an overall increase in the total energy consumed by the building due to the low thermal performance of glazing systems. High levels of insulation and careful placement of fenestration are central to energy conservation in cold climates. The use of alternative-energy systems to compensate for uninformed design decisions demonstrates poor design and resource planning. The decisions regarding the envelope design are central to reducing building energy loads and take precedence over alternative-energy-related power generation as a priority.

Alternative-Energy Potential and Prioritization of Strategies

The development of alternative-energy strategies begins with examining the magnitude and availability of solar and/or wind energy to the site. The use or application of alternative-energy technologies should be paired with the high-load demand conditions, as well as cost considerations. If a building type, such as a hospital, has a high domestic hot-water demand load and is also located in a climate with strong solar radiation conditions, then evacuated tubes would be a prudent choice to reduce fossil fuel use for heating water. If a building has a need to be shaded from excessive solar radiation and high electrical loads, using photovoltaic panels as part of shading system on a southern wall would be a possible solution. Solar-air and water-based panels, evacuated tubes, photovoltaic panels and wind turbines are all alternative-energy systems which can be used to reduce fossil fuel use. Alternative-energy costs can be made affordable for clients when tax incentives, rebates, or low interest loans are available and should be part of initial feasibility studies. The reader is referred to Chapter 12 on alternative-energy source for more information on applying alternative-energy sources.

Although not an alternative-energy source, the use of geothermal heat pump systems is another way to reduce fossil fuel use.

PRIORITIZATION AND STUDIES OF EXISTING BUILDINGS' ENERGY RESOURCE UTILIZATION

Evidence-based design or performance-based outcomes require definable problems based on data. The increasing levels of adoption of performance-based metrics by municipalities, states, and corporations (e.g., those used for building labeling) will eventually place responsibility on designers to validate design proposals with appropriate data. Once climatic and site conditions and the alternative-energy potential have been identified, the energy demands based on building type should be known and design alternatives can be theoretically evaluated based on building energy simulation studies and other supporting data.

Use of Case Studies of Existing Buildings

Referenced buildings can be used as benchmarks to predict typical energy consumption for a building's various purposes (i.e., energy used for heating, cooling, lighting, domestic hot water, and specialized equipment). Case studies are readily available from professional organizations such as the ASHRAE, AIA, and USGBC. When choosing benchmarks, the buildings selected should be similar in

- building function,
- building size,
- expected occupancy, and
- climate zone.

Using benchmarks along with microclimatic data should narrow and define the priorities and lead the designer to identifying the most viable design strategies for resource conservation. Benchmarking with high-performing reference buildings will provide the designer with exemplary targets.

Use of Computer-Simulation Studies and Energy

The evaluation of alternative sustainable designs can be augmented by the use of whole-building energy simulation software. Ease of use is now possible with interoperable databases linking BIM 3D models with object-oriented software and embedded information where building data is exported directly into the simulation software. Alternatives and various design options for form, shading, massing, insulation, fenestration, alternative-energy potential, etc., can be explored with direct and rapid results indicating the impact on energy use. Unfortunately, not all building system types, use conditions, and environmental characteristics can be modeled using such software. See the section titled, "The Role of Energy Modeling During Conceptual Design" in Chapter 8, "Conceptual Engineering Design-Load Determination," for an in-depth discussion of this topic.

INTENTIONS IN ARCHITECTURE AND BUILDING SYSTEM DEVELOPMENT

There are several philosophies and approaches that may be applied by building designers for sustainable architecture. These vary from theoretical and abstract to those with more measurable, defined outcomes and several may be combined to meet sustainability goals. All used must include a concern for meeting human as well as ecological needs and rights. After all, the architect and the engineer hold both the safety and welfare of building occupants in their hands and in striving for sustainability, the welfare of the environment with its effect on humans and humans' effect on the earth. The philosophies and approaches which follow are offered for consideration by designers.

Ecosystem/Bio-Regionalism. Understand and design for local ecological systems first and paramount (i.e., preservation of ecologically sensitive areas/endangered species). Knowledge of the ecosystem should precede human development actions and prevent harmful effects to it.

Human Vitality. The intent is to design for human needs based on performance and productivity but constrained by a proenvironmental outlook. When logging efforts in the underdeveloped countries lead to soil erosion, flooding, and other negative effects on the ecosystem and the human occupants, constraints to development are needed.

Circle of Life Principle. Minimize waste as is accomplished in nature, where all is recycled (i.e., waste in one process becomes food in another). The *Cradle-to-Cradle* concepts by William McDonough exemplify this approach. The consequences of human development should be thought of as part of, or akin to, natural processes of plant waste which renew the earth in some manner.

Biomimicry. Consider nature as an example and model with its symbiotic relationships between organisms, protective skins, regenerative nature, and continuous nurturing processes. Today's thermal solar collectors imitate the polar bear. The dark or selective surfaced absorber plate absorbs solar energy just like the polar bear's black skin (yes, it is really black). The bear's transparent and clear hollow core hair is replaced by an insulating air space and glazing, which assists in allowing relatively short wave sunlight through while reducing long wave infrared radiative losses. The open source "Ask Nature" through the Biomimicry 3.8 Institute explores several ideas for applying biomimicry (www.asknature.org). In architecture, examples of applying biomimicry can be found in organic architecture initiated by Frank Lloyd Wright in the late nineteenth and used by the Japanese Metabolists in the 1970s. A recent example is in the literal organic growing of houses by Mitchell Joachim. He has a molecular biology lab in his architecture office and prints three-dimentional, test-tube-grown products. The reader can view videos of his lectures on the Internet by searching his name with "growing houses."

Conservation and Renewable Resources. Energy is a critical resource. We must promote the use of clean, renewable, and non-carbon-emitting energy sources. The basic tenants of the environmental movement are reduce, reuse, and recycle materials and use renewable, rather than fossil, energy.

Holistic Theory. Collaborate with other disciplines and promote integrated design practice to guide the process. Proponents attempt to challenge the rules, think outside of the box, set aside conventional solutions, and use innovation in design.

Sustainable Design Intent. Use basic sustainable processes to conserve water, energy, etc. This may best be accomplished using a metrics-based approach that clearly defies sustainable criteria, measures, and outcomes. Implement sustainability standards (e.g., ANSI/ASHRAE/USGBC/IES Standard 189.1) and apply building labels to measure success (e.g., ASHRAE Building Energy Quotient [bEQ], LEED, Green Leaf, Green Globes, Go Green, GB Tools, ENERGY STAR, etc.) Refer to Chapter 7 for building energy label information.

The reader may also consider using the ten "Design Strategies for Sustainability" defined by the man who coined the term *sustainability*, John Tillman Lyle, in his book, *Regenerated Design for Sustainable Development*.

These strategies overlap those stated above and include the following:

1. Let nature do the work.
2. Consider nature as both model and context.
3. Aggregate rather than isolate.
4. Match technology to the need.
5. Seek common solutions to disparate problems.
6. Shape the form to guide the flow.
7. Shape the form to manifest the process.
8. Use information to replace power.
9. Provide multiple pathways.
10. Manage storage (as a key to sustainability).

BUILDING-TYPE GREENTIPS

The GreenTips in this chapter are called Building-Type GreenTips. The intent is to give the design engineer a very general idea of concepts that are specific to certain building types. References are included for more detailed study and analysis of design options.

Performing Arts Spaces

GENERAL DESCRIPTION

Performing arts spaces include dance studios, black-box theaters, recital halls, rehearsal halls, practice rooms, performance halls with stages and fixed seating, control rooms, back-house spaces, and support areas.

HIGH-PERFORMANCE STRATEGIES

Acoustics

1. Place less acoustically sensitive spaces around more sensitive spaces as acoustic buffers.
2. Work closely with the acoustic consultant, structural engineer, architect, and construction manager to integrate strategies that eliminate the distribution of vibration and equipment noise from the HVAC systems to the performance spaces.
3. Understand the different conditions for noise criteria.
4. Locate equipment as far away from acoustically sensitive spaces as practical. This includes rooftop air-handling units, which should not be installed above acoustically sensitive rooms.
5. The architect needs to provide enough space for the HVAC engineer to design duct systems to minimize noise transfer to acoustically sensitive spaces. Sound levels can be attenuated by the following: lowering air turbulence through proper duct design with lowered velocities, smooth bends, preferred duct aspect ratios of 3:1 or lower; gradual diameter transitions; placing of terminal devices as far from elbows and takeoffs as possible; and designing appropriately long duct length leads to and from fans and bends. Absorptive liners can be added to attenuate high-frequency noise. These measures require space. Fan vibration transmission is reduced by flexible duct connectors, applying dampening lagging to ducts, low-pressure

drop sound attenuators and silencers, or fiberglass lined plenums at least ten times the duct cross-sectional area. For high-frequency noise, noise-cancellation technology is used.
6. Do not route piping systems through or above spaces that are acoustically sensitive.

Energy Considerations

1. Consider using demand-controlled ventilation for high-occupancy spaces.
2. Consider using heat recovery for spaces served by air-handling units (AHUs) with 100% outdoor air capability or over 50% outdoor air component.
3. Consider strategies that allow the significant heat gain from the theatrical lighting equipment to stratify, rather than handle all of the equipment heat gain within the conditioned space zones in the building.
4. Because of the significant variation in the cooling load throughout the day, incorporating a thermal energy storage system into the central plant design will reduce the size of the chiller plant equipment, saving capital costs and energy and operational costs.
5. Consider the use of desiccant wheels to remove moisture to avoid the need for reheat in performance halls.

Occupant Comfort

1. Consider low-velocity underfloor-air-distribution (UFAD) strategies for large halls with fixed seating. High-velocity air entering under the feet of occupants causes discomfort due to convective heat losses, whereas low-velocity air should not provide such discomfort.
2. Consider separating stage-area air distribution from seating-area air distribution in halls.
3. Consider humidification control for all spaces where musicians and vocalists will practice, store musical instruments or sensitive equipment, and perform.

KEY ELEMENTS OF COST

1. If properly integrated, an underfloor air distribution system (UFAD) should not add significant capital costs to the project.
2. Heat recovery strategies should be assessed using life-cycle analyses. All components of the strategy must be taken into account, including the negative aspects such as adding fan static pressure and, therefore, using more fan energy when enthalpy wheel, desiccant wheel, or heat pipe strategies are considered.

SOURCES OF FURTHER INFORMATION

ASHRAE. 2013. *ASHRAE Handbook—Fundamentals,* Chapter 8, Sound and Vibration. Atlanta: ASHRAE.

ASHRAE. 2011. *ASHRAE Handbook—Applications,* Chapter 48, Noise and Vibration Control. Atlanta: ASHRAE.

ASHRAE 2013. *UFAD Guide: Design, Construction and Operation of Underfloor Air Distribution Systems.* Atlanta: ASHRAE.

ASHRAE Building-Type GreenTip #5-2

Health Care Facilities

GENERAL DESCRIPTION

Healing patients is the primary mission of health care organizations and the facilities where they practice or deliver those services must share and support that same mission. Health care facilities represent a wide range of space types that can consist of some doctors' offices that involve little to no special infrastructure, to research and specialty hospitals and diagnostic centers that involve an array of specialized medical equipment and supporting infrastructure components that cost into the tens of millions of dollars. Certainly, energy use and the other sustainability areas will be significantly impacted by the type of health care facility.

Health care facilities and their systems must be flexible to accommodate remodeling and changes to the delivery of health care services. That is true more than ever looking forward to the anticipated changes to reimbursement programs. Changes are also driven by the advancement of medical technology and practices. Sustainable practices that are selected need to be capable of changing with the changes in the facility. The degree of flexibility is often an economic consideration. Another key health care consideration is system reliability and uptime. When sophisticated systems are evaluated for installation, care should be taken to ensure excellent equipment reliability and ease of maintenance.

The tips provided here are based on the common inpatient hospital-type facility considerations. Many outpatient facilities are similar to other office building types, and their green HVAC tips are likely to apply for some of these heath care facilities. Certainly, for some facilities, such as outpatient surgery facilities, it will be helpful to also review the office buildings GreenTip.

Design teams must have extensive inpatient facility experience in addition to having the green/sustainable experts. Many of the sustainable certification programs, guidebooks, and references promote an integrated design and delivery approach to incorporate the sustainable features. By using an integrated design team of knowledgeable and experienced professionals who will carefully plan and execute sustainable options, time, cost, and redesign will be minimized.

HIGH-PERFORMANCE STRATEGIES

The suggestions provided are focused on the uniqueness of inpatient health care facilities. They do not represent the vast array of sustainable ideas and tips that apply generally to all building types. In addition, the recommendations are more focused on those that impact HVAC. For recommendations and tips outside of HVAC, a few sources include U.S. Green Building Council's LEED for Healthcare, *Green Guide for Healthcare* (www.gghc.org/), and proposed ASHRAE Standard 189.3P, *Design, Construction, and Operation of Sustainable, High-Performance Health Care Facilities*. Another organization to investigate is Practice Greenhealth (https://practicegreenhealth.org/), and other publications to consider are *GreenSource: The Magazine of Sustainable Design* (www.construction.com/greensource/), *Health Facilities Management* (www.hfmmagazine.com/hfmmagazine/), and *HERD (Health Environments Research and Design Journal)* (https://www.herd-journal.com/).

Safety and Infection Control

1. Consider high-efficiency particulate air (HEPA) filtration for all air-handling equipment serving the facility. Double HEPA filters may be used to compensate for lower airflow rates.
2. Consider air distribution strategies in operating rooms and trauma rooms that zone the spaces from most clean to least clean. Start with the cleanest zone being the operation/thermal plume location at the patient (the volume of air directly above the patient which is warmed due to the patient's body

temperature), then the zone around the doctors, then the zone around the room, and finally, the zone outside the room.

3. Pressurize rooms consistent with American Institute of Architects (AIA) and/or ASHRAE guidelines (e.g., pressurized operating rooms, reduced pressure in rooms where patients are contagious, and installing adjacent vestibules to reduced pressure rooms to further control transmission of airborne pathogens).

4. Consider providing air exchange rates in excess of AIA guidelines in operating rooms, intensive care units, isolation rooms, trauma rooms, and patient rooms.

5. Provide redundancy of equipment for fail-safe operation and optimal full- and part-load energy-efficient operation.

6. Model intake/exhaust location strategies to ensure no reintroduction of exhaust into the building.

7. If adding atria containing plants, ensure that soils and plants are free of potential contaminates.

8. Take steps to prevent contamination from Legionella, especially for patients with compromised immune systems, to include the following.

    - Carefully design potable water systems to prevent Legionella growth in piping. Biofilms and scale can develop in piping where water may sit in dead legs or in piping of areas closed off for some reason. Limit dead legs to five pipe diameter lengths. Maintain hot water tanks at temperatures of at least 140°F (60°C). Legionella thrive and multiply when held at temperatures ranging from 95°F to 115°F (35°C to 46°C).
    - Maintain decorative fountains and cold water storage tank water temperatures below 68°F (20°C).
    - Use potable, noncirculated water at or below 68°F (20°C) for cold-water humidifiers.
    - Design piping systems to allow complete system flushing.
    - Specify point-of-use fittings to minimize bacteria growth and to regulate for safe, nonscalding shower temperatures.
    - Caution owners and operators of the importance of maintaining safe levels of water treatment chemicals for cooling towers, hot tubs, whirlpool baths, fountains, and swimming pools.

Energy Considerations

Energy conservation and high efficiency is arguably the most or one of the most significant areas to implement sustainability. Hospitals use over 836 trillion Btu (245,000 kWh) of energy annually. Hospitals have an Energy Utilization Index (EUI) that is about 2.5 times that of typical office buildings. According to the World Health Organization, the use of electricity in the U.S. adds over $600 million in direct health costs and over $500 billion in indirect costs. U.S. hospitals spend over $500 billion on energy that typically represents 1% to 3% of their operating budget or around 15% of profits or net operating income (DOE 2009). Reducing energy costs can play a significant role in a hospital's organization profit or net operating income. Keep in mind, for a hospital operating on a 4% margin, it takes $25 of gross revenue to generate $1 of profit or net operating income. Every dollar of energy savings is a dollar added to the profit or net operating income. According to Target 100 and *Advanced Energy Design Guide for Large Hospitals,* it is possible to reduce a new hospital's energy use by 50% or 60% over the minimum requirements to meet ANSI/ASHRAE/IESNA Standard 90.1-2004. Typically, the amount of energy savings comes down to an economic decision. Life-cycle economic cost analysis is recommended, including maintenance of the various components and systems. Often, the highest savings are achieved with more sophisticated equipment with more complicated operating control sequences. Appropriate assumptions must be applied to the cost of maintenance. Equally important, the design team needs to clearly outline to the building operators the expected maintenance requirements as part of the system selection process. This is an area commonly omitted from of the systems selection and economic analysis.

With the amount of energy savings potential, there are a significant number of options to consider. Energy modeling is recommended to help sort out the interplay of building envelope, lighting, HVAC, and service hot-water energy-saving options. Design teams and the health care operational staff should consider developing energy usage targets and goals at the onset of the project. Those will be useful in guiding the expectations and decision making.

For facilities to realize the potential of any design, three key steps are needed. First, ensure the systems are operating as intended. The design and construction team needs to functionally test and prove all systems are operating per the intent. Following occupancy, it should be an expectation that additional system shakedown will be needed. The actual operating conditions, both in terms of occupancy and facility operations as well as varying weather conditions, will uncover additional issues from that of the functional testing preoccupancy that will need to be addressed. Second, a real commitment to monitoring, benchmarking, and having a continuous improvement process in place is equally important to optimize the energy-saving potential. Third, which is really a subpart of the second idea, is a commitment to budget and fund proper maintenance over the life of the buildings and systems.

Two health care-specific resources for energy-saving ideas for building envelope, lighting, HVAC, and service water heating are outlined for the various climate zones in the United States are part of the Advanced Energy Design Guides (AEDG) series:

- *Advanced Energy Design Guide for Small Hospitals and Healthcare Facilities* (30% better than ANSI/ASHRAE/IES Standard 90.1-1999)
- *Advanced Energy Design Guide for Large Hospitals* (50% better than ANSI/ASHRAE/IES Standard 90.1-2004)

The AEDGs provide case studies and how-to tips for implementing the various suggested strategies. Two other excellent studies that have been referenced already are from the work done through the University of Washington's Integrated Design Lab. See the report "Target100! Envisioning the High Performance Hospital: Implications for A New Low Energy, High Performance Prototype" (2010) and review the website Targeting 100! website (http://idlseattle.com/t100/). There are two critical observations from these reports that are consistent from the modeling results used in the two AEDG for health care facilities:

- Space heating (reheat) due to the mandatory air exchange requirements is the largest, single energy use area.
- There is a significant amount of heat rejection from a variety of processes.

Energy recovery strategies that capture the heat which would otherwise be lost to use for space heating and water heating are great opportunities to explore. For example, consider using heat recovery for spaces served by AHUs with 100% outdoor air capability. The references provide a variety of ways that can be done. Other recommendations include the use of variable-air-volume (VAV) systems in noncritical spaces working in conjunction with lighting occupancy sensors and using dedicated outdoor air systems (DOAS).

Occupant Comfort

Energy considerations are important, but an overall well-designed green health care facility will also include focus on other factors. These factors are addressed with the following recommendations for occupant comfort:

1. Acoustics of systems and spaces must be designed with patient comfort in mind.
2. Daylight and views should be provided, while minimizing the HVAC load impact of these benefits.
3. Provide individual temperature control of patient rooms with the capability of adjustment by patient.
4. Building pressurization relationships/odor issues should be carefully mapped and addressed in the design and operation of the building.

KEY ELEMENTS OF COST

Costs are the bottom line as well.

1. HEPA filtration costs are significant in both first cost and operating costs. The engineer should work closely with the infection control specialists at the health care facility to determine cost/benefit assessment of the filtration strategies.
2. Heat recovery strategies should be assessed using life-cycle analyses. All components of the strategy must be taken into account, including the negative aspects, such as adding fan static pressure and, therefore, using more fan energy when heat wheel or heat pipe strategies are considered.

SOURCES OF FURTHER INFORMATION

AHA. 2012. Business Case Cost-Benefit Worksheet. Sustainability Roadmap for Hospitals, the American Society for Healthcare Engineering, American Hospital Association, Chicago, IL. www.sustainabilityroadmap.org/strategies/businesscase.shtml.

AIA. 2006. *Guidelines for Design and Construction of Health Care Facilities*. Washington, DC: American Institute of Architects.

American Society of Healthcare Engineering. *Green Guide for Health Care*
www.gghc.org.

ASHRAE. 2012. *50% Advanced Energy Design Guide for Large Hospitals*. Atlanta: ASHRAE.

ASHRAE. 2011. *50% Advanced Energy Design Guide for Small to Medium Office Buildings*. Atlanta: ASHRAE.

ASHRAE. 2008. ANSI/ASHRAE/ASHE Standard 170, *Ventilation of Health Care Facilities*. Atlanta: ASHRAE.

ASHRAE. 2000. Guideline 12, *Minimizing the Risk of Legionellosis Associated With Building Water Systems*. Atlanta: ASHRAE.

ASHRAE. 2013. *HVAC Design Manual for Hospitals and Clinics*, 2nd Ed. Atlanta: ASHRAE.

ASHRAE. 2008. *ASHRAE Handbook–HVAC Applications*. Chapter 8, Health Care Facilities. Atlanta: ASHRAE.

ASHRAE. 2009. *ASHRAE Handbook—Fundamentals*, Chapters 10 and 11. Atlanta: ASHRAE.

ASHRAE. 2013. Standard 188P, *Prevention of Legionellosis Associated with Building Water Systems*. Atlanta: ASHRAE. (in review at the time of this writing).

CDC. 2003. *Guideline for Environmental Infection Control in Health-Care Facilities*. 2003 Recommendations of CDC and the Healthcare Infection Control Practices Advisory Committee (HICPAC). www.cdc.gov/ncidod/dhqp/gl_environinfection.html.

DOE. 2012. EERE Network News. New Hospital Energy alliance to Promote Clean Energy in Healthcare. http://apps1.eere.energy.gov/news/news_detail.cfm?news_id=12485.

George, R. 2012. (unpublished ASHRAE Chapter Presentation). Preventing legionella associated with building water systems. Legionella Website: www.legionellaprevention.org

NFPA. 2012. NFPA 99, *Health Care Facilities Code*. Quincy, MA: National Fire Protection Agency.

The Center for Health Design
www.healthdesign.org.

WHO. 2009. *WHO Guideline 2009–Natural Ventilation for Infection Control in Health-Care Settings*. World Health Organization, Geneva, Switzerland.

ASHRAE Building-Type GreenTip #5-3

Laboratory Facilities

GENERAL DESCRIPTION

Laboratory facilities are infrastructure-intensive and include many different types of spaces. The HVAC systems for these different types of spaces must be designed to address the specific needs of the spaces being served. The first considerations should always be safety and system redundancy to ensure the sustainability of laboratory studies. Life-cycle cost analysis for different system options is critical in developing the right balance between first and operating costs.

HIGH-PERFORMANCE STRATEGIES

Safety

1. Design fume hoods and associated air distribution and controls to protect the users and the validity of the laboratory work.
2. Pressurize rooms to be consistent with the *ASHRAE Laboratory Design Guide* (McIntosh et al. 2002) and any other code-required standards. Use building pressurization mapping to develop air distribution, exchange rate, and control strategies.
3. Optimize air exchange rates to ensure occupant safety, while minimizing energy usage.
4. Design storage and handling exhaust and ventilation systems for chemical, biological, and nuclear materials to protect against indoor pollution, outdoor pollution, and fire hazards.
5. Model intake/exhaust location strategies to ensure that lab exhaust air is not reintroduced back into the building's air-handling system.

Redundancy

1. Consider a centralized lab exhaust system with a redundant (n + 1) exhaust fan setup.
2. Redundant central chilled-water, steam, or hydronic heating, air-handling, and humidification systems should be designed for fail-safe operation and to optimize full-load and part-load efficiency of all equipment.

Energy Considerations

1. Consider energy recovery for areas served by AHUs with more than a 50% outdoor air component. Carefully design heat recovery for hazardous spaces.
2. Use VAV systems to minimize air exchange rates during unoccupied hours.
3. Consider the use of computational fluid dynamics (CFD) and other modeling software to establish safe air change rates for particular laboratory needs and follow up with monitoring to verify its effectiveness. Consider low-flow fume hoods with constant volume controls where this concept can be properly applied.
4. Look for opportunities for sharing spaces to reduce conditioned space volume, (e.g., preparation areas and laboratory support spaces could be shared between labs). The number of laboratories can also be reduced if work schedules can be shifted and the same space can be used through the day and night.
5. Isolate and minimize the size of laboratory space for those processes requiring greater airflow for safety.

Occupant Comfort

1. Air systems should be designed to allow for a collaborative working environment. Acoustic criteria should be adhered to in order to maintain acceptable levels of noise control.
2. Daylight and views should be considered where lab work will not be adversely affected. When appropriate, glazed walls to other spaces or hallways will permit views without requiring the laboratory to be on an outside wall.

KEY ELEMENTS OF COST

1. Heat recovery strategies should be assessed using life-cycle analyses. All components of the strategy must be taken into account, including the negative aspects, such as adding fan static pressure and, therefore, using more fan energy when strategies using enthalpy wheels, desiccant wheels or heat pipes are considered.
2. Low-flow fume hoods should be evaluated considering the impact of reducing the sizes of air-handling, heating, cooling, and humidification systems.

SOURCES OF FURTHER INFORMATION

ASHRAE. 2011. *ASHRAE Handbook—HVAC Applications*, Chapter 16, Laboratories. Atlanta: ASHRAE.

Labs 21, Environmental performance criteria www.labs21century.gov.

McIntosh, I.B.D., C.B. Dorgan, and C.E. Dorgan. 2002. *ASHRAE Laboratory Design Guide*. Atlanta: ASHRAE.

NFPA. 2011. NFPA 45, *Standard on Fire Protection for Laboratories using Chemicals*. Quincy, MA: National Fire Protection Association.

Tschudi, W., Dale Sartor, Evan Mills and Tim Xu. 2002. *High-Performance Laboratories and Cleanrooms: A Technology Roadmap*. Berkeley, CA: Lawrence Berkeley National Laboratory.

VanGeet, O. and S. Reilly. 2006. Ventilation heat recovery for laboratories. *ASHRAE Journal* 48, no. 3: 44–53.

ASHRAE Building-Type GreenTip #5-4

Student Residence Halls

GENERAL DESCRIPTION

Student residence halls are made up primarily of living spaces (e.g., bedrooms, living rooms, kitchen areas, common spaces, study spaces, data/communications closets). Most of these buildings also have central laundry facilities, assembly/main lobby areas, and central meeting/study rooms. Some of these spaces also include classrooms, central kitchen and dining facilities, etc. Many of the strategies outlined below can also be applied to hotels and multiunit residential complexes, including downtown luxury condominium developments.

HIGH-PERFORMANCE STRATEGIES

Energy Considerations

1. Use heat recovery for spaces served by AHUs with 100% outdoor air capability serving living units (exhaust from toilet rooms/supply air to occupied spaces).
2. Use VAV systems with economizers or induction systems for public/common spaces and consider VAV also for sleeping areas.
3. Specify that the static pressure setpoint for the central supply fans be based on the most open VAV box position.
4. Specify CO_2 demand-controlled ventilation for areas with variable use (i.e., study spaces, lounges, meeting rooms, etc).
5. Use natural-ventilation (operable windows) and hybrid natural ventilation strategies.
6. Use variable-speed drives for chilled-water pumps and HVAC fans, including fan-coil units.
7. Consider using a ground-source heat pump (GSHP) system.
8. Consider using skylights in corridors for daylighting. Also integrate electrical lighting with daylighting through controls.

Occupant Comfort

1. Systems should be designed to appropriately control noise in occupied spaces.
2. Daylight and views should be optimized, while minimizing load impact on the building.
3. Consider providing occupant control in all dorm rooms. However, provide overrides for the case when a window is open.

KEY ELEMENTS OF COST

1. While there is a premium to be paid in first cost for energy control measures (ECM), many utility companies have energy rebate programs that make this concept acceptable, even on projects with tight budgets.
2. Heat recovery strategies should be assessed using life-cycle analyses. All components of the strategy must be taken into account, including the negative aspects, such as adding fan static pressure and, therefore, using more fan energy when enthalpy wheel, desiccant wheel, or heat pipe strategies are considered.
3. Hybrid natural ventilation strategies could be used, such as operable windows, properly designed vents, using the Venturi effect to optimize natural airflow through the building, and the shutdown of mechanical ventilation and cooling systems during ambient temperature ranges between 60°F and 80°F (16°C and 27°C). These measures will significantly reduce operating costs. The costs of the operable windows and vents will need to be weighed against the energy savings.

REFERENCES AND RESOURCES

Friedman, G. 2010. Energy-Saving Dorms. *ASHRAE Journal* 52, no. 5: 20–24.

SOURCES OF FURTHER INFORMATION

ASHRAE. 2013. *ASHRAE Handbook—Fundamentals*. Chapter 16, Ventilation and Infiltration. Atlanta: ASHRAE.

BRESCU, BRE. 1999. *Natural Ventilation for Offices* Guide and CD-ROM. ÓBRE on behalf of the NatVent Consortium, Garston, Watford, UK, March.

CIBSE. 2005. *Natural Ventilation in Non-Domestic Buildings.* London: Chartered Institution of Building Services Engineers.

Svensson C., and S.A. Aggerholm. 1998. Design tool for natural ventilation. ASHRAE IAQ 1998 Conference, October 24–27, New Orleans, LA.

Athletic and Recreation Facilities

GENERAL DESCRIPTION

Athletic and recreational spaces include natatoriums, gymnasiums, cardio rooms, weight-training rooms, multipurpose rooms, courts, offices, and other support spaces, etc.

HIGH-PERFORMANCE STRATEGIES

Energy Considerations

1. Consider using demand-controlled ventilation for high-occupancy spaces.
2. Consider using heat recovery for spaces served by AHUs with more than 50% outdoor air component.
3. Consider strategies that allow the significant heat gain in high-volume spaces to stratify, rather than handling all of the heat gain within the conditioned space zones in the building.
4. Consider heat recovery/no-mechanical-cooling strategy for the pool area in moderate climates.
5. Consider occupied/unoccupied mode for large locker room and toilet room areas to set back the air exchange rate in these spaces during unoccupied hours and save fan energy.
6. Consider heating pool water with waste heat from pool dehumidification system.
7. Consider using a water-based geothermal heat pump system (GHPS).

Occupant Comfort

1. Consider CO_2 sensors in all spaces that have infrequent, dense occupancy.
2. Consider high-occupancy and low-occupancy modes for air-handling equipment in gymnasiums using a manual switch and variable-frequency drives.

3. Consider hybrid natural ventilation strategies in areas that do not have humidity control issues (e.g., pools, training rooms).

KEY ELEMENTS OF COST

1. The pool strategy described above should reduce first cost and operating costs.
2. Heat recovery strategies should be assessed using life-cycle analyses. All components of the strategy must be taken into account, including the negative aspects, such as adding fan static pressure and, therefore, using more fan energy when strategies using enthalpy wheels, desiccant wheels or heat pipes are considered.
3. Demand-controlled ventilation adds minimal first cost and often provides paybacks in one to two years.

SOURCES OF FURTHER INFORMATION

ASHRAE. 2011. *ASHRAE Handbook—HVAC Applications*, Chapter 5, Places of Assembly, and Chapter 7, Educational Facilities. Atlanta: ASHRAE.

ASHRAE. 2012. *ASHRAE Handbook—HVAC Systems and Equipment*, Chapter 24, Desiccant Dehumidification and Pressure-Drying Equipment, and Chapter 25, Mechanical Dehumidifiers and Related Components. Atlanta: ASHRAE.

Crowe, Derek. 2011. State-of-Art School. *ASHRAE Journal*. 53(5):36–42.

Lawrence, Tom. 2004. Demand-Controlled Ventilation and Sustainability. *ASHRAE Journal* 46 (12), pp. 117, 120–121.

ASHRAE Building-Type GreenTip #5-6

Commercial Office Buildings

GENERAL DESCRIPTION

Commercial office buildings are made up primarily of office spaces, meeting rooms, and central core facilities such as toilet rooms, storage space, and utility rooms (including telephone and data). Some of these spaces also include central kitchen and dining facilities. The strategies outlined below can also be applied to most large-scale commercial office buildings.

HIGH-PERFORMANCE STRATEGIES

Energy Considerations

1. Consider a dedicated outdoor air system (DOAS) with total energy recovery (TER).
2. Use high-efficiency fan-coil units or chilled-beam/induction systems in conjunction with the DOAS concept.
3. Use natural ventilation and hybrid natural ventilation strategies.
4. Use energy conservation measures (ECM) for fan-coil units.
5. Use ground source heat pumps (GSHP) where feasible.
6. Consider using underfloor air distribution (UFAD).
7. Incorporate energy-efficient lighting strategies that share integrated occupancy controls with the HVAC system.

Occupant Comfort

1. Systems should be designed to appropriately control noise in occupied spaces.
2. Daylight and views should be optimized, while minimizing heating and cooling load impact on the building.

KEY ELEMENTS OF COST

1. While there is a premium to be paid in first cost for ECMs, many utility companies have energy rebate programs that make this concept acceptable, even on projects with tight budgets.
2. Energy recovery, UFAD, and DOAS strategies should be assessed using life-cycle analyses. All components of the strategy must be taken into account, including the negative aspects, such as adding fan static pressure and, therefore, using more fan energy when heat wheel or heat pipe strategies are considered.
3. Hybrid natural ventilation strategies could be used such as using operable windows, properly designed vents, the Venturi effect to optimize natural airflow through the building, and the shutdown of mechanical ventilation and cooling systems during ambient temperature ranges between 60°F and 80°F (16°C and 27°C). This will significantly reduce operating costs. The costs of the operable windows and vents will need to be weighed against the energy savings.

REFERENCES AND RESOURCES

Mumma, S. A. 2011. DOAS misconceptions. *ASHRAE Journal* 53, no. 8: 76–79.

SOURCES OF FURTHER INFORMATION

ASHRAE. 2011. Advanced Energy Design Guide for Small to Medium Office Buildings: Achieving 50% Energy Savings. Atlanta: ASHRAE.

ASHRAE. 2009. *ASHRAE Handbook—Fundamentals*. Chapter 16, (Natural Ventilation). Atlanta: ASHRAE.

ASHRAE. 2011. *ASHRAE Handbook—HVAC Applications*. Chapter 3, Commercial and Public Buildings. Atlanta: ASHRAE.

BRESCU, BRE. 1999. *Natural Ventilation for Offices* Guide and CD-ROM. ÓBRE on behalf of the NatVent Consortium, Garston, Watford, UK, March.

Svensson C., and S.A. Aggerholm. 1998. Design tool for natural ventilation. ASHRAE IAQ 1998 Conference, October 24–27, New Orleans, LA.

ASHRAE Building-Type GreenTip #5-7

K–12 School Buildings

GENERAL DESCRIPTION

K–12 school buildings are made up primarily of classrooms, gymnasiums, libraries, an auditorium, a central kitchen, and dining facilities, etc. (Also see GreenTip #5-6: Commercial Office Buildings and GreenTip #5-5: Athletic and Recreation Buildings.)

HIGH-PERFORMANCE STRATEGIES

Energy Considerations

1. Use heat recovery for spaces served by AHUs with more than 50% outdoor air capability serving occupied spaces.
2. Use dedicated outdoor air systems (DOASs) with fan coils or chilled-beam/induction systems for spaces.
3. Use natural ventilation and hybrid natural ventilation strategies.
4. Use ECMs for fan-coil units.
5. Use GSHPs where feasible.

Occupant Comfort

1. Systems should be designed to appropriately control noise in occupied spaces.
2. Daylight and views should be optimized while minimizing load impact on the building.

KEY ELEMENTS OF COST

1. While there is a premium to be paid in first cost for energy conservation measures (ECM), many utility companies have energy rebate programs that make this concept acceptable, even on projects with tight budgets.
2. Energy recovery strategies should be assessed using life-cycle analyses. All components of the strategy must be taken into

account, including the negative aspects, such as adding fan static pressure and, therefore, using more fan energy when heat wheel or heat pipe strategies are considered.
3. Hybrid natural ventilation strategies could be used, such as using operable windows, properly designed vents, using the Venturi effect to optimize natural airflow through the building and the shutdown of mechanical ventilation and cooling systems during ambient temperature ranges between 60°F and 80°F (16°C and 27°C). These actions will significantly reduce operating costs. The costs of the operable windows and vents will need to be weighed against the energy savings.

SOURCES OF FURTHER INFORMATION

ASHRAE. 2009. *ASHRAE Handbook—Fundamentals*. Chapter 16, (Natural Ventilation). Atlanta: ASHRAE.

BRESCU, BRE. 1999. *Natural Ventilation for Offices* Guide and CD-ROM. ÓBRE on behalf of the NatVent Consortium, Garston, Watford, UK, March.

Mumma, S. A. 2011. DOAS misconceptions. *ASHRAE Journal* 53, no. 8: 76–79, www.scopus.com.

Svensson C., and S.A. Aggerholm. 1998. Design tool for natural ventilation. ASHRAE IAQ 1998 Conference, October 24–27, New Orleans, LA.

ASHRAE Building-Type GreenTip #5-8

Existing Buildings

GENERAL DESCRIPTION

Renovating an existing building can provide an excellent opportunity to reduce the environmental impact of that building. Significant improvements can be made in building energy and water efficiency. And, of course, extending the life of an existing building via a renovation project obviates the need to develop a new site and uses fewer materials than would be required by a new building.

HIGH-PERFORMANCE STRATEGIES

Energy Considerations

1. Perform an audit of the building to determine the potential to reduce energy use and to identify specific opportunities to improve energy efficiency.
2. Retrocommission control and HVAC systems that will remain in place after completion of the renovation project.
3. Consider installing energy recovery systems for existing or new HVAC systems, particularly for AHUs providing 100% outdoor air.
4. Use air- and/or water-side free cooling (AHU economizer controls and plate and frame heat exchangers for existing or new cooling towers) where feasible.
5. Consider installing variable-speed drives on new or existing fan and pump motors to improve the match between air and water flows on the one hand and heating and cooling loads on the other.
6. Replace existing HVAC equipment and systems with more efficient equipment and systems where necessary to meet new owner requirements or where cost-justifiable.
7. Retrofit or replace existing lighting fixtures using current energy-efficient lamp and ballast technologies.

8. Consider the use of occupancy sensors to switch lamps off when not needed and dimming controls to take advantage of daylighting where feasible.
9. Consider incorporating renewable energy technologies into the renovation project, such as solar thermal, photovoltaics, and small-scale wind turbines.

Water Considerations

1. Address water use and efficiency concerns in the building audit.
2. Consider replacing existing plumbing fixtures (i.e., water closets, urinals, lavatory faucets, and shower heads) with new, more efficient plumbing fixtures to reduce both water and energy use.
3. Consider modifying landscapes to reduce outdoor water use (e.g., replace grass with native plants, pavers, or gravel).
4. Retrocommission or install new landscape irrigation controls where feasible.
5. Consider alternatives to storm drains (e.g., create rain gardens, bioswales, permeable surfaces allowing rain water to sink into the ground in place of hard surfaces, and use stormwater on-site for irrigation and toilet flushing).

Occupant Comfort

1. Be sure that existing or new control and HVAC systems are capable of meeting the requirements of ANSI/ASHRAE Standard 55-2010, *Thermal Environmental Conditions for Human Occupancy* (ASHRAE 2010) and ANSI/ASHRAE Standard 62.1-2013, *Ventilation for Acceptable Indoor Air Quality* (ASHRAE 2013).
2. Improve occupant comfort and performance with daylighting, new lighting technologies employing high-efficacy lamps/ballasts, and integrated lighting controls.
3. Provide occupant control over HVAC and lighting systems (i.e., operable windows, thermostats, and light fixture switches), where feasible.

KEY ELEMENTS OF COST

1. Energy and water cost savings are typically used to justify the installation costs of energy and water efficiency upgrades. Establishing cost savings requires a good understanding of baseline operating conditions, proposed operating conditions, utility rates, and interactions among conservation opportunities. It may be necessary to develop whole-building simulation models in order to estimate potential savings accurately.
2. Often, there are ancillary benefits to installing energy and water efficiency upgrades, such as improved comfort, improved maintainability, lower cooling loads, and extended system life. It is important to discuss these benefits thoroughly with the building owner or owner representative to prioritize energy and water-conservation opportunities properly.
3. Alternative sources of funding may be available for energy and water efficiency upgrades. These include utility company rebates, state and federal grants and tax credits, and renewable energy credits. Obtaining access to such alternative funding sources may allow for an increase in renovation project scope.

REFERENCES AND RESOURCES

Published and Software

ASHRAE. 2009. *ASHRAE Handbook—Fundamentals*. Atlanta: ASHRAE.

Evans, B. 1997. Daylighting design. In *Time-Saver Standards for Architectural Design Data*. New York: McGraw-Hill.

Heerwagen, J. 2001. Do green buildings enhance the well-being of workers? www.edcmag.com.

Hilten, R.N. 2005. An analysis of the energetics and storm water mediation potential of green roofs. Master's thesis, University of Georgia, Athens, GA.

Savitz, A.W., and K. Weber. 2006. *The Triple Bottom Line*. San Francisco: John Wiley & Sons, Inc.

Spiegel, R. and D Meadows 1999. *Green Building Materials: A Guide to Product Selection and Specification*, John Wiley & Sons, Inc., New York.

Online

www.calrecycle.ca.gov/GreenBuilding/Materials/
www.buildinggreen.com/auth/article.cfm/2012/2/2/What-Makes-a-Product-Green/
"Tips for Daylighting with Windows"
 http://windows.lbl.gov/pub/designguide/designguide.html
Cool Roof Rating Council
 www.coolroofs.org
Green Roof Industry Information Clearinghouse and Database
 www.greenroofs.com
Regional Climate Data
 www.wrcc.dri.edu/rcc.html
National Oceanic and Atmosphere Administration
 www.noaa.gov
U.S. Environmental Protection Agency, Building Energy Analysis Tool, eQuest
 http://energydesignresources.com/resources/software-tools/equest.aspx/

SOURCES OF FURTHER INFORMATION

ASHRAE. 2010. ANSI/ASHRAE Standard 55-2010, *Thermal Environmental Conditions for Human Occupancy*. Atlanta: ASHRAE.

ASHRAE. 2006. ANSI/ASHRAE/IESNA Standard 100-2006, *Energy Conservation in Existing Buildings*. Atlanta: ASHRAE.

ASHRAE. 2011. *ASHRAE Handbook—HVAC Applications*. Chapter 36. Atlanta: ASHRAE.

ASHRAE. 2013. ANSI/ASHRAE Standard 62.1-2013, *Ventilation for Acceptable Indoor Air Quality*. Atlanta: ASHRAE.

ASHRAE Building-Type GreenTip #5-9

Data Centers

GENERAL DESCRIPTION

Data centers are energy-intensive buildings using 20 W/ft² (215 W/m²) to greater than 1000 W/ft² (10 764 W/m²). They support information technology (IT) spaces, electronic equipment, and electronic communications. These facilities are known by many different names: data centers, computer rooms, server rooms, telecom rooms, datacom, or information technology equipment (ITE) rooms. (The term *ITE* will be used in the remainder of this article to designate these spaces.) The HVAC systems for data centers can be as diverse as the data centers themselves and include various forms of air or liquid cooling. While the first considerations must be reliability of the infrastructure and IT systems to minimize the risks and losses associated with outages, it is also necessary to minimize energy consumption. To this end, many strategies of redundancy, operational practice, and service diversity are used, along with green practices such as outdoor air, "free cooling," air containment, and automatic control. In the end, the purpose of a data center is always to provide a suitable environment (power, cooling, and security) for the electronic equipment in an energy-efficient, scalable, and reliable manner.

HIGH-PERFORMANCE STRATEGIES

Reliability and Scalability

1. Install only the necessary redundancy to meet the owner's reliability requirements; optimize redundancies.
2. Use zoning strategies for different types of ITE, based on power densities and preferred cooling strategies for each type.
3. Rightsize the cooling and power infrastructure for the IT loads with sufficient stages to operate at best efficiency points as loads change.
4. Design scalable infrastructure that can change with the rapid changes in ITE and its software (data center occupants).

Energy Considerations

1. Use air management best practices (e.g., separate hot/cold airstreams, maximize temperature differences across HVAC heat exchangers, eliminate air leaks, use preferred ITE equipment cooling classes, etc.) to achieve ITE inlet temperature uniformity as described in ASHRAE's third edition of *Thermal Guidelines for Data Processing Environments*.
2. Install monitoring and control equipment (building management systems) to measure efficiency and performance of HVAC as well as IT systems.
3. Install data center infrastructure management (DCIM) systems to measure and manage ITE and infrastructure systems more holistically.
4. Use cooling plant best practices (e.g., VFDs, optimize chiller/condenser supply temperatures, chilled water vs. DX, etc.).
5. Consider liquid cooling when heat loads or total cost of ownership (TCO) requirements demonstrate energy savings.
6. Avoid high-energy-consuming humidity controls such as reheat. Control humidity using dew-point sensing. Consider alternatives such as desiccant dehumidification and adiabatic humidifying.
7. Use air or water economizers where TCO and risks are acceptable.
8. Improve power distribution efficiencies (e.g., evaluate transformer efficiency, reduce number of power conversions, use higher voltages, consider DC power, minimize power run distances, optimize redundancy strategy).
9. Capture and reuse waste heat where possible.
10. Determine optimal operating temperature and humidity ranges using the steps defined in ASHRAE's third edition of *Thermal Guidelines for Data Processing Environments* (The recommended range should be used as the starting point if no optimization has been performed.)

Electronic Equipment and Environmental Requirements

1. Implement server virtualization where applicable.
2. Collaborate with IT department on ITE purchases. When prudent, select the appropriate higher ASHRAE Classes of hardware (A2, A3, and A4) to permit higher operating temperatures, balance TCO, reliability, locale, and other owner objectives. The reader is referred to the 3rd Edition (2012) of *Thermal Guidelines for Data Processing Environments*.
3. Enable power management features on ITE.
4. Specify high-efficiency power supplies and eliminate unnecessary power conversions. Consider DC-powered hardware if available.
5. Remove "comatose" ITE hardware.
6. Perform technology refreshes to increase computer performance per watt.
7. Optimize IT equipment supply air conditions per *Thermal Guidelines for Data Processing Environments*.
8. Control HVAC systems based on entering air conditions at the ITE.
9. Determine appropriate gaseous and particulate contamination control measures for the data center location.

KEY ELEMENTS OF COST

1. Infrastructure redundancy comes with a significant capital cost as well as energy-efficiency penalties. The engineer should work to optimize the required redundancy without exceeding the owner's business requirements and risk models.
2. TCO is an important concept for data centers that requires careful evaluation of IT hardware classes, deployment and technology refresh strategies, system redundancies, energy efficiency, cooling transport media (air, water, refrigerant), scalability, etc.

SOURCES OF FURTHER INFORMATION

ASHRAE. 2012. ANSI/ASHRAE Standard-127R, *Method of Testing for Rating Computer and Data Processing Room Unitary Air Conditioner*. Atlanta: ASHRAE.

ASHRAE. 2009. *Thermal Guidelines for Data Processing Environments*, 3rd Edition, ASHRAE Datacom series, Book #1. Atlanta: ASHRAE.

ASHRAE. 2012. *Power Trends and Cooling Applications*, 2nd Edition, ASHRAE Datacom series, Book #2. Atlanta: ASHRAE.

ASHRAE. 2005. *Design Considerations for IT Equipment Centers*, ASHRAE Datacom series, Book #3. Atlanta: ASHRAE.

ASHRAE. 2009. *Best Practices for Datacom Facility Energy Efficiency*, 2nd Edition, ASHRAE Datacom series, Book #6. Atlanta: ASHRAE.

ASHRAE. 2008. *High Density Data Centers, Case Studies and Best Practices*, ASHRAE Datacom series, Book #7. Atlanta: ASHRAE.

ASHRAE. 2012. *Particulate and Gaseous Contamination in Datacom Environments*, ASHRAE Datacom series, Book #8. Atlanta: ASHRAE.

ASHRAE. 2007. *ASHRAE Handbook—HVAC Applications*, Chapter 17, Data Processing and Electronic Office Areas. Atlanta: ASHRAE.

OTHER REFERENCES AND RESOURCES

Sources for climate data and sustainable design strategies based on climate include, but not limited to the following plus historical and current publications by authors such as Victor Olgyay and Murray Milne (UCLA):

Published

ASHRAE. 2009. *ASHRAE Handbook—Fundamentals*, Ch. 14, Climatic Design Information. Atlanta: ASHRAE.

ASHRAE. 2011. ANSI/ASHRAE/USGBC/IES Standard 189.1-2011, *Standard for the Design of High-Performance Green Buildings*. Atlanta: ASHRAE.

ASHRAE. 2011. *ASHRAE Handbook—HVAC Applications*, Ch. 44, Building Envelopes (includes "Quick Design Guide for High-Performance Building Envelopes"). Atlanta: ASHRAE.

ASHRAE. 2012. *Thermal Guidelines for Data Processing Environments*. Atlanta: ASHRAE.

R.J. Morris. 1987. Climatic design data for engineers in Canada: A 20-year update. *ASHRAE Transactions*, 93(2), Nashville, TN

Lechner, Norbert. 2001. *Heating Cooling and Lighting*. John Wiley.
Said, S.A.M., H.M. Kadry, and B.I. Ismail. 1996. Climate conditions for Saudi Arabia. *ASHRAE Transactions*, 102(1), 37–44.
Wimmers, Guido and the Light House Sustainable Building Center. "Passive Design Toolkit for Homes." 2009. City of Vancouver, British Columbia, available online at: http://vancouver.ca/files/cov/passive-home-design.pdf.

Online

ASHRAE Advanced Energy Design Guides
www.ashrae.org/standards-research--technology/advanced-energy-design-guides.

Climate Zone, for international climate data
http://climate-zone.com/.

Energy Efficiency & Renewable Energy (EERE) DOE Buildings Performance Database
https://www.buildingenergy.com/.

Energy Efficiency & Renewable Energy (EERE) (international Weather Data compatible with EnergyPlus simulation software format). http://apps1.eere.energy.gov/buildings/energyplus/weatherdata_about.cfm?CFID=1411744&CFTOKEN=2be19baca70de793-AD7C57A3-D289-9D16-321EAADEDE6F7D96.

Intergovernmental Panel on Climate Change (IPCC)
www.ipcc-data.org/index.html.

National Oceanic and Atmospheric Administration (NOAA) National Climatic Data Center
www.ncdc.noaa.gov/.

World Meteorological Organization (WMO) (international climate data in several languages)
http://worldweather.wmo.int/.

Ask Nature
www.asknature.org.

Software

ASHRAE. 2009. *Weather Data Viewer*, Version 4.0 2009. Atlanta: Atlanta: ASHRAE.

CHAPTER SIX

COMMISSIONING

The commissioning process, administered by the commissioning authority (CxA) is a quality control process that provides continuous oversight from the predesign phase of the project throughout occupancy. The commissioning process also provides an essential foundation for integrated design and delivery, eases adapting to changing needs over the facility's lifetime, and establishes benchmarks for evaluating achievement of the owner's defined goals and objectives as they change over time. Commissioning is applied to new building projects and existing facilities to ensure that a facility meets the needs of the occupants to effectively and efficiently deliver the owner's purpose for that building and their organization's and financial goals. The true measure of a high-performance building is how it performs over its lifetime.

Initial guidance on commissioning was produced by ASHRAE in ASHRAE Guideline 0-2005, *The Commissioning Process* (2005), which describes the overall process and provides guidance on implementation for commissioning, and ASHRAE Guideline 1.1, *HVAC&R Technical Requirements for the Commissioning Process* (2007), which contains additional information specifically focused on HVAC&R applications. The new ASHRAE Standard 202, *Commissioning Process for Buildings and Systems*, will help fill the need for these concepts to be written in code language.

Commissioning for new construction projects can be broken down into five phases: predesign, design, construction, acceptance, and warranty/ongoing commissioning. Commissioning of existing buildings also has four phases: investigation, analysis, recommendations, and implementation. Distinct commissioning activities occur during each of these phases. Achieving desired building performance starts with clearly defining the Owner's Project Requirements (OPR) at predesign (before development of the architectural program). During design, the CxA provides checklists to designers to assist them in their design quality control process and remind the designers about the specifics the CxA will be focusing on during commissioning design reviews. Chapter 4 of this guide provides further information on commissioning activities in the design phase. During construction, the CxA conducts site visits and provides construction checklists to the contractors to assist them in their quality control process and, as in the design reviews, verifies that the contractor's quality control process is functioning well. These efforts significantly improve the chances that the systems being commissioned will only need minor modifications during performance testing and will reduce the delivery team's efforts. The acceptance phase verifies through testing that the systems perform as intended and helps resolve issues prior to occupancy. During the warranty phase, the CxA monitors system performance and verifies that training provided was understood by operators. The CxA assists operators in better understanding their systems and maintaining maximum performance, which helps prevent inappropriate modifications by the operators (due to lack of understanding). During the new construction process, the CxA identifies specific deliverables required from the project team needed for the systems manual. The systems manual is the repository for the OPR, Basis of Design, commissioning reports, O&M manuals, equipment operation instructions, and as-built documentation including any modifications made to the facilities systems and assemblies, along with the reasoning behind the modifications. Detailed information to be included in the systems manual is defined in the ASHRAE guidelines and Standard 202.

The OPR includes sustainable development goals that are in harmony with the owner's mission. Many teams focus on green rating system credits without knowing how these credit selections will affect the owner's needs. For example, on a recent project, the design team selected use of native and adaptive plants to achieve a green rating system credit but neglected to understand that this conflicted with the military facility's mission and requirements for no vegetation that was 4 in. (100 mm) in height above the ground.

The commissioning process focuses upon verifying and documenting that the building and all of the systems and assemblies are planned, designed, installed, tested, operated, and maintained to meet the owner's goals and objectives for the project. Commissioning is not only an important part of successful project delivery, but it's an essential and, in many cases, required part of green-building design and construction. Commissioning is not an exercise in blame. It is, rather, a collaborative effort to identify and reduce potential design, construction, and operational problems by resolving them early in the process, at the least cost to everyone. This chapter will include the selection and role of the commissioning provider, a discussion of various commissioning models, the choice of building systems for commissioning, and the long-term benefits afforded by implementing the commissioning process.

Commissioning provides benefits to everyone: the owner, the design and construction teams, building occupants, and building operators. What owner, particularly a long-term owner, does not want reduced risk, fewer change orders (and the resulting cost avoidance), improved energy efficiency, lower operating costs, satisfied tenants/occupants, and a building that operates as intended over the lifetime of the building?

What contractor or designer does not dream of a project with few or no problems or callbacks and the additional costs that come with such problems? The commissioning process should begin early in the predesign phase to ensure maximum benefits from commissioning. Starting early improves designer and contractor quality control processes, makes the CxA part of the process, and identifies and helps resolve problems during design (when corrective action is the least expensive). During construction, commissioning can also provide benefit when the contractor has the materials and resources on site for efficient corrective action (minimizing postoccupancy rework and repairs).

How, and to what extent, an owner incorporates commissioning into their project generally depends on the owner's understanding of commissioning. Commonly, an owner might start with construction phase commissioning but may soon see how much more they would have benefited by starting in the predesign phase. Other factors that determine the extent of commissioning incorporated in any given project are how long the owner holds the property, the owner's staff capabilities and funding mechanisms for design, construction, and operation, the project schedule, and ownership experience.

One of the most beneficial attributes of sustainable development principles contained in U.S. Green Building Council's (USGBC) Leadership in Energy and Environmental Design (LEED®) rating system is the inclusion of commissioning. If a building's high performance design features do not function as intended, there is little benefit in incorporating them into the design. Commissioning provides verification that the building systems will and do operate as intended.

COMMISSIONING PHASES

The role of the CxA varies according to the phase of the project. Because of the nature of the delivery process, the further along the team is in the design and construction process, the more difficult and expensive changes become (see Figure 4-1). Waiting until after predesign to define project end goals, occupant requirements, and team roles and responsibilities can lead to increased project cost. The easiest time to revaluate goals, objectives, and criteria (and make changes) is during the schematic design phase, and this is also the best time to reduce the control cost of any changes. A steep decline in the feasibility to implement sustainability features occurs after the schematic design phase. The ability to make significant cost-efficient changes ends when the project moves into design development. The earlier a design team addresses and implements sustainable features into a design, the more likely these will be included without significant cost. This is a foundation of integrated design and critical to cost-effective green buildings. Historically, owners and contractors set up a contingency fund intended to cover the unpredictable cost of changes in a project. If the design does not meet the owner's needs, the owner may be forced to accept the project as is, because changes needed to meet his or her requirements would be too costly at that point. If the CxA is engaged as late as the construction phase, there is some, but very limited, opportunity to address potential design problems. The longer an owner waits to engage in the commissioning process, the less influence the CxA has on resolving problems cost effectively.

Predesign

Commissioning is a team effort performed within a collaborative framework. The entire project team is part of the commissioning process. The commissioning plan defines the project team's roles and responsibilities, and execution of the commissioning process. It is essential that the owner clearly defines (contractually) the project team's commissioning roles and responsibilities as well as prepare the OPR. The value of the quality and quantity of information provided by owners is often related to their development experience. Institutional owners who have developed many buildings and who have held properties for extended periods of time have over time developed the information that design teams need in order to understand an owner's basic needs. One of the greatest values of involving a CxA in the predesign phase is to develop a comprehensive OPR document that serves as the project benchmark, guiding all project team members.

The CxA's Role in OPR Development

It is the CxA's role to lead the collaborative team effort required to balance competing interests so that the owner's needs and mission are not lost. To accomplish this task, a benchmark is needed. This benchmark is the OPR. The OPR is a written document that details the functional requirements of a project and expectations of how it will be used and operated. The OPR includes project and design goals, measurable performance criteria, budgets, schedules, success criteria, and supporting information (i.e., specific information that should be included and can be found in ASHRAE Guideline 0 [ASHRAE 2005] and ASHRAE Standard 202 [ASHRAE 2013]). It also includes information necessary for all disciplines to properly plan, design, construct, operate, and maintain systems and assemblies.

Ideally, it is in the predesign phase of a project, prior to engaging the design team, that the owner would engage a CxA to develop a draft commissioning plan. The draft commissioning plan defines the team's roles and responsibilities, suggested communication protocols, commissioning activities, and the schedule of the activities. Project success is dependent on each team member's understanding of what is expected of them and obtaining their buy-in. The commissioning plan provides the owner with clearly defined roles and responsibilities for each team member for inclusion in contractual agreements and for improved team efficiency. It is more economical to define these requirements early, before selecting and contracting the various project team members. After the project team selection, an owner will state the basic requirements of the project that form the starting point for the OPR and architectural program. This information will typically include justification for the project, programming needs, intended building use, basic construction materials and methods, proposed systems, project schedule, and general information (such as attaining LEED certification).

Many confuse the OPR with an architectural program. The OPR is different from the typical architectural program, which focuses on project floor area needs, adjacencies, circulation, cost, and structural predesign test results. The OPR documents how the owner intends for this building to function and fulfill the needs of the owner and the occupants. For instance, an architectural program does not contain requirements for how the building will be operated. It also does not contain operation and maintenance (O&M) training requirements, or post-construction documentation requirements for the building, whereas the OPR does. Developing the OPR in conjunction with the architectural program reduces programming effort and provides valuable information to the designers that they typically do not have. In combination with the architectural program, the OPR provides a strong foundation for a successfully integrated design and delivery process and the building's operational criteria. The OPR, however, should not duplicate information contained in the architectural program. The OPR forms the basis from which the commissioning provider verifies that the developed project meets the needs and requirements of the owner. An effective commissioning process depends on a clear, concise, and comprehensive OPR with benchmarks for each of the objectives and criteria. This written document details the functional requirements of the building and the expectations of how it will be used and operated.

If no formal program exists, the OPR document can be used to assist with identifying the criteria the design team is tasked with meeting. However, the main purpose of the OPR is to document the owner's objectives and criteria. The designer's Basis of Design documents, the assumptions the designers made to meet the OPR, and a summary of this information is provided to the operators of the project long after the design and construction team have left the project. As such, the OPR document provides the benchmark against which the design, construction, and project operating performance can be measured.

Design Phase

During the design phase commissioning, the CxA often will develop design checklists and specifications that incorporate commissioning into the project. The design checklists reflect objectives and criteria that designers should check during their quality control process, the amount and type of information to be provided at each stage of the commissioning design review, and the designer's assertion that items in the commissioning checklists are complete. Typically, two or three reviews occur in most design phases. The specifications identify the roles and responsibilities of the construction phase project team, the systems that will be commissioned, and the criteria for acceptance of the commissioned systems.

Commissioning design reviews focus on assessment of the design meeting the OPR. Typically, the OPR provides the project vision; expected service life; energy and water efficiency goals; maintainability; training of operational and maintenance personnel; infrastructure for monitoring-based, ongoing commissioning; expectations of the building envelope's ability to resist weather, air, and water intrusion; and other functional requirements needed for occupants to effectively and efficiently deliver their daily mission. Commissioning design reviews can also randomly check the designer's quality control process by identifying systemic design issues that lead to potential change orders, including missing or misleading information and insufficient detail needed for accurate pricing and competitive bids. Commissioning design reviews can reduce owner and project team risk, allow corrections to be made at the lowest possible cost, and reduce requests for information, supplemental instructions, and schedule impacts. Third-party commissioning provides great benefits to the project team and owner. (See further discussion in the "Commissioning Models" section later in this chapter.)

Design phase commissioning promotes communication, identifies disconnects, questions design elements that appear incorrect, and shares experience to produce a better set of contract documents and a better building.

Several engineering trade magazines and long-term owners, along with insurance companies who provide errors and omissions (E&O) coverage to the design community, have all voiced their concerns about the quality of construction documents and have charted how E&O premiums are affected as a result of judgments or settlements. Design phase commissioning reduces the risks of change orders, accompanying construction delays, and E&O claims, and helps clarify construction documents. Design phase commissioning, if correctly implemented, is a seamless process that provides benefits to the entire project team.

One role of the CxA is to assemble a review team experienced in the type of building being designed. Generally, the CxA has a team of reviewers with specific background and experience to review the disciplines selected for design phase commissioning. This process often requires the most senior individuals as part of the design phase commissioning provider team. (See "Selection of a CxA" later in this chapter.) A typical design review process is as follows:

- Written comments from the reviewers are provided to the design team and owner.
- Comments are reviewed, and the design team returns written responses.
- Meetings are scheduled with the review team and the designers to adjudicate comments as necessary, allowing the owner to understand the issues and have an opportunity to provide direction as needed.
- Design concerns, comments, and actions taken are recorded in the design review document. Changes are made as agreed and the commissioning review team verifies the change and closes the issue as appropriate.

Using a best-practices approach, the design phase commissioning process could occur four different times during the design: at 100% of schematic design, 100% of design development, 50% complete construction documents, and approximately 95% of construction documents. (Combining the review of the first two phases can be done on smaller projects.) An advantage of four reviews, however, is that the design is evaluated based on the OPR document goals before the design development phase starts. Although four reviews is ideal, having at least two (as required for commissioning by ANSI/ASHRAE/USGBC/IES Standard 189.1) provides at least some level of checks and balances.

Changes to optimize building performance, daylighting considerations, system selection, and stacking/massing synergies can best be addressed during schematic design review. Review during design development allows the team to identify potential problems and constructability perspectives early enough to resolve many issues before the construction documents phase starts.

The quantity of design concerns typically increases during the 95% construction document phase because more detail is provided about each building system and component. The concerns identified at that point typically revolve around details—finishes, coordination conflicts, etc. Resolving these concerns provides clearer direction to the contractors, resulting in better cost and schedule predictions.

It is not uncommon that concerns identified at the 50% construction documents phase will go unaddressed by a design team. This is especially true in fast-track projects when designers, responding to owner and contractor demands for documents, struggle to finish and deliver their work product. This is why the later 95% review is so important.

Construction Phase

The CxA's role during the construction phase is to review the 95% complete construction documents and submittals; develop and integrate contractor construction checklists; identify and track issues to resolution; develop, direct, and verify functional performance tests; observe construction of commissioned systems; review the O&M manuals provided under the contract; and oversee development of a systems manual. The purpose of these activities is to verify that the requirements of the OPR document have been met, commissioned systems are serviceable, commissioned systems perform as intended, and operational personnel receive the training and documentation necessary for operating and maintaining building performance.

The review of final construction documents is used to verify that the concerns and issues raised during the design reviews have been resolved, reducing the risk of contractors building flaws into the project. Sometimes, agreement between the designers and the commissioning design review team is not reached during the design phase. An example of this would be a disagreement over building pressurization control: the designer may feel that the design provides adequate control, and the CxA may disagree. The CxA must verify whether or not the building is correctly pressurized through performance tests. If the designer is correct, the CxA closes the issue after verification. If pressurization is an issue, then the team has the opportunity to correct the concern before more serious and costly ramifications can occur. Systems failing to perform are identified and the project team works to resolve the issue while the entire team is still engaged in the project.

The CxA reviews product submittals to look for potential performance problems and to verify that what was proposed is in fact delivered. An example of this would be contractor's ductwork shop drawings showing high-pressure loss fittings that would increase energy consumption. The CxA review activity does not take the place of the designer's review, nor is it meant to. The CxA review should reveal whether the contractors are following the designer's intent. Several purposes can be combined by the CxA in his or her review of the submittals, depending on the role the owner defines for the CxA. For instance, if the CxA is also assisting the team with LEED certification, the CxA can verify that the sustainable development goals identified in the OPR document are being met.

At the start of construction, a CxA may choose to be part of prebid conference meetings with the general contractor or construction manager to provide a description of commissioning activities, to describe the general roles and responsibilities the contractor will be asked to fulfill in the commissioning process, and to answer questions. Clear communication with the contractors during prebid has proven important in preventing high bids that reflect a fear factor from contractors who are unfamiliar with the commissioning process.

In the early stages of construction, the CxA develops a commissioning plan that defines the commissioning process, the roles and responsibilities of the project team, lines of communication, systems being commissioned, and a schedule of commissioning activities. The CxA conducts an initial commissioning scoping meeting where the commissioning plan is reviewed by the project team and, based on this information, a final commissioning plan is developed and implemented.

Prior to the testing of systems and equipment, the CxA will ensure that the necessary prefunctional tests and observations have been completed. The prefunctional tests are those where the equipment installation is confirmed as ready to receive power. This includes checks such as correct phasing of electrical connections, correctly installed mounting fasteners, accessible valve operators, and secure pipe and duct connections. Often, the prefunctional test procedures are developed by the manufacturer and verified by the contractor. All prefunctional test reports should be retained and made part of the systems manual.

Throughout construction, the CxA observes the work to identify conditions that would impair preventive maintenance or repair, hinder operation of the system as intended and/or compromise useful service life, and to verify other sustainability goals (such as indoor air quality [IAQ] management) during construction. In recent years, there has been a growing recognition of the importance for building envelope commissioning, and it is during the construction process that envelope commissioning is most valuable. Typically, envelope commissioning is provided by third-party specialists and is often a separate contract from systems commissioning. The CxA may help develop activities to help perfect installation procedures at the start of the specific construction activity and coordinate activities to help ensure that the contractor's quality control process is working through verification of construction checklists, start-up procedures, and testing and balancing. When contractors have completed their construction checklist for a specific system being commissioned, they are stating that their systems will perform as intended. The CxA verifies this by directing and witnessing functional performance testing.

Acceptance Phase

The functional testing phase of the commissioning process is often referred to as the *acceptance phase* and is the one phase that is most commonly associated with commissioning. With designer and contractor input, the CxA develops system tests (functional tests) to ensure that the systems perform as intended under a variety of conditions. Contractors under the CxA's direction execute the test procedures, while the CxA records results to verify performance. The tests should verify performance at the component level through inter- and intrasystem levels. Another practice is to have the contractors simulate failure conditions to verify alerts and alarms, as well as system reaction and interaction with associated systems. Problems identified are resolved while contractors and materials are still on site and the designers are engaged.

The functional tests are preestablished by the CxA and submitted to the contractor early in the construction phase. The contractor is aware at an early stage of all the tests that must be observed. Testing and balancing (TAB) typically provided by a third party (not the CxA), should include a dry run of the functional tests before the formal CxA witness to ensure that the systems are performing as intended. Often, the systems are not checked in advance of the formal CxA-witnessed test and the systems fail, causing time delays and additional expenses for all parties involved. The contractor must correct the discrepancy, and the CxA must return for a rewitness. It is important that the CxA have clear contractual contingencies in place in order to adequately compensate for retests.

Warranty Phase

The first year of a project is critical to finding and resolving issues that arise, and the CxA plays an important role in helping ensure that a building performs at its optimum. The warranty period is also the period when the contractors and manufacturers are responsible for the materials and systems installed, and is the only time the owner has to identify warranty repairs without additional construction costs. As such, the CxA takes on a vital role to assist the owner and operational staff in identifying and resolving problems and assisting with resolution that gives the least interruption to the occupants during this critical period. This phase may also be coordinated with a postoccupancy evaluation, particularly if one is being done as part of the LEED green-building rating program.

During the first several months, the CxA verifies that systems are performing as intended through monitoring of system operation. Many systems cannot be fully tested until the building is occupied or the systems are called to be operated during near-peak cooling or heating situations. There might be a small percentage of system components that pass functional testing but, under actual load, fail to perform as intended. These components must be identified in the warranty period and replaced or repaired as necessary. The CxA's role in conjunction with the operational staff is to search out problems that only become evident under actual load. To accomplish this, the commissioning provider performs several specific tasks.

The commissioning provider may also help identify system points to trend, verifying efficient system operation; installs independent data loggers to measure parameters beyond the capabilities of the building automation system (BAS); and monitors utility consumption. Using the trend system data from the selected BAS input points, the CxA can help in the analysis of the information, looking for operational sequences that consume resources unnecessarily and conditions that could compromise occupant satisfaction within the working environment. In addition, the CxA also looks for conditions that could result in building failure, such as high humidity in interstitial spaces in the building's interior, hot spots in the electrical distribution system, or analysis of electrical system harmonics where power quality is essential to the owner. These functions can only be tested after occupancy.

Some systems, such as the heating and cooling equipment, can only be fully tested when the season allows testing under design load conditions. The CxA works closely with the operational staff to identify and help resolve issues that become apparent in the warranty period and verifies that the operational staff fully understands and meets their warranty responsibilities. In addition, the CxA should provide operational staff with the specific functional test procedures developed for their use in maintaining building performance for the life of the building. By having the CxA work with the operational staff during the warranty period, the operators gain valuable insight into how the building should operate and what to look for to ensure continued performance. This helps to overcome a typical industry problem: the bypassing of system components and controls because of a lack of understanding of how the systems are intended to operate.

Ongoing Commissioning Phase

An important part of green design is verification that the goals defined by the owner are integrated by the design and construction team to achieve the owner's objectives for the lifetime of the building, as defined in the OPR. The OPR is a living document that changes over time as the owner requires change. Once the building is occupied, changes in the owner/occupant mission can occur that impact and modify the original project requirements. The original OPR is maintained for its historical value in marking the original project requirements. Using the original OPR as a starting point, the OPR is converted to the current facility requirements (CFR), which documents the changes the owner needs to match the changes in his or her daily mission. The CFR provides the foundation to guide the future project teams in modifications needed to meet the changing needs of the owner/occupant.

Ongoing commissioning in both new and existing projects provides the owner and operators with the tools necessary to efficiently manage financial and human resources to achieve desired returns on their investment, building performance, and reduction of environmental impact. Owners and operators cannot manage what is not measured. During occupancy, the CxA monitors the building's performance and identifies deterioration of performance. This deterioration typically falls into one of two categories: operator error or system malfunction. Monitoring-based, ongoing commissioning identifies the cause of the deterioration and recommendations can be made for correction. Operator errors resulting in deterioration of building performance provide training opportunities to improve O&M staff understanding of correct building operation, which helps prevent future operational mistakes. Monitoring-based, ongoing commissioning identification of degradation of building performance due to system malfunction provides in-depth failure information, which lessens the O&M staffs' troubleshooting efforts and can verify that the repairs performed resolve the problem. Monitoring-based commissioning is easily implemented by the CxA when integrated into the OPR. Implementation in existing buildings does require more effort and some additional cost, but it provides significant financial benefits immediately and over the lifetime of the building. Commissioning authorities can also help integrate measurement and verification (M&V) plans and procedures that can be used to identify when a building begins operating outside of allowable tolerances, signaling the owner that corrective action is needed to maintain performance.

The operational staff's knowledge and understanding of how the building should be operated, and their methods of operation in practice, typically determine the actual building system's performance. The most common reason facilities fail to meet performance expectations or experience deterioration of performance is tied directly to how the building is operated. Changes in operational personnel often result in loss of the institutional knowledge needed to maintain the building's performance and meet occupants' needs. The system manual is the repository for essential information needed by the operator to maximize occupant satisfaction and building performance. Without this information, the operator is running blind, without reference points for correct system operation. The result is that it may be difficult to meet the owner's goals and objectives and operation costs may increase. Monitoring-based, ongoing commissioning provides regular monitoring and analysis of utility usage, system interaction, and operator performance. Performed on a regular basis, monitoring-based commissioning improves financial and operator performance by changing the culture from constantly responding to problems and replacing it with routine maintenance. It also helps to ensure that institutional knowledge is not lost through changes in personnel, occupant mission, and building modifications. For owners to get maximum performance from their facilities, they must know when systems fall outside of allowable performance tolerances or be aware of when operational personnel make changes that could negatively impact the facilities' performance. This is best done through monitoring-based, ongoing commissioning.

For ongoing commissioning activities, it is not necessarily required that a third party CxA be retained to oversee the day-to-day activities. Ideally, the facilities staff can perform the function and integrate ongoing CxA into the management of the facility. Modern facility management is a highly skilled and high-tech enterprise and some major universities are now offering doctoral programs in this field.

ASHRAE Guideline 14, *Measurement of Energy and Demand Savings* (ASHRAE 2002) provides several methods to establish operational benchmarks for energy consumption. Sustainable, healthy, high-performing facilities often measure other parameters including water consumption, waste generation, recycling, pesticide use, etc. The operational tracking of these parameters reduces total cost of ownership, impact on the environment, building occupants' quality of life, and can increase building value. An owner can obtain guidance on integrating sustainable operation practices by adopting USGBC's LEED for Existing Buildings Operation and Maintenance program at this stage of the building's life.

Additional information and insight into the application of these concepts in the operation of existing buildings can be found in Chapter 18 of this book, "Operation/Maintenance/Performance Evaluation."

Selection of the CxA

As with finding a doctor, lawyer, contractor, or other professional, the key element is that the commissioning provider should have experience in the types of systems an owner wants commissioned. In other words, an owner must match the experience with the job. A good CxA generally has a broad range of knowledge: hands-on experience in O&M, design, construction, and investigation of building/system failures. CxAs must also be detail-oriented, good communicators, and able to provide a collaborative approach that engages the project team.

ASHRAE provides Commissioning Process Management Professional certification for owners implementing commissioning and engaging commissioning authorities. People who pass this certification understand how to apply the commissioning process and select commissioning providers, which helps ensure the owner gets what they paid for.

SELECTION OF SYSTEMS TO COMMISSION

Although numerous commissioning service models exist, commissioning should be performed by a third-party provider or the owner's own commissioning team whenever possible. Commissioning of all systems using the whole-building approach has proven to be beneficial. However, due to budget constraints, owners may want to look at commissioning systems that will yield the greatest benefit to them. In other cases, the systems selected for commissioning may be specified by code or standard, such as projects that are being conducted under the requirements of ANSI/ASHRAE/USGBC/IES Standard 189.1. Long-term owners have an advantage and can apply their experience of where they have historically encountered problems and elect to commission only those systems. Others who do not have that depth of experience may wish to talk with long-term owners or insurance providers to gain perspective.

Commissioning was originally developed in areas where energy efficiency was a prime driver. The commissioning process has expanded beyond the original commissioning of HVAC systems to include building envelope, electrical, plumbing, lighting, security, etc. This expansion from HVAC is often referred to as *whole-building commissioning*.

There are many factors that define which building systems should be commissioned, but there are no published standards yet to help guide owners through such selection (ASHRAE has produced guidelines, but not standards, in this area.) It often depends on the associated risk of not commissioning. E&O insurance providers publish graphs of claims against design professionals by discipline. Interestingly enough, 80% of the claims against architects are for moisture intrusion, thus indicating a need for considering building envelope commissioning. In addition, some organizations have suggested that up to 30% of new and remodeled buildings worldwide may be the subject of excessive complaints related to IAQ. The reasons for sick buildings include inadequate ventilation, chemical pollutants from both indoor and outdoor sources, and biological contaminates.

Commissioning provides several benefits, two of which are risk reduction and generally an overall lower total cost of ownership. Based on a specific climate such as Phoenix, Arizona, the risk of not commissioning the building envelope is much less than in Atlanta, Georgia (due to humidity concerns). Based on functional requirements, declining to commission the security systems in a conventional office building may have minimum risk compared to a federal courthouse. So, what should be commissioned?

The best time for determining what systems should be commissioned is during the development of the OPR document. Generally, there are three main system categories that, as a minimum, should be commissioned: building envelope, mechanical and plumbing systems, and electrical systems. If the building project includes an on-site renewable energy system, that should also be considered a definite candidate for commissioning. Depending on the functional requirements of a building and the complexity of systems, additional systems that may be commissioned include security, voice/data, selected elements of fire and life safety, irrigation and/or process water systems, energy or water monitoring systems, and daylighting controls. In some cases, these systems may be commissioned by the manufacturer's installer, such as is the case with fire and life safety systems. ANSI/ASHRAE/USGBC/IES's Standard 189.1 is more specific on the systems that should be commissioned and could be considered a good general guideline to follow. Some building systems are installed with oversight and field testing by the manufacturer (elevators, for example) and these systems do not require a third party commissioning agent.

USGBC's LEED rating system recognizes the benefits of commissioning and its importance to green-building design, construction, and operation. As such, the LEED reference guide implies that building systems that affect energy consumption, water usage, and indoor environmental quality should be commissioned. Each element of green design needs verification to ensure that the design, construction, and operation of the high-performance building meet the expectations of the team and realize the financial return envisioned by the owner.

COMMISSIONING MODELS

Independent third-party commissioning is the preferred commissioning approach, because it reduces the potential for conflict of interest and puts a quality advocate squarely in the owner's corner. It also allows for integration of commissioning professionals specialized to meet a project's specific needs concerning the building envelope, security systems, labs, etc.

Having commissioning be part of the general contractor's responsibility has many of the same problems as the model using the design professional. Contractors, by the very nature of the construction business, are focused on schedule and budget. This focus is not always in the owner's best long-term interest. Most contractors are quality-minded and do their best to identify problems and assist with resolutions (though often to their detriment because they inadvertently take responsibility for the design in doing so). If a constructability issue arises that will adversely affect the schedule or budget, the contractor may choose to fix the issue and hope that it does not create a warranty callback. The main problem is that many of the issues are discovered too late in the process, again resulting in change orders, construction delays, and additional costs.

To truly be an owner's advocate, the CxA must owe allegiance to no one but the owner. A third-party CxA will verify that the goals defined by the owner and integrated by the design and construction team are achieved as intended, from the first day of occupancy. If the CxA is separate from the design professional or contractor, he/she will provide unbiased reporting of issues to the team and will guide them toward timely solutions without finger-pointing, delays, and liability.

THE GROWTH OF COMMISSIONING

Commissioning of projects has been growing in importance during the past decade. This is partly due to the LEED mandatory requirement for commissioning and to the growing awareness of the benefits commissioning brings to a project. In addition, ANSI/ASHRAE/USGBC/IES Standard 189.1-2011 includes a requirement for commissioning for all buildings with floor area greater than 5,000 ft^2 (500 m^2). There is some debate on whether all high-performance green buildings should be required to undergo a full commissioning process. Certainly any building that includes complex systems or a larger number of energy and water consuming devices should be considered for commissioning.

ONE FIRM'S COMMISSIONING CHECKLIST

- Begin the commissioning process during the design phase; carry out a full commissioning process from lighting to energy systems to occupancy sensors.
- Verify and ensure that fundamental building elements and systems are designed, installed, and calibrated to operate as intended.
- Engage a CxA that is independent of both the design and construction team.
- Develop an OPR document and review designer's basis of design to verify requirements have been met.
- Incorporate commissioning requirements into project contract documents.
- Develop and use a commissioning plan.
- Verify installation, functional performance, training, and O&M documentation.
- Complete a commissioning report.
- Perform additional commissioning:
  - Conduct a focused review of the design prior to the construction documents phase.
  - Conduct a detailed review of the construction documents when these are considered nearly complete and prior to their issuance for construction.
  - Conduct reviews of contractor submittals that are relevant to systems being commissioned.
  - Provide information required for recommissioning systems in a single document to the owner.
- Have a contract in place to review with operations staff the actual building operation and any outstanding issues identified during commissioning, and to provide assistance resolving these issues within the warranty period.

> - Encourage long-term energy management strategies.
>   - Provide for the ongoing accountability and optimization of building energy and water consumption performance.
>   - Design and specify equipment to be installed in base building systems to allow for comparison, management, and optimization of actual vs. estimated energy and water performance.
>   - Use M&V functions where applicable.
>   - Tie contractor final payments to documented M&V system performance and include in the commissioning report.
>   - Provide for an ongoing M&V system maintenance and operating plan in building O&M maintenance manuals.
> - Operate the building ventilation system at maximum fresh air for at least several days (and ideally several weeks) after final finish materials have been installed before occupancy.
> - Provide for the ongoing accountability of waste streams, including hazardous pollutants.
> - Use environmentally safe cleaning materials.
> - Train O&M workers.

REFERENCES AND RESOURCES

Published

ASHRAE. 2002. ASHRAE Guideline 14-2002, *Measurement of Energy and Demand Savings*. Atlanta: ASHRAE.

ASHRAE. 2005. ASHRAE Guideline 0-2005, *The Commissioning Process*. Atlanta: ASHRAE.

ASHRAE. 2007. ASHRAE Guideline 1.1-2007, *HVAC&R Technical Requirements for the Commissioning Process*. Atlanta: ASHRAE.

ASHRAE. 2013. ANSI/ASHRAE Standard 202, *Commissioning Process for Buildings and Systems*. Atlanta: ASHRAE.

CHAPTER SEVEN

GREEN RATING SYSTEMS, STANDARDS, AND OTHER GUIDANCE

Rapid growth in interest in green buildings over the past two decades has occurred with corresponding growth in the number, depth, and breadth of green-building resources available.

There are three general types of programs or resources that exist to encourage green-building design. The first type comprises green-building rating systems (sometimes referred to as *building label* programs), such as the LEED program. Second are general guidelines or resources, such as this guide, that have created and published to encourage and assist designers in achieving green-building design. Third is the recent adoption of green-building practices as part of building design standards and the code enforcement process. This chapter provides a brief summary of each type and cites several specific examples.

GREEN-BUILDING RATING SYSTEMS

Various rating systems that have been developed by organizations around the world that strive to indicate how well a building meets prescribed requirements and to determine whether a building design is green and to what level. They all provide useful tools to identify and prioritize key environmental issues. These tools incorporate a coordinated method for accomplishing, validating, and benchmarking sustainably designed projects. As with any generalized method, each has its own limitations and may not apply directly to every project's regional, political, and owner design-intent-specific requirements. Some rating systems concentrate on energy efficiency, whereas, others place higher values in other areas of sustainability, such as "greening" brown fields, making use of existing structures or increasing the indoor environmental quality (IEQ) (Grondzik 2001).

It should not be implied that this guide advocates the exclusion of these elements. By advocating both green and good design in any building design endeavor, this exclusion is avoided. Thus, while this guide does not endorse or recommend use of any one particular green-building rating system or program, it does encourage their use when the application will produce an exceptional green design and encourages the building operators to maintain and operate the building in a manner which provides its occupants a continuing healthy and energy-efficient living/work space.

It could easily be said that the green-building movement really started in earnest with the initial establishment of the BREEAM® rating system in 1990. BREEAM is a creation of the Building Research Establishment (BRE) in the UK and the acronym stands for BRE Environmental Assessment Method. This is a voluntary, consensus-based, market-oriented assessment program. With one mandatory and two optional assessment areas, BREEAM encourages and benchmarks sustainably designed office buildings. The mandatory assessment area is the potential environmental impact of the building; the two optional areas are design process and operation/maintenance. Several other countries and regions have developed or are developing related spinoffs inspired by BREEAM, and BREEAM has been adopted in other countries. Although initially focused on specific building types, it is been adapted to include a wider range of different types of buildings. Similar to what has happened with other green rating systems, BREEAM has been adapted to various type of programs (called schemes), such as BREEAM for new construction, domestic refurbishment, communities, in-use (existing buildings), and for homes (EcoHomes).

The rating method primarily used in the United States is the Leadership in Energy and Environmental Design (LEED®) program, created by U.S. Green Building Council (USGBC). USGBC started offering this system in 1998 and describes it as a voluntary, consensus-based, market-driven green-building certification system. It evaluates environmental performance from a "whole-building" perspective over a building's life cycle, providing a numerical standard for what constitutes a green building. USGBC's goal has been to raise awareness of the benefits of building green, and it has transformed the marketplace. LEED has been applied to numerous projects over a range of project certification levels, and its use has grown rapidly over the past several years. The LEED rating system started out with a basic program for new construction, but because a large majority of buildings already exist, a LEED for existing buildings was released in 2004 and has been revised several times since. LEED rating systems have also been developed for a variety of specific building types. These include, for example, building core and shell and commercial interiors for project developers and tenants (respectively), as well as schools, retail, hospitality, data centers, warehouses and health care, and homes. Growing in importance is the role of LEED for Existing Buildings Operations and Maintenance (EBOM). The LEED program and registered building projects have already been, or are being, established in other countries including India, Australia, Canada, and China, and new project registrations in countries outside the U.S. make up about 40% of the total, according to recent trends. Many other countries are developing their own green-building rating systems based on the criteria in LEED and other rating systems. A brief synopsis of the LEED program is included at the end of this section.

Another rating method that was originally developed in Canada and is being used in the United States is the Green Globes program. Green Globes is an online auditing tool that includes many of the same concepts as LEED. While both aim to help a building owner or designer develop a sustainable design, Green Globes is primarily a self-assessment tool (although third-party assessment is an option) and also provides recommendations for the project team to follow for improving the sustainability of the design. In the UK, Green Globes is known as the Global Environmental Method program.

Other major green-building rating programs exist in countries throughout the world, such as Australia's Green Star, Japan's CASBEE (Comprehensive Assessment System for Built Environment Efficiency, Hong Kong's BEAM (Building Environmental Assessment Method), and the Estidama program in the United Arab Emirates, among others.

The procedures used by those organizations and governmental providing building rating systems vary. Many building rating programs use static building labels, that is, the building's energy (or other green attributes) is evaluated once, the label is applied, and the providing organization does no reassessment to determine if the building continues to actually meet the original specifications. The application of dynamic labeling is preferred. A few of the building rating programs actually do include a reassessment (e.g., every two years), and buildings not continuing to perform lose their building label (Means and Walters 2010).

THE LEED RATING SYSTEM

Since its development and introduction in the late 1990s, the LEED program has become a major factor in the advancement of green buildings, as well as an influence on how all buildings are thought of in the design and construction process.

LEED is a voluntary program that uses a point-based rating system for a given building project. In 2009, USGBC released a significant revision to the program that, among other things, reworked the point ratings in order to provide a more effective focus to drive positive environmental and health benefits. Part of this revision was based on using the Tool for the Reduction and Assessment of Chemical and Other Environmental Impacts (TRACI) developed by the U.S. Environmental Protection Agency (EPA). An illustration of the impact of the various credit point option weightings is given in Figure 7-1.

Another 2009 change to the LEED program (which came to be known as LEED for Version 3) was the introduction of regional credits, which are used to help put additional emphasis on design features that are particularly important in terms of the climate and area where the project is to be located. For example, in areas with known stormwater problems, the Sustainable Sites Credit 6.1 for

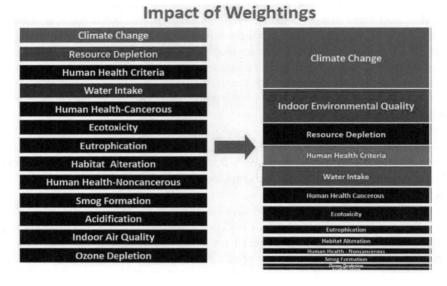

Image courtesy of Brendan Owens, U.S. Green Building Council.

Figure 7-1 The impact of weightings on revising the LEED credit point system for LEED Version 3.

Stormwater Design—Quantity Control may be included in this priority list. A full list of the Regional Priority credits can be found on the USGBC website (see the "References and Resources" section at the end of the chapter).

USGBC went through an extensive process for another significant revision to LEED, which will be known as LEED v4. The final draft of this revision will be voted upon during the summer of 2013 (just after the time of this writing), so a final decision on the fourth version is not known. However, the notes below provide a summary of some of the changes in the draft version available as of this writing.

The breakdown of credits and points available under the LEED v4 for the new construction rating system is given below. One significant point is that the Sustainable Sites topic has been broken up into Sites and Location and Transportation. These areas can generally be described as the impact of the building project design on a given site and area (Sustainable Sites) and the external factors to the building design (Location and Transportation):

- Integrative Process   1 credit/1 point
- Location and Transportation   8 credits/16 points
- Sustainable Sites   6 credits/10 points

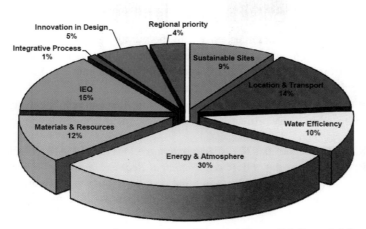

Image courtesy of IBrendan Owens, U.S. Green Building Council.

Figure 7-2 Credit point distribution for LEED for new construction, v4.

- Water Efficiency   4 credits/11 points
- Energy and Atmosphere   7 credits/33 points
- Materials and Resources   5 credits/13 points
- Indoor Environmental Quality (IEQ)   9 credits/16 points

Total core points   100
Innovation/design   6
Regional priority   4
TOTAL POSSIBLE   **110**

Achieving 40 points will earn LEED Certified status, and the higher levels of silver, gold, and platinum can be achieved with 50, 60 and 80 points, respectively.

The percentage distribution of these credits is illustrated in Figure 7-2 for the most recent version of LEED. "Energy and Atmosphere" is the category with the highest number of points available, and this is in the direct purview of the HVAC&R engineer. Note that the LEED energy prerequisites and credits have been updated to use the more stringent Standard 90.1-2013 as the baseline reference.

USGBC also produces a reference guide to go along with the corresponding program. This reference provides guidance on complying with the credits, but also can be considered a good general resource for any project. The latest reference guide available (as of the writing of this *GreenGuide* edition) is for version 3, and is listed in the References section below.

GUIDELINES AND OTHER RESOURCES

Other programs and resources exist with the purpose of guiding or encouraging green-building design but without assigning a specific rating. This section outlines just some of those resources that the green-building designer can access, starting with those that have come from ASHRAE.

ASHRAE has produced a number of publications that can help the inexperienced and experienced industry professional alike. These include the following:

- ASHRAE Guideline 0, *The Commissioning Process* (ASHRAE 2005) and ASHRAE Guideline 1.1, *HVAC&R Technical Requirements for the Commissioning Process* (ASHRAE 2007). (Note that commissioning is the subject of a new ASHRAE Standard 202, *Commissioning Process for Buildings and Systems* [2013].)
- The Advanced Energy Design Guide series. (The 30% Guides were developed initially and were intended to offer a 30% savings compared to Standard 90.1-1999, while the more recent 50% Guides are designed for 50% savings compared to the 2004 version of Standard 90.1.
- *Indoor Air Quality Guide: Best Practices for Design, Construction and Commissioning* (ASHRAE 2009)
- *The ASHRAE Guide for Buildings in Hot and Humid Climates*, 2nd Edition (ASHRAE 2009)
- ASHRAE also developed user's manuals for many of its most widely used standards, such as Standards 62.1, 62.2, 90.1 and 189.1.
- *High Performing Buildings* magazine. This publication was created by ASHRAE several years ago to provide real-world, case study examples for reference.

Many of these are also discussed elsewhere in this guide but are mentioned here as a reminder.

The Advanced Energy Design Guide series is a series of books that provide a set of prescriptive technical approaches to achieve significant energy savings. The documents are focused on specific building types, typically smaller building projects that may not have resources available for much engineering study and analysis of energy saving technologies (e.g., small retail stores). Recommendations are provided based on the climate zone the project is located in. The initial series of guides was targeted toward achieving 30% energy savings compared to Standard 90.1-1999, and the intent is to repeat the series with increasing efficiency levels leading to net zero designs. The more recent 50% Guides are designed for 50% savings compared to the 2004 version of Standard 90.1. Additional guides with a higher efficiency target are in the planning process, and funding considerations will determine how far they go (e.g., a net zero target by 2020 for the AEDG series is one consideration).

These energy guides were produced in collaboration with the American Institute of Architects (AIA), the Illuminating Engineering Society of North America (IES), and USGBC, with assistance provided by the U.S. Department of Energy (DOE). These guides are available for free from the ASHRAE website and have been widely distributed since their release.

The *Indoor Air Quality Guide* was released in January of 2010, and was developed in conjunction with AIA, the Building Owners and Managers Association (BOMA), the Sheet Metal and Air Conditioning Contractors National Association (SMACNA), the U.S. EPA and USGBC. A summary of the guidance offered is available for free download from the ASHRAE website, while the detailed guidance document is available for purchase.

BUILDING ENERGY QUOTIENT (bEQ)

In 2009, ASHRAE unveiled a building energy labeling program known as the Building Energy Quotient (bEQ) program. The program provides a method to rate a building's energy performance both "As Designed" (Asset Rating) and "As Operated" (Operational Rating).

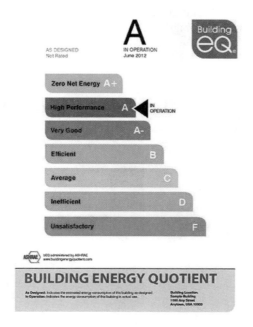

Figure 7-3 Sample Building Energy Quotient plaque.

ASHRAE's Building Energy Quotient program provides the general public, building owners and tenants, potential owners and tenants, and building operations and maintenance staff with information on the potential and actual energy use of buildings. This information is useful for the following reasons:

- Building owners and operators can see how their building compares to peer buildings to establish a measure of their potential for energy performance improvement.
- Building owners can use the information provided to differentiate their building from others to secure potential buyers or tenants.
- Potential buyers or tenants can gain insight into the value and potential long-term cost of a building.
- Operations and maintenance staff can use the results to inform their decisions regarding maintenance activities, influence building owners and managers to pursue equipment upgrades, and demonstrate the return on investment for energy efficiency projects.

Beyond the benefit received by individual building owners and managers, the increased availability of building data (specifically the relationship between the design and operation of buildings) will be a valuable research tool for the building community.

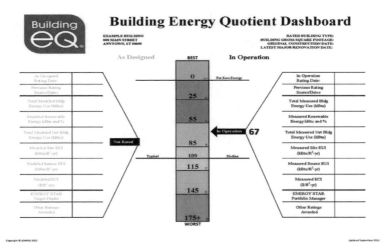

Figure 7-4 Sample Building Energy Quotient dashboard.

Other Resources

Further information regarding these resources can be found in the References and Resources section at the end of the chapter. Other guides and methods include the following:

- *The Whole Building Design Guide*
- *The Living Building Challenge* V2.0 (An updated version 2.1 of this program was released in May 2012)
- Green Building Advisor
- California Collaborative for High Performance Schools (CHPS)
- *Minnesota Sustainable Design Guide*
- New York High Performance Building Guidelines (Released in 1999 by the nonprofit Design Trust for Public Space, but still has relevant information for today)

Work referred to by architects includes:
- *The Hannover Principles* (1992)
- GreenSpec® Product Guide
- Information from The Natural Step (a nonprofit organization)
- International Organization for Standardization (ISO 14000 family of standards)
- Building for Environmental and Economic Sustainability (BEES). (This software tool to analyze the impact of a building uses a life-cycle assessment approach as specified in ISO 14040. An online version was released in 2010.)
- Tool for the Reduction and Assessment of Chemical and Other Environmental Impacts (TRACI)

IMPLEMENTATION IN THE FORM OF STANDARDS AND BUILDING CODES

Since the middle part of the past decade, there has been a movement to make green-building practices a more mandatory part of the normal building code process. Several cities in the United States now require LEED certification for building projects above a certain size or classification (such as a government building). In addition, ASHRAE has initiated a process to create a series of new standards for high-performance green buildings, releasing in early 2010 the initial version of Standard 189.1, *Standard for the Design of High-Performance Green Buildings* (ASHRAE 2009). Also in 2010, California became the first state to issue a green-building code (known as CAL-Green). See the accompanying Sidebar for more information on CALGreen.

ASHRAE/USGBC/IES STANDARD 189.1

In 2006, ASHRAE (in conjunction with USGBC and the Illuminating Engineering Society) began a process to create a standard that would address a growing need within the industry for a code-language document for green buildings suitable for adoption as part of building codes. ANSI/ASHRAE/USGBC/IES Standard 189.1-

CALGREEN CODE: AMERICA'S FIRST STATEWIDE GREEN-BUILDING CODE

In 2010, the California Green Building Standards Code (CALGreen) was adopted to promote the design of efficient and environmentally responsible residential and nonresidential buildings in California. The CALGreen code is part of the overall California Building Standards Code and is the first statewide green code established in the U.S. It was developed in an effort to meet the provisions of Assembly Bill (AB) 32, which requires a cap on greenhouse gas emissions by 2020, with mandatory reporting. The 2010 CALGreen Code became effective January 1, 2011 and a modified version takes effect in January 2014.

To reduce the overall environmental impact of new buildings constructed in California, the CALGreen Code adopts many green-building practices as mandatory building code requirements. The CALGreen Code includes requirements (divisions) for planning and design, energy efficiency, water efficiency and conservation, material conservation and resource efficiency, and environmental quality. The code also requires building commissioning to verify and ensure that all building systems operate as designed to meet their maximum energy efficiency targets.

Some similarities to LEED programs include standards for stormwater pollution prevention, light pollution reduction, indoor and site water savings, construction waste management, energy performance, outdoor air delivery, carbon monoxide monitoring, and materials selection. In some cases, CALGreen has stricter targets than LEED; in others, LEED is stricter, and in many others, the requirements are identical. There are several CALGreen requirements not found in LEED, such as installing water meters on buildings that are greater than 50,000 square feet, providing weather-resistant exterior walls and foundation envelopes, defining the type of fireplace that can be installed, and employing acoustical control (interior and exterior).

In addition to the mandatory statewide CALGreen requirements, a city or county may adopt local ordinances to require more restrictive standards that go above and beyond the mandatory measures. These packages of voluntary measures, called Tier 1 and Tier 2, include a set of provisions from each code division. These provisions are additional measures which, in some cases, are stricter than the mandatory codes. For instance, building energy performance must exceed the California Energy Code (Title 24) by 15% and 30% for Tier 1 and Tier 2, respectively. Additionally, Tier 1 includes one additional elective from the water efficiency division, whereas Tier 2 includes 3. Some of the cities that have adopted Tier 1 include Burlingame, Napa, and Santa Rosa. As of the writing of this edition, only Palo Alto has adopted Tier 2.

2009, *Standard for the Design of High-Performance Green Buildings* (ASHRAE 2009) was developed during a more than three-year process with extensive public review and was initially published in early 2010. This standard is in a continuous maintenance process, and an updated version was released in 2011. Another update is planned for release in 2014.

The standard differs from LEED or other products in that it is not a rating system, nor is it a design guideline per se. The purpose of Standard 189.1 is to provide minimum requirements for the siting, design, construction and plans for operation of high-performance green buildings, while attempting to balance environmental responsibility, resource efficiency, occupant comfort and well-being, and community sensitivity. One key point of this is that Standard 189.1 is not targeted for any building project, but rather for high-performance building projects. This document is intended to help fill a perceived gap in the evolving building codes in this area, as localities begin to adopt green-building designs as a requirement. Standard 189.1 is also a compliance option of the *International Green Construction Code*™ (IgCC) (ICC 2010).

While many of the topics and criteria may overlap or seem similar to LEED for new construction, Standard 189.1 differs in that it establishes mandatory, minimal requirements across all topical areas. Besides the obvious intent of providing a vehicle for adoption into building codes, this standard may also be used by developers, corporations, universities, or governmental agencies to set requirements for their own building projects.

Standard 189.1 is not intended to do away with other ASHRAE standards. Rather, it builds upon key ASHRAE standards and adopts these with modifications when considered necessary to develop a document that deals with high-performance green buildings, as illustrated in Figure 7-5.

This standard includes mandatory criteria in all topical areas (for example, water or energy) and provides for two compliance paths. The prescriptive path includes simple compliance criteria; simple in the sense that they are more like a checklist of technologies or system requirements. The performance path is more complicated in that it requires more analysis to verify that compliance is indeed achieved.

A brief overview of some key criteria in Standard 189.1 that the typical ASHRAE member or design professional should be aware of is given below. This is just a brief overview and is not intended to be an all inclusive summary.

Sustainable Sites

In addition to a cool-roof requirement for cooling-dominated climates (climate zones 1–3), the standard also has provisions for building walls to be shaded or to have a minimum solar reflective index value for the opaque wall materials in all but the colder climate zones.

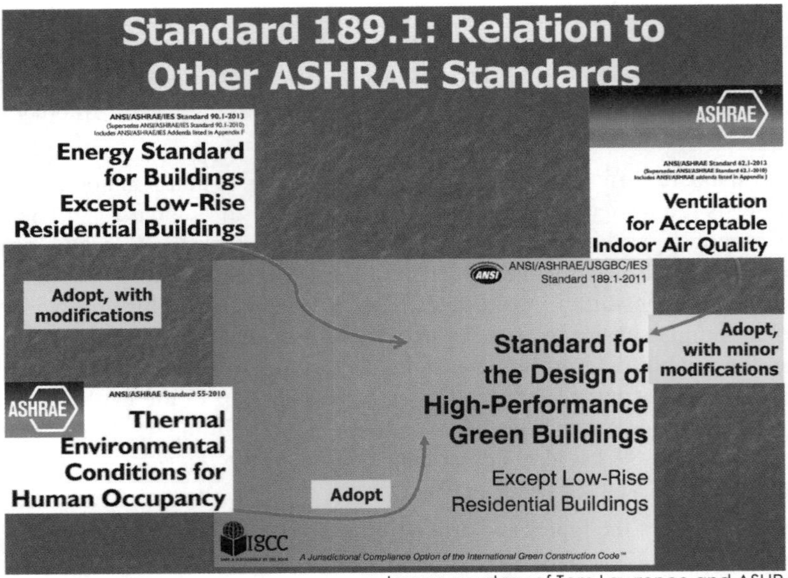

Image courtesy of Tom Lawrence and ASHRAE.

Figure 7-5 Relation of Standard 189.1 (ASHRAE 2009) to other key ASHRAE standards.

Water Use Efficiency

Standard 189.1 puts limits on the number of cycles of water through a cooling tower. It also requires condensate collection on air-handling units above 5.5 tons (19 kW) of cooling capacity in more humid regions (areas with a climate-zone designation of humid).

Standard 189.1 is also unique in that it requires the installation of meters with data storage and retrieval capability on systems and areas above a given threshold in water usage. Similar provisions are included in the energy section for energy use and IEQ section for monitoring of outdoor airflow.

Energy Efficiency

While the LEED program, through EA Credit 1, provides for a sliding scale of points awarded for energy efficiency improvements above Standard 90.1, Standard 189.1 began with the intended goal of providing mandatory measures that would result in buildings using 30% less energy than what is designed according to Standard 90.1-2013, including process loads.

During the development of Standard 189.1, many concepts and requirements were considered for improvements. At the same time, addenda to Standard 90.1-2013 were proposed and approved that increased the overall efficiency levels of Standard 90.1. The net result is that the overall average energy utilization index (EUI) for buildings designed according to Standard 189.1 and Standard 90.1-2013 are not that great. It is ASHRAE's intent to have the energy efficiency levels for Standard 189.1 improve at a faster rate than those for Standard 90.1, with an ultimate goal of having Standard 189.1 reach nearly net zero status by the year 2020 (although the definition of an optimum nearly net zero value has yet to be established).

One of the key considerations when developing the energy requirements for Standard 189.1 was whether to include on-site renewable energy, and if so, to what extent. Many renewable energy systems are not yet fully cost competitive with conventional energy sources, particularly when excluding incentive programs that may go away at any time. Therefore, the standard only includes the provisions for being renewable ready as a mandatory requirement. Exceptions are included for areas with low solar incidence or local shading. In the prescriptive path, on-site renewable energy is included, but a project can still comply with this standard using other methods that would have equivalent benefits or the engineers and designers may elect to go with the performance compliance path.

Energy metering is required for key systems (e.g., HVAC) above certain thresholds. Even when a building is initially designed to be energy efficient, it can quickly slip into being less than stellar in energy efficiency if not continuously monitored.

Standard 189.1 makes numerous modifications to the requirements in Standard 90.1-2013 regarding HVAC systems. Here is a summary of key points:

- The threshold for occupancy levels requiring demand-controlled ventilation is lowered.
- The minimum size requirement for economizers is reduced. Other specific exception and requirement changes are included, but a description is beyond the scope of this *GreenGuide*.
- Fan power limits (per volume of air moved) are lowered.
- The requirements for energy recovery from exhaust air are expanded and the minimum effectiveness of the energy recovery device is set at 60%.
- Levels of duct insulation are increased.
- Unoccupied hotel/motel guest room controls are included.

Depending on the project approach, additional requirements for equipment efficiency beyond Standard 90.1 may also be incorporated. In addition, requirements are set for automated peak demand reduction of the building.

The performance path for showing compliance with the energy section of ANSI/ASHRAE/USGBC/IES Standard 189.1 includes demonstrating equivalent performance in terms of both energy cost and CO_2 equivalent emissions, compared to if the building project had been designed strictly to the criteria in the prescriptive path.

IEQ

There are several criteria for indoor environmental quality that are relevant to ASHRAE members: tobacco smoke control, outdoor air monitoring, filtration /air cleaning, and determination of outdoor airflow rate.

The minimum ventilation design for outdoor airflow is to be according to ASHRAE Standard 62.1 (ASHRAE 2013) using the Ventilation Rate Procedure. Outdoor air monitoring is to be done using permanently mounted, direct outdoor airflow measurement devices. In contrast to LEED, CO_2 monitoring in densely occupied zones is not included as part of Standard 189.1.

Tobacco smoke control is achieved by simply banning smoking within the building and near entrances, outdoor air intakes, or operable windows.

Materials and Resources

A number of requirements included in this section parallel those included with the LEED program for new construction. Items of particular note to an ASHRAE member would be a construction waste management provision to divert a minimum of 50% of nonhazardous waste and demolition debris from being sent to a landfill, a ban on CFC-containing equipment, and for fire suppression systems to contain no ozone-depleting substances.

Construction and Plans for Operation

Standard 189.1 includes provisions for not only how a building should be constructed, but also for planning for how it should be operated once occupied. Since it is written and intended for adoption into building codes, only items that would be expected to be developed and in place at the time a certificate of occupancy is issued could reasonably be considered for inclusion in this standard. The approach taken within Standard 189.1 is to set requirements for the development of plans for operation in critical areas.

This standard includes requirements for building acceptance testing and/or commissioning, erosion control, IAQ, moisture control and idling of construction vehicles to be implemented during construction. Commissioning is to be done according to requirements that in essence parallel those in ASHRAE guidelines. IAQ requirements during construction and before occupancy are similar to those in the LEED program but not identical and when different are generally more stringent.

Plans for operation are required in key areas that would be needed to help ensure the building performs as would be expected for a high-performance, green building. These include criteria in setting up long-term monitoring and verification of water and energy use, as well as IAQ through provisions such as outdoor air monitoring.

Maintenance and service life plans are required as well, and these involve equipment and systems relevant to the HVAC&R or MEP engineer.

INTERNATIONAL GREEN CONSTRUCTION CODE (IgCC)

Soon after the initial release of Standard 189.1, ASHRAE and the International Code Council reached an agreement whereby the standard would be included as an appendix to *International Green Construction Code*™ (IgCC). The IgCC was released in March 2012; it specifies Standard 189.1 as a compliance option. The project team has a choice for compliance: they can comply with the IgCC or with Standard 189.1.

Each of these two (IgCC and Standard 189.1) address the key topical issues for a high-performance green building but with significant differences in the approach. The IgCC specifically lists certain requirements, such as electrical peak demand limitations, to be selected by the local jurisdiction. Two other key differences to note with the IgCC on the topic of energy are (1) the IgCC requirements for on-site renewable energy are more detailed and based on a percentage of the building's total annual energy use, rather than a production rate based on roof area as done in Standard 189.1; and (2) the energy performance-based compliance includes modeling to compare energy consumption to a standard baseline reference building and the determination of an energy performance index based on the ratio of energy consumption to the baseline building.

Space limitations do not allow for a more detailed discussion of the IgCC in this *GreenGuide*.

REFERENCES AND RESOURCES

Published

AIA. 1996. *Environmental Resource Guide*. Edited by Joseph Demkin. New York: John Wiley & Sons.

ASHRAE. 2007. ANSI/ASHRAE Standard 52.2-2007, *Method of Testing General Ventilation Air-Cleaning Devices for Removal Efficiency by Particle Size*. Atlanta: ASHRAE.

ASHRAE. 2010. ANSI/ASHRAE Standard 55-2010, *Thermal Environmental Conditions for Human Occupancy*. Atlanta: ASHRAE.

ASHRAE. 2005. ASHRAE Guideline 0-2005, *The Commissioning Process*. Atlanta: ASHRAE.

ASHRAE. 2007. ASHRAE Guideline 1.1-2007, *HVAC&R Technical Requirements for the Commissioning Process*. Atlanta: ASHRAE.

ASHRAE. 2009. *The ASHRAE Guide for Buildings in Hot and Humid Climates*. Atlanta: ASHRAE.

ASHRAE. 2009. *Indoor Air Quality Guide: Best Practices for Design, Construction and Commissioning*. Atlanta: ASHRAE.

ASHRAE. 2011. ANSI/ASHRAE/USGBC/IES Standard 189.1-2011, *Standard for the Design of High-Performance Green Buildings*. Atlanta: ASHRAE.

ASHRAE. 2013. ANSI/ASHRAE Standard 62.1-2013, *Ventilation for Acceptable Indoor Air Quality*. Atlanta: ASHRAE.

ASHRAE. 2013. ANSI/ASHRAE/IES Standard 90.1-2013, *Energy Standard for Buildings Except Low-Rise Residential Buildings*. Atlanta: ASHRAE.

Deru, M., and P. Torcellini. 2007. *Source Energy and Emissions Factors for Energy Use in Buildings*. Technical Report NREL/TP_550-38617, National Renewable Energy Laboratory, Golden, CO.

Grondzik, W.T. 2001. The (mechanical) engineer's role in sustainable design: Indoor environmental quality issues in sustainable design. HTML presentation available at www.polaris.net/~gzik/ieq/ieq.htm.

Jarnagin, R.E., R.M. Colker, D. Nail and H. Davies. 2009. ASHRAE Building EQ. *ASHRAE Journal* 51(12):18–21.

Means, J.K. and F.H. Walters. 2010. Do High Performance-Labeled Buildings Really Perform at the Promised Levels? CLIMA 2010 REHVA World Congress 10, Antalya, Turkey. ISBN 978-0975-6907-14-6: Proceedings cd-rom.

Taylor, S.T. 2005. LEED and Standard 62.1. *ASHRAE Journal Sustainability Supplement*, September.

USGBC. 2009. *LEED 2009 Green Building Design and Construction Reference Guide*. Washington, D.C.: U.S. Green Building Council.

Online

The American Institute of Architects
  www.aia.org.
ASHRAE Building Energy Quotient
  www.buildingenergyquotient.org/.
ASHRAE Advanced Energy Design Guides
  www.ashrae.org/standards-research--technology/advanced-energy-design-guides.
BuildingGreen (for purchase)
  www.greenbuildingadvisor.com.

Building Research Establishment Environmental Assessment Method
 (BREEAM®) rating program
 www.breeam.org.
Center of Excellence for Sustainable Development, Smart Communities Network
 www.smartcommunities.ncat.org.
Collaborative for High Performance Schools
 www.chps.net.
Estidama (United Arab Emirates)
 www.estidama.org.
Green Building Advisor
 www.greenbuildingadvisor.com.
Green Globes
 www.greenglobes.com.
GreenSpec® Product Guide
 www.buildinggreen.com/menus/.
Green Star (Australia)
 www.gbca.org.au/green-star/.
The Hannover Principles
 www.mindfully.org/Sustainability/Hannover-Principles.htm.
International Living Building Institute, The Living Building Challenge
 http://living-future.org/lbc.
International Organization for Standardization family of 14000 standards,
 www.iso.org/iso/iso14000.
Intergovernmental Panel on Climate Change
 www.ipcc.ch.
International Performance Measurement and Verification Protocol
 www.evo-world.org.
Lawrence Berkeley National Laboratories, Environmental Energy Technologies Division
 http://eetd.lbl.gov.
Minnesota Sustainable Design Guide
 www.sustainabledesignguide.umn.edu.
National Institute for Standard and Technology, Building for Environmental and Economic Sustainability (BEES)
 www.nist.gov/el/economics/BEESSoftware.cfm.
National Renewable Energy Laboratory, Buildings Research
 www.nrel.gov/buildings.
Natural Resources Canada, EE4 Commercial Buildings Incentive Program
 http://canmetenergy-canmetenergie.nrcan-rncan.gc.ca/eng/software_tools/ee4.html.
The Natural Step
 www.naturalstep.org.

New Buildings Institute
www.newbuildings.org.
Oikos: Green Building Source
www.oikos.com.
Rocky Mountain Institute
www.rmi.org.
Sustainable Building Challenge
www.iisbe.org.
Sustainable Buildings Industry Council
www.sbicouncil.org.
Tool for the Reduction and Assessment of Chemical and Other Environmental Impacts (TRACI)
http://link.springer.com/article/10.1007/s10098-010-0338-9#page-1.
U.S. Green Building Council, Leadership in Energy and Environmental Design, Green Building Certification System
www.usgbc.org/leed.
U.S. Green Building Council, Regional Priority Credit Listing
www.usgbc.org/rpc.
The Whole Building Design Guide
www.wbdg.org.

CHAPTER EIGHT

CONCEPTUAL ENGINEERING DESIGN—LOAD DETERMINATION

Economic benefits can be realized by focusing your efforts on reducing building loads. Reducing loads allow the designers to select smaller, less expensive HVAC equipment, which saves money. Plus, the reduced HVAC system's energy consumption leads to lower operational costs throughout the lifetime of the building.

The traditional load determination methods, such as the cooling load temperature difference method or rules of thumb, are rough, first approximations that are based on old correlations or simplified heat transfer calculations. Designing low-energy buildings requires the engineer to have a thorough understanding of the dynamic nature of the interactions of the building with the environment and the occupants. ASHRAE has published the *Load Calculation Applications Manual* (Spitler 2009), which focuses on the most current load calculation methods: the Heat Balance Method, and the Radiant Time Series Method. To optimize the design, detailed computer simulations that use these methods allow the engineer to accurately model the major loads and interactions.

Loads can be divided into those stemming from the envelope and those from internal sources. Envelope loads include the impact of the architectural features; heat and moisture transfer through the walls, roof, floor, and windows; and infiltration. Internal loads include lights, equipment, people, and process equipment. Examine all of the loads in two ways: separately, to determine their relative impact, and together, to determine their interactions.

The engineer must also understand the energy sources and flows in the building and their location, magnitude, and timing. Once the engineer understands these sources and flows, he or she can be creative in coming up with solutions. The charts in Figure 8-1 show how average energy use breaks down in commercial and residential buildings. As designs are developed, breakdowns such as these should be kept in mind so that the energy-using areas that matter most are given priority in the design process.

When trying to minimize energy use in buildings, the first step is to identify which aspects of building operation offer the greatest energy-saving opportunities. For example, as shown in Figure 8-1, space heating and cooling, lighting, and water heating make up a combined 78% of energy use in the average commercial building in the United States.

Another important aspect of this exercise in the design process is that it allows the design team, along with the owner, to set the appropriate goal for energy usage. Many organizations and institutions have adopted the Architecture 2030 Challenge. Conceptual load calculations help to determine if a specific building design concept meets these requirements.

To find energy end-use statistics for many types of buildings (e.g., office, education, health care, lodging, retail, etc.), based on building location, age, size, and principal energy sources, consult the Commercial Building Energy Consumption Survey data published by the U.S. Energy Information Administration at www.eia.doe.gov/emeu/cbecs/.

THE ROLE OF ENERGY MODELING DURING CONCEPTUAL DESIGN

To achieve the greatest possible energy reductions, energy modeling should begin during the early conceptual stage of the design process. Decisions about the building's form, orientation, percentage of glazing on each façade, and construction materials are generally made during this stage and can have a profound impact on how the building will perform.

Often, the architectural designers will develop several design options to present to the client. Perhaps one option has an elongated shape, another may be L-shaped, and a third may be rectangular with an interior courtyard. The options give the client the opportunity to provide feedback on the proposed design and interior space planning. Of course, if an energy use assessment of each option can be provided at this stage, then that will be another very important factor for the team to consider.

Commercially available energy simulation software can now give design teams the capability to directly read geometry from 3D design models. This is a great time saver, and allows the design team to get feedback quickly on the performance of proposed designs rather than having to re-create geometry each time the design changes. For designs that have very complex geometry, this feature becomes especially important.

Once the geometry is created, the energy model can be populated with information on the loads. At the conceptual design phase, exact information is not yet likely to be determined, so engineering judgment is important to provide a realistic assessment.

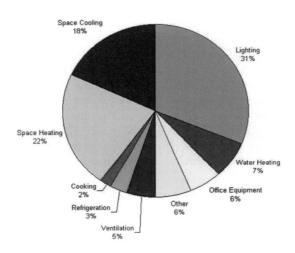

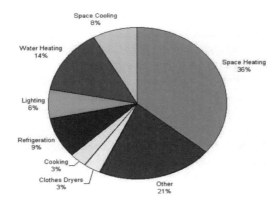

Image courtesy of CTG Energetics, Inc.

Figure 8-1 Average commercial and residential building energy use in the United States.

Suppose that the design team is working on a new office building. A massing model may be developed, however it is unlikely that specific office locations have been determined, so at this stage it would be reasonable for the energy model to be divided into simple block zones (for each exterior exposure) and interior zones. ANSI/ASHRAE Standard 62.1, *Ventilation for Acceptable Indoor Air Quality*. (ASHRAE 2013) provides data on typical occupancy densities and ventilation requirements, and ANSI/ASHRAE/IES Standard 90.1, *Energy Standard for Buildings Except Low-Rise Residential Buildings* (ASHRAE) provides data on typical lighting and equipment power densities for a wide range of different building types. Many energy simulation software tools have this information stored in a library (or allow creation of custom libraries) to allow users to quickly apply it to the energy model. Once this information is applied to the model, the design team can perform iterative simulations to determine the best performing massing scheme, the optimal orientation, the impact of different construction materials, and the impact of different window-to-wall ratios.

Since some assumptions must be made at this stage, the energy simulations may not predict energy consumption to the exact Btu (joule). However, if our assumptions are sound, we can get a good sense of the impacts of design decisions at an appropriate order of magnitude. Moving forward, we can be confident that the right concepts are in place and can then assess them in greater detail as the design progresses.

DETERMINING THE LOAD DRIVERS WITH PARAMETRIC SIMULATIONS

To gain an understanding of how the proposed design of the building affects energy use, the engineer should perform a series of parametric simulations. This process involves sequentially removing each load from the energy balance individually to determine the effect on overall energy use. The simulations that show the greatest energy reductions identify which loads are driving the building's energy consumption.

For instance, to effectively remove wall conductive heat transfer from the energy balance, set the wall thermal resistance to a very high value (such as R-100 [R-17.6]). Run the simulation model, and note the building energy requirements. Then, reset the wall insulation back to the actual R-value and proceed to the next parameter. Continue to do this individually with the floors, roofs, and windows. Then remove the window solar gain, daylighting, and other site-specific shading that may be involved. Assess the impact of infiltration and the effect of ventilation air by setting them sequentially to zero.

Parametric simulations should be completed for the internal loads in a process similar to that described in Chapter 5. Set the lighting load to zero, set the equipment load to zero, and then take all the people out of the building. In this manner, the engineer can understand what is driving the energy use in the building.

Plot the results for each case by annual energy use or by peak load to compare the impacts, as illustrated in Figure 8-1.

Look for creative solutions to minimize the impact of each load, starting with those with the largest impact, or those that are the easiest to implement. From the parametric analysis example in Figure 8-2, adding underfloor air distribution results in the largest impact on annual energy use, so it should be one of the first solutions considered. Alternatively, a solution such as high-efficiency lighting may not result in as large of an impact, but may be easier to implement.

ENERGY IMPACTS OF ARCHITECTURAL FEATURES

It is important to collaborate with the architects to help guide decisions that will improve the thermal envelope's performance. Determine the greatest sources of heat gains and losses through the building's skin, and look for opportunities to minimize the effect.

Solar radiation through the windows can be one of the largest gains during the summer cooling season. Providing external shading can effectively minimize these gains. Horizontal shading, or overhangs, on the south face of the building will block the high sun during the day for much of the year, and vertical shading on the east and west can reduce gains early and late in the day as the sun rises and sets. However, the microclimate around the building should be completely analyzed in

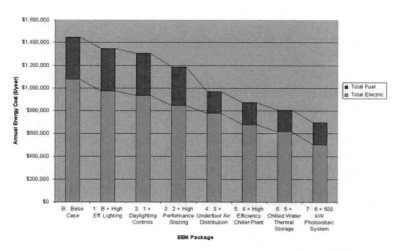

Image courtesy of CTG Energetics, Inc.

Figure 8-2 Sample parametric analysis to determine relative impacts of building envelope.

order to make sure any external shading actually works as intended. For example, there are a number of areas where the summer months are mostly foggy (e.g., Santa Barbara, California). Providing external shading for high angle summer sun may not be very effective in that area.

Another means to control solar gains is to specify high-performance glazing with a low solar heat gain coefficient. These glazing options can be either tinted glass, or spectrally selective glass. The benefit of the spectrally selective option is that they have a clear appearance and a high visible light transmittance, allowing for greater daylighting potential and views to the exterior. However, they are generally the more costly option.

In climates where heating the building is of primary concern, it can often be beneficial to allow solar heat gain into the building as it acts as a free heating source. In climates where heating and cooling are both significant concerns, use a building simulation software program to understand the trade-offs associated with various solar heat gain coefficient values.

Conduction gains through the building skin are also significant in the overall heat balance equation. Where possible, try to reduce the window-to-wall ratio. Even the best-performing glass selections cannot compare to the thermal characteristics of an insulated wall.

Glass selections with low U-factors will reduce the heat gains and losses. Also, try to minimize the U-factors of the walls and roof by using insulation with high R-values. Where thermal bridging may occur due to structural elements of the building, provide a continuous interior layer of insulation to minimize this effect.

It is very important for designers to look at the entire window assembly properties, including the frame and the spacers and not only for the glazing. Good glazing U-factors can be severely compromised by having nonthermally broken frames.

Lastly, try to minimize infiltration gains from leakage of untreated air into the building. Positively pressurizing the building with the HVAC systems will minimize this impact; however, stack-driven and wind-driven pressure differentials will still cause exterior air to infiltrate the building. Opening and closing of doors to the exterior will also contribute to infiltration gains. Specify a continuous air barrier to minimize air infiltration through the building skin, and install entry vestibules or revolving doors to minimize airflow into the building through open doors.

THERMAL/MASS TRANSFER OF ENVELOPE

Basic, steady-state energy transfer through building envelopes is well known, well understood, and easy to calculate. Increased R-values are certainly beneficial in heating climates and can also help in cooling-load-dominated climates. Use as a minimum the recommended amounts spelled out in ASHRAE's energy standard (Standard 90.1 [ASHRAE 2013], latest approved edition). While values above those recommended can be beneficial in simple structures, high values can be coun-

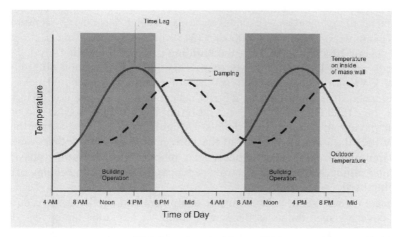

Image courtesy of U.S. Department of Energy Efficiency and Renewable Energy.

Figure 8-3 The effects of thermal mass.

terproductive over a heating/cooling season in more complex structures. It is always wise to evaluate R-value benefits through the application of load and energy simulation programs.

The effects of thermal mass are sometimes not as easy to gauge intuitively. (See Figure 8-3.) Therefore, the above-mentioned simulation programs, when properly applied, are very useful in this regard. If thermal mass is significant in the building being planned (or if increased mass would be easy to vary as optional design choices), then such programs are essential for evaluating the flywheel effects of thermal mass on both loads and longer-term energy use.

GreenTip #8-1 describes a technique for combining nighttime ventilation with a building's thermal mass to achieve load and/or energy savings.

ENGINEERING INTERNAL LOAD-DETERMINING FACTORS

Internal loads are significant contributors to the energy balance, but can be effectively reduced during the design phase. For office buildings and other buildings in mild climates, lights are usually the major culprits. Therefore, optimize the daylighting and electric lighting design. Refer to Chapter 13 for a detailed discussion on lighting systems.

Likewise, evaluate office equipment loads. Make recommendations about the effects of the choices of computers, monitors, printers, and other types of equipment. In many offices, this equipment is left on all night. An office building should have very few loads when the building is unoccupied. Leaving an office full of equipment powered up all night and on weekends can easily add up to large energy consumption.

For example, assume one 34,000 ft^2 (3159 m^2) office building has nighttime plug loads of 10 kW, and assume the building is unoccupied for 14 h/day during the week and 24 h/day on the weekend. This adds up to around 6000 h/yr and 60,000 kWh—or $4200 (at $0.07/kWh)—of electricity that could have easily been reduced. It is important for the engineer to bring up these issues during the design stage (and later, during the operation) of the building, because no one else may be paying attention to such details. Designing and building a great building is only half the job; operating it in the correct manner is the other—and often more important—half. The engineer can make better operation possible by designing piping, wiring, and controls capable of easily turning things off when not being used.

Educate owners about efficient office equipment and appliances to reduce plug loads, covering such things as flat screen computer monitors, laptops vs. desktop central processing units, copy machines, refrigerators, and process equipment (e.g., lab equipment, health care machinery, etc.). Consider measuring usage in one of the clients' existing buildings to get an accurate picture of load distribution and population profiles. Attempt to develop a total building electric load profile for every minute of one week.

SYSTEM/EQUIPMENT EFFICIENCIES

It is important to use cooling and heating equipment of the correct size. The old rule of thumb of 250 to 350 gross ft^2/ton (20 to 35 gross m^2/ton) cooling load does not apply to sustainable buildings. Recent high-performance building projects operate between 600 and 1000 gross ft^2/ton (55 to 90 gross m^2/ton). Set cooling equipment and system performance targets in terms of kW/ton (kW/kWR) as indicated in Chapter 10.

Industry standards such as Standard 90.1 (ASHRAE 2013) give minimum requirements for equipment efficiencies and system design and installation. Understand that these standards represent the least-efficient end of the spectrum of energy-conserving buildings that should be built. To be considered green, a building must exceed them.

In addition to ASHRAE, there are other sources of information on energy efficiency as related to green-building design (see the "References and Resources" section at the end of this chapter). This guide does not endorse any of them; the information is presented for informational purpose only. Readers should be aware that the sources use various methods to arrive at their final recommendations and that some of the guidance offered may have a hidden (or not so hidden) agenda. Some may use economics as a basis; however, the underlying economic assumptions should be understood prior to using the information. Others attempt to push energy efficiency to its technical limits. Therefore, before using any of these sources for guidance, investigate the premises used, the methods of analysis, and the background of the author.

LIFE-CYCLE COST ANALYSIS (LCCA)

As we can see, there are a wide range of strategies to achieve efficiency improvements, but which are the most cost effective? Life-cycle cost analysis (LCCA) can be used to compare various options to determine which will achieve the greatest net monetary savings over the building's life.

All of the costs associated with each option are considered including first cost, fuel costs associated with the building's operation, maintenance costs, and any other recurring costs. Escalation rates and increases in fuel prices are accounted for over the length of the study, and the totals are adjusted to calculate the total cost of each option in terms of today's dollars. This total is generally referred to as the *net present value*.

Interest rates and discount factors for performing these calculations are widely available. Additionally, there are numerous software programs available to perform LCCA studies. For additional resources, refer to the following websites:

- www1.eere.energy.gov/library/default.aspx?Page=3#lifecycle
- www.wbdg.org/resources/lcca.php

Refer to Chapter 1 for more discussion on LCCA.

Digging Deeper

KEY CONSIDERATIONS IN THE HVAC DESIGN PROCESS

Design Intent

- Set goals for performance:
  - Energy performance (what code and what energy savings percentage should the project achieve? Should the comparison be an energy cost comparison or an energy consumption comparison?)
  - Environmental performance
  - Comfort
  - Operating cost
  - Determine how to achieve the goals
- System by system:
  - Integrated design

Verify that Design Intent is Met

- In design:
  - Verification of commissioning goals in design
  - Coordination between design disciplines
  - Include commissioning in design documents
- In construction:
  - Procurement of equipment and materials
  - Installation
- At start-up and testing
- In operations

Design Integration

- Integration with other disciplines
  - Architecture, lighting, interiors, and structural
  - Daylighting
  - Underfloor air distribution
  - Form-follows-function design
- Increased emphasis on HVAC performance
  - Thermal comfort
  - Indoor air quality (IAQ)
  - Energy efficiency

HVAC Systems

- High-efficiency equipment
- Systems responsive to partial loads
  - 80% of year, system operates at <50% of peak capacity.
- Emphasis on free cooling and heating
  - Economizers (i.e., air or water)

- Evaporative cooling (cooling towers or precooling)
- Heat recovery
- Emphasis on IAQ
- Underfloor air distribution is new wave

Load Reduction

- Reduce envelope loads
  - Solar loads
- Reduce lighting loads
  - Standard 90.1 (ASHRAE 2013) or local energy codes as a design maximum
- Reduce power loads
  - Site- and building-type specific, perhaps 1.0 to 1.5 W/ft^2 (16 W/m^2) as a maximum
- Reduced air-conditioning tonnage
  - Can provide higher air-conditioning efficiency for same cost

Cooling and Heating Load Reduction

- Envelope loads
  - Shading
  - Glass selection
  - Glass percentage
- Internal loads
  - Lighting power density
  - Equipment loads (advocate for ENERGY STAR®)
  - Controls/occupancy sensors

ASHRAE GreenTip #8-1

Night Precooling

GENERAL DESCRIPTION

Night precooling involves the circulation of cool air within a building during nighttime hours with the intent of cooling the structure. The cooled structure is then able to serve as a heat sink during the daytime hours, reducing the mechanical cooling required. The naturally occurring thermal storage capacity of the building is thereby used to smooth the load curve and for potential energy savings. More details on the concept of thermal mass on building loads are included in Chapter 5, "Architectural Design Impacts."

There are two variations on night precooling. One, termed *night ventilation precooling*, involves the circulation of outdoor air into the space during the naturally cooler nighttime hours. This can be considered a passive technique, except for any fan power requirement needed to circulate the outdoor air through the space. The night ventilation precooling system benefits the building IAQ through the cleansing effect of introducing more ventilation air. With the other variation, *mechanical precooling*, the building mechanical cooling system is operated during the nighttime hours to precool the building space to a setpoint that is usually lower than that of normal daytime hours.

Consider these key parameters when evaluating either concept:

- Local diurnal temperature variation
- Ambient humidity levels
- Thermal coupling of the circulated air to the building mass

The electric utility rate structure for peak and off-peak loads also is important when determining cost-effectiveness, in particular for a mechanical precooling scheme.

A number of published studies show significant reductions in overall operating costs by the proper precooling and discharge of building thermal storage. The lower overall costs result from load shifting from the day to the nighttime with its associated off-peak utility rates. For example, Braun (1990) showed significant energy cost savings of 10% to 50% and peak power requirements of 10% to 35% over a traditional nighttime setup control strategy. The percent savings were found to be most significant when lower ambient temperatures allowed night ventilation cooling to be performed.

For a system incorporating precooling to be considered a truly green design concept, the total energy used through the entire 24-hour day should be lower than without precooling. A system that uses outdoor air to do the precooling only requires the relatively lower power needed to drive the circulation fans, compared to a system that incorporates mechanical precooling. Electrical energy provided by the utility during peak demand periods also may be dirtier than that provided during normal periods, depending on the utility and circumstances.

The system designer needs to be aware of the introduction of additional humidity into the space with the use of night ventilation. Thus, the concept of night ventilation precooling is better suited for drier climates. A mechanical nighttime precooling system will prevent the introduction of additional humidity into the space by the natural dehumidification it provides, but at the expense of greater energy usage compared to night ventilation alone.

Both variations (i.e., night ventilation and night mechanical precooling) are not 100% efficient in the thermal energy storage in the building mass, particularly if the building is highly coupled (thermally) with the outside environment. Certain building concepts used in Europe are designed to increase the exposure of the air supply or return with the interior building mass (see, for example, Andersson et al. [1979]). This concept will increase the overall efficiency of the thermal storage mass.

For either type of system, the designer must carefully analyze the structure and interaction with the HVAC system air supply using transient simulations. A number of techniques and commercially available computer codes exist for this analysis (Balaras 1995).

WHEN/WHERE IT'S APPLICABLE

Night precooling would be applicable in the following circumstances:

- When the ambient nighttime temperatures are low enough to provide sufficient opportunity to cool the building structure through ventilation air. Ideally, a low ambient humidity level would also occur. A hot, dry environment, such as the southwestern U.S., is an ideal potential area for this concept.

- When the building occupants would be more tolerant of the potential for slightly cooler temperatures during the morning hours.
- When the owner and design team are willing to include such a precooling system concept and to commit to (1) a proper analysis of the dynamics of the building's thermal performance and (2) the refinement of the control strategy (upon implementation) to fine-tune the system performance.
- More massive buildings, or those built with heavier construction materials such as concrete or stone, have a greater potential for benefits. Just as important, is the interaction of the building mass with the building internal and HVAC system circulating air. This interaction may allow for more efficient transfer of thermal energy between the structure and the air space.

PRO

- Night ventilation precooling has good potential for net energy savings because the power required to circulate the cooler nighttime air through the building is relatively low compared to the power required to mechanically cool the space during the daytime hours.
- Mechanical precooling could lead to net energy savings, although there will likely be a net increase in total energy use due to the less-than-100% thermal energy storage efficiency in the building mass.
- Both variations require only minor, if any, change to the overall building and system design. Any changes required are primarily in the control scheme.
- Night ventilation can provide a better IAQ environment, due to increased circulation of air during the night. A greater potential exists with the ventilation precooling concept. Both will be better than if the system was completely shut off during unoccupied hours.

CON

- Temperature control should be monitored carefully. The potential exists for the building environment to be too cool for the occupants' comfort during the early hours of the occupied period. This will result in increased service calls or complaints and may end with the night precooling being bypassed or turned "off."
- The increased run time on the equipment could lead to lower equipment life expectancy or increased frequency in maintenance. Careful attention should be given to the resulting temperature profile through the day during the commissioning process. Adjustments to the control schedule may be necessary to keep the building within the thermal comfort zone.
- Proper orientation must be given to the building operator so he or she can understand how the control concept affects the overall system operation throughout the day.
- Future turnovers in building ownership or operating personnel could negatively affect how successfully the system performs.
- Occupants would probably need at least some orientation so that they would understand and be tolerant of the differences in conditions that may prevail with such a system. Future occupants may not have the benefit of such orientation.

KEY ELEMENTS OF COST

The following provides a possible breakdown of the various cost elements that might differentiate a nighttime precooling scheme from a conventional one and gives an indication of whether the net cost for the precooling option is likely to be lower (L), higher (H), or the same (S). This assessment is only a perception of what might be likely, but it may not be correct in all situations. There is no substitute for a detailed cost analysis as part of the design process. The listings below may also provide some assistance in identifying the cost elements involved.

First Cost

- Mechanical ventilation system elements   S
- Architectural design features   S
- System controls   H
- Analysis and design fees   H

Recurring Costs

- Energy for mechanical portion of system
  - Ventilation precooling   L
  - Mechanical precooling   S/H
- Total cost to operate cooling system   L
- Maintenance of mechanical ventilation and cooling system   S/H
- Training of building operators   H
- Orientation of building occupants   H
- Commissioning cost   H
- Occupant productivity   S

SOURCES OF FURTHER INFORMATION

Andersson, L.O., K.G. Bernander, E. Isfält, and A.H. Rosenfeld. 1979. Storage of heat and cool in hollow-core concrete slabs. Swedish experience and application to large, American style buildings. Second International Conference on Energy Use and Management, Lawrence Berkeley National Laboratory, LBL-8913.

Balaras, C.A. 1995. The role of thermal mass on the cooling load of buildings. An overview of computational methods. *Energy and Buildings* 24(1):1–10.

Braun, J.E. 1990. Reducing energy costs and peak electrical demand through optimal control of building thermal storage. *ASHRAE Transactions* 96(2):876–88.

Keeney, K.R., and J.E. Braun. 1997. Application of building precooling to reduce peak cooling requirements. *ASHRAE Transactions* 103(1):463–69.

Kintner-Meyer, M., and A.F. Emery. 1995. Optimal control of an HVAC system using cold storage and building thermal capacitance. *Energy and Buildings* 23:19–31.

Ruud, M.D., J.W. Mitchell, and S.A. Klein. 1990. Use of building thermal mass to offset cooling loads. *ASHRAE Transactions* 96(2):820–29.

ASHRAE GreenTip #8-2

Night-Sky Cooling

GENERAL DESCRIPTION

When using chilled-water-based cooling systems, peak load reduction opportunities exist through the use of thermal storage systems. Many thermal storage systems use installed chiller capacity to charge the storage system during off-peak periods. However, while this takes advantage of cheaper energy during off-peak periods, it can actually increase overall energy consumption. Yet, as night-time utility generation capacity is generally from cleaner sources than during peak periods, thermal storage systems can be an effective way to reduce greenhouse gas emissions as well as save operating costs.

Another good way to create a source of cooling is to use a roof spray cooling system. This system would spray water on a roof that is cooled by evaporation and radiation heat transfer. Such a system would work best in areas with limited cloud cover at night and with lower humidity, thus limiting where it would be effective. The consumption of water is also a concern that needs to be balanced with the energy effectiveness.

The cooled water from the roof spray system is collected from the roof surface, filtered and stored for use the following day. Researchers at UC Davis' Western Cooling Efficiency Center have reported Annual Energy Efficiency Ratios (EERs) of 40 to 150 for night-sky radiant cooling roof spray systems, versus EERs of 13 to 20 for ground-source heat pump systems (generally considered a highly efficient system). One manufacturer of these systems claims EERs of 50 to 100.

The roof spray cooling system consists of a piping grid and conventional water spray nozzles located on the roof surface and connected to a water pumping and filtration system. During the cooling season, water is sprayed over the roof surface at night. This water is first cooled by evaporation during the spray process and then further cooled by radiation to the night sky. The water is recirculated through this spray-cooling process until it is cooled and then stored for later use in providing cooling to the building. Studies have shown that the minimum water temperature reached during operation can be 5°F–10°F (3°C–5°C) less than the minimum dry-bulb air temperature reached at night.

WHEN/WHERE IT'S APPLICABLE

Research data suggests that, where summer dry-bulb temperatures fall below 65°F (18°C) at night, system capacities can be expected to be approximately 25 ton-hours per 1,000 square feet of roof area (300 Btu per square foot [0.95 kWh/m^2]). Another report places capacities at approximately 290 Btu per square foot (0.9 kWh/m^2).

Cooling via night-sky radiation relies on the fact that the effective temperature of the night sky can be significantly cooler than the ambient air temperature. Thus, an object set outside at night will radiate more heat to the night sky than it will absorb from the night sky and the net loss of heat will cause the object to cool below the surrounding air temperature.

Most of the cooling via night-sky radiation occurs as the water film left on the roof from the spray process exchanges radiant heat with the night sky. The typically large roof areas available allow for substantial radiative cooling at relatively little additional building cost. The effective nighttime sky temperature is a function of both the dry-bulb air temperature and the humidity content of the air, with higher humidity reducing the differential between the effective sky temperature and dry-bulb air temperature. Although not an easily measured parameter, sky temperature data are available in Typical Meteorological Year (TMY) weather data files or can be estimated from other weather parameters using established algorithms.

PROS AND CONS

Pro

- Very high efficiency cooling system.
- Enhances redundancy of systems.

Con

- Higher first cost, unless chiller capacity can be reduced by the peak tons available from the chilled-water storage system.
- Sensitive to quality of the potable water used for make-up (e.g. clogging of spray nozzles).

- Requires good design of stratified thermal storage tank.
- Available capacity subject to weather patterns.

KEY ELEMENTS OF COST

The following provides a possible breakdown of the various cost elements that might differentiate a night sky radiant cooling system from a conventional one and an indication of whether the net cost for an option is likely to be lower (L), higher (H), or the same (S). This assessment is only a perception of what might be likely and it obviously may not be correct in all situations. There is no substitute for a detailed cost analysis as part of the design process. The listings below may also provide some assistance in identifying the cost elements involved.

First Cost

- Chilled-water storage system   H
- Interface of chilled-water storage to cooling system   H
- Roof spray system   H
- Controls and instrumentation   H

Recurring Cost

- Energy consumption   L
- Testing and balancing (TAB)   S
- Maintenance   S/H
- Commissioning   H

SOURCES OF FURTHER INFORMATION

Bourne, R., C. Carew. 1996. Design and Implementation of a Night Roof-Spray Storage Cooling System. Proceedings of the 1996 ACEEE Summer Study on Energy Efficiency in Buildings, Washington, D.C.

California Energy Commission, 1992. Cool Storage Roofs ETAP Project, Consultant Report. Prepared by the Davis Energy Group. CEC P500-92-014, Davis, California.

Martin, M., P. Berdahl. 1984. Characteristics of Infrared Sky Radiation in the United States, *Solar Energy*, 33:321–326.

Pacific Northwest National Laboratory. December, 1997. Technology Installation Review, "WhiteCap™ Roof Spray Cooling System: Cooling Technology for Warm, Dry Climates".

Chen, Bing, John Kasher, John Maloney, and Girgis A. Girgis, and David Clark. Passive Solar Research Group, University of Nebraska. Determination of the Clear Sky Emissivity for Use in Cool Storage Roof and Roof Pond Applications.

Allen, C.P. The Estimation of Atmospheric Radiation for Clear and Cloudy Skies, M.S. thesis at Trinity University, 1977.

Berdahl, P., and R. Fromberg. 1982. The Thermal Radiance of Clear Skies, *Solar Energy*, 29(4):299–314.

Bourne R. and B. Chen. 1989. Full Year Performance Simulation of a Direct-Coupled Thermal Storage Roof (DCTSR), *Proceedings of the annual meeting of the American Solar Energy Society, Denver, CO.*

ASHRAE GreenTip #8-3

Plug Loads

GENERAL DESCRIPTION

As the efficiency of various energy-using systems (i.e., HVAC, lighting, or building envelopes) improves, the impact of plug loads on the total annual building energy usage increases. This is accelerated by the rapidly growing use of computers and other electronic equipment in most building types. The purpose of this GreenTip is to identify some key issues to watch for that are relative to plug loads.

TYPICAL PLUG LOAD ENERGY BUDGETS

Historically, plug loads in office buildings have been designed to support connected loads in the 2 to 5 W/ft^2 (20 to 50 W/m^2) range. For laboratories or computing-intensive buildings, this can go up to 10 to 15 W/ft^2 (100 to 150 W/m^2). The electrical distribution system must be designed to support this level of load, in compliance with the *National Electrical Code*© (NFPA 2011). Yet the actual diversified load exerted on the building electrical demand and upon the HVAC system will more likely be in the range of 1 to 1.5 W/ft^2 (10 to 15 W/m^2) or even less for highly efficient installations, as described below.

DIFFERENCE BETWEEN CONNECTED LOAD AND EFFECTIVE LOAD

The reason for this disparity is the fundamental difference between connected load for electrical equipment and the effective load imposed on the HVAC systems and electrical systems by the actual equipment when diversity of use is taken into account. The nameplate ratings of equipment are not representative of the actual energy usage of that equipment, which is typically a fraction of nameplate on average. Thus, if energy usage and cooling loads from office equipment and other plug loads are based upon nameplate, they can be overestimated by 200% to 500%.

There is debate about how much further plug loads are likely to drop in the future, as ever-improving efficiencies of equipment are being offset by increasing amounts of equipment being used. Programs like the U.S. EPA's ENERGY STAR program will continue to improve the energy efficiency of individual pieces of equipment.

CONTROLS OF PLUG LOADS

One of the key factors in reducing effective plug loads is the impact of controls on the equipment loads. The primary issue is the power management controls now built into all computers and most other office equipment—putting the equipment in standby mode or completely shutting it off after preset intervals of no activity. Another way to accomplish this is through the use of occupancy sensors connected to plug strips that control equipment at a workstation, so it can be shut off when not occupied. This also works well for task lighting, which is not otherwise equipped with power management.

SOURCES OF FURTHER INFORMATION

M. Piette et al. June 1995 and subsequent updates. Office technology energy use and savings potential in New York. Lawrence Berkeley National Laboratory, Berkeley, CA.

REFERENCES AND RESOURCES

Published

ASHRAE. 1998. *Fundamentals of Water System Design*. Atlanta: ASHRAE.

ASHRAE. 2011. *ASHRAE Handbook—HVAC Applications*. Atlanta: ASHRAE.

ASHRAE. 2013. ANSI/ASHRAE Standard 62.1-2013, *Ventilation for Acceptable Indoor Air Quality*. Atlanta: ASHRAE.

ASHRAE. 2013. ANSI/ASHRAE/IES Standard 90.1-2013, *Energy Standard for Buildings Except Low-Rise Residential Buildings*. Atlanta: ASHRAE.

NFPA. 2011. *National Electrical Code®*. Standard 70-2014. Quincy, MA: National Fire Protection Association.

Spitler, J. 2009. *Load Calculation Applications Manual*. Atlanta: ASHRAE.

Trane. 2000. *Multiple-Chiller-System Design and Control Manual*. Lacrosse, WI: Trane Company.

Online

Air-Conditioning, Heating and Refrigeration Institute
www.ahrinet.org/.

Alliance to Save Energy
http://ase.org.

American Council for an Energy-Efficient Economy
www.aceee.org.

Architecture 2030 Challenge
www.architecture2030.org/.

California Energy Commission
www.energy.ca.gov.

CoolTools Chilled Water Plant Design Guide
www.taylor-engineering.com/downloads/cooltools/
EDR_DesignGuidelines_CoolToolsChilledWater.pdf.

ENERGY STAR®
www.energystar.gov.

Geoexchange
www.geoexchange.org.

Heschong Mahone Group
www.h-m-g.com/projects/daylighting/projects-PIER.htm.

Savings by Design
www.savingsbydesign.com.

U.S. Department of Energy, Federal Energy Management Program
www.eere.energy.gov/femp.

U.S. Energy Information Administration,
   Commercial Building Energy Consumption Survey
   www.eia.doe.gov/emeu/cbecs/.
U.S. Environmental Protection Agency
   www.epa.gov.

CHAPTER NINE

INDOOR ENVIRONMENTAL QUALITY

INTRODUCTION

Indoor environmental quality is inarguably one of the most important characteristics of green buildings intended for human occupancy. While it is challenging as well as important to provide good IEQ in an energy efficient manner, in some cases the most effective means to improve IEQ can also save energy. No sacrifice of IEQ can be justified in order to obtain energy use reductions. After all, the purpose of such buildings is to support the activities for which the building exists and to do so in a manner that does the least harm to the environment while enhancing the health and well-being of the human occupants.

The four major aspects of indoor environmental quality are indoor air quality, thermal conditions, illumination, and acoustics. Perhaps even more important are the interactions within and among the factors and between the building, the occupants, and the indoor and outdoor environment.

Indoor Air Quality (IAQ)

In the context of ASHRAE *GreenGuide*, IAQ refers to the quality of the air in buildings intended for nonindustrial use. Good IAQ is defined by the absence of harmful or unpleasant constituents. The major means for achieving good IAQ are eliminating or reducing the sources of pollution in building materials and furnishings and from appliances, equipment, and consumer products. After vigorous efforts to eliminate pollutant sources, ventilation is applied to remove unavoidable pollutants, including the metabolic products from human and other living occupants, as well as pollutants from materials and furnishings used.

Thermal Conditions

The thermal environment within a building is important not only for comfort but also for the health and well-being of occupants and for the long-term durability of the building fabric and its contents. Moisture and chemical pollutants associated with poor control of thermal conditions are associated with a deteriorating building envelope as well as uncomfortable or even unhealthy occupants.

Illumination

Maintaining proper light levels is essential for virtually all human occupancy and, although not commonly thought of in terms of the health consequences (except eyesight), insufficient or inappropriate illumination can result in other health and safety hazards as well as poor performance on tasks requiring good illumination.

Acoustics

Noise and vibration are commonly combined into one term, *acoustics*. Acoustics involve far more than the control of noise emitted from mechanical equipment. Sound intensity, often characterized in decibels, can be too strong or too weak for building occupants to successfully use buildings for the intended purpose. Individuals have different sensitivity to different noise frequencies, so control of noise spectrum also matters. Speech recognition requires sufficient volume without competing background noise, while concentration during activity or sleep and relaxation may require significantly quieter conditions.

Interactions among indoor environmental factors and occupants' health and well-being are also important. All of the indoor environmental factors interact with each other as well as with human occupants and the building itself. Considerable attention to these interactions is required for a truly successful building. ASHRAE Guideline 10, *Interactions Affecting the Achievement of Acceptable Indoor Environments*, is devoted to discussion of the interactions among indoor environmental conditions and the considerations important for designers as well as building operators (2011).

GREEN-BUILDING DESIGN AND INDOOR ENVIRONMENTAL QUALITY

Various definitions or conceptions of green building exist, and there is no single definition that is universally accepted. ASHRAE considers the quality of the indoor environment an essential component of green buildings, and a substantial amount of guidance is available from ASHRAE.

Some of the texts and much of the content of the present chapter is based on *Indoor Air Quality Guide*, a document that is the most comprehensive single guide to design for good IAQ in commercial buildings (2009a). Readers of this chapter are encouraged to obtain a copy of *Indoor Air Quality Guide*, which is available for free download. More information on obtaining this resource is found in the references section at the end of this chapter.

INTEGRATED DESIGN APPROACHES AND SOLUTIONS

Building design professionals understand that the design of virtually every building element affects the performance of other elements, so it makes sense to integrate various design elements of a building. Energy efficiency and IEQ strategies often affect each other. Close interactions also exist among the different IEQ design factors.

Thermal comfort and good IAQ are intricately bound together both in the characteristics of the indoor environment and in the way building occupants respond to the indoor environment. Ventilation and thermal control solutions affect and are affected by the acoustic and illumination requirements and solutions. Noise from mechanical systems, waste heat from electrical illumination sources, or heat loss or gain through glazing are as important to the selection of ventilation solutions as the pollutant loads coming from building materials, occupant activities, building equipment, appliances, or any other sources. Air-cleaning technologies can be used to remove target compounds and help reduce the overall ventilation rate required for pollutant dilution. These technologies also lead to energy savings for heating and cooling. Daylighting helps reduce artificial lighting energy, but also requires more careful design and control of the lighting system for illumination quality. The easiest and most effective way to accomplish integrated design is to assemble the entire design team at the beginning of the project and to brainstorm siting, overall building configuration and zoning for intended building use, ventilation, thermal control, and illumination concepts as a group. These concepts should also be considered in conjunction with energy supply, water utilization, material selection, and aesthetics along with concepts of green design.

Once the initial design concept is agreed upon, then the evolution of the design through its various stages can occur with a shared concept and the potential for direct interaction among team members as challenges arise later in the process. The integrated design process is discussed in more detail in Chapter 4 of this book.

Commission to Ensure that the Owner's IAQ Requirements Are Met

No manufacturer today can stay in business without an effective quality control program. Yet many buildings are still designed and built without such a process. Commissioning can meet this need for a well defined and run quality-control process. Commissioning is "a process for achieving, verifying, and documenting that the performance of facilities, systems, and assemblies meets defined objectives and criteria" (ASHRAE 2005). Commissioning for IAQ may need to address the site, building enclosure, HVAC systems, and other elements of the design. Many IAQ-related examples are given in the *Indoor Air Quality Guide* (ASHRAE 2009a).

Contrary to popular misconception, commissioning is not just a postconstruction inspection process. Commissioning starts at project inception. During predesign, the commissioning authority (CxA) should help the owner articulate the key functional requirements for the project. For IAQ, this could include such things as ensuring that the building enclosure keeps out rain and humid air that may cause mold and moisture problems or ensuring that outdoor air intakes are located to minimize introduction of outdoor contaminants. While these criteria may sound basic, failure to meet such requirements has been responsible for serious IAQ problems in many buildings. During predesign the CxA should also review and provide input to the project schedule to ensure that the key steps required to achieve the owner's objectives are incorporated. For example, commissioning of the building enclosure may require that mock-ups be built and tested or that assemblies be inspected while they are still open and visible. Additional discussion on commissioning is found in Chapter 6 of this book.

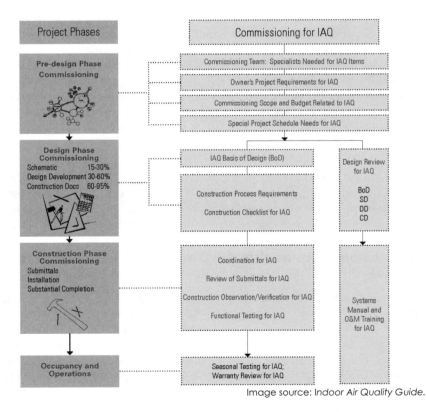

Image source: *Indoor Air Quality Guide.*

Figure 9-1 Commissioning for IAQ.

Select HVAC Systems to Improve IAQ and Reduce the Energy Impacts of Ventilation

When selecting HVAC systems, their ability to meet IAQ objectives should be carefully considered. All too often, the type of HVAC system is decided by the architect before the engineering firm is even retained. In other cases, several types of HVAC systems may be compared, but their implications for IAQ are not fully considered.

Design for optimal IAQ requires a fully integrated design process. In general, the more green the design, the more this is true. For example, buildings that use natural ventilation and passive cooling require close coordination of the architectural and mechanical design to meet IAQ and comfort objectives.

For more conventional designs, the type of HVAC system selected can also affect IAQ in critical ways, and thus it is essential to consider IAQ as part of system selection. For example, in mechanically cooled buildings in humid climates, it is essential to maintain positive building pressurization to reduce the risk of condensation and mold growth in the building enclosure. Some types of HVAC systems are less able or even unable to maintain such positive pressurization. Similarly, HVAC systems may vary in their ability to control indoor humidity levels, the flexibility they allow in the location of outdoor air intakes, their ability to provide higher efficiency particle filtration, time and effort required to perform IAQ-related maintenance, and other factors.

Building codes and standards set minimum ventilation requirements for indoor air quality. Different types of HVAC systems differ in the amount of energy they consume in delivering that ventilation. Both the IAQ and energy impacts of system selection are discussed in greater detail in Strategy 1.3 in the *Indoor Air Quality Guide* (2009a).

Use Project Scheduling and Manage Construction Activities to Facilitate Good IAQ

The schedule and the sequencing of construction activities can have a significant impact on IAQ in the finished building. As mentioned previously, the schedule must allow time for key quality control tests related to IAQ. For example, it is much better to inspect or test drainage planes, sliding glass doors, or shower pans after a few have been installed and correct any problems then, rather than to discover problems after hundreds of feet or hundreds of units have been installed. Therefore, the inspection of mock-ups or early work, and suspension of additional work until these are approved, should be incorporated into the schedule.

Compressed scheduling, improper sequencing of work, and poor construction management can cause IAQ problems. For example, an unrealistic timeline or a bonus for early completion can create pressure to install materials or equipment that are not water resistant before the building is sealed up and while the materials may still be exposed to moisture. Ductwork stored outdoors can accumulate dirt that facilitates future growth and circulation of microorganisms, even though high efficiency filtration is used in the finished project. Operating HVAC equipment for temporary heating or cooling during construction or for occupied areas in a partially completed building can cause widespread system contamination, transfer of construction dust and volatile organic compounds (VOCs) to the occupied area, or other problems. Schedule and construction management issues that affect IAQ are discussed in more depth in the *Indoor Air Quality Guide* (2009a).

Limit Entry of Outdoor Contaminants

Make an Airtight Enclosure. Designers should understand the importance and benefits of making the enclosure of a building airtight. There are also some caveats to consider.

Among the benefits are reduced chances of thermal comfort complaints, amount of power used to condition the building, chances of condensation inside enclosure assemblies in both heating and air conditioning mode, and chances of ice dam problems in snowy climates.

If done correctly, a tight enclosure forms the basis for effective pest exclusion (See Section 2.7 of the *Indoor Air Quality Guide*). The benefits include making it easier to manage the building pressurization relative to the outdoors and, in multistory buildings with multiple tenants or uses, air sealing suites, units, or whole floors, it allows airflow between zones to be managed and ventilation systems to perform more effectively.

Many Things Could Go Wrong. There are a few potential problems that can arise from air-sealing buildings:

- The accidental ventilation rate goes down. Whether fan-powered or wind- and stack-powered, mechanical ventilation systems must provide adequate intentional ventilation (See ASHRAE Standard 62.1 and Standard 62.2). While operable windows may be a part of passive ventilation strategy, they are not a complete ventilation system in most U.S. climates.

- As mentioned above air pressure relationships between zones and between indoors and outdoors can be managed with smaller airflows in a compartmentalized or air-sealed building. That means that smaller, accidental airflows can induce significant pressure differences in a building. For example, in an air-sealed building how big an indoor-outdoor pressure difference will be induced if the fire alarm system turns on smoke control fans? What unexpected pressure differences might be created when a test and balance (TAB) contractor tests the HVAC system in failure modes?

Defining Airtight. Most building airtightness measurements are conducted using fan pressurization methods. This test is simple in concept.

The test building has the following:

- All exterior doors and windows closed and latched, and all outdoor air intakes and exhausts and relief outlets off sealed by dampers or masking.
- As many interior doors open as it takes to create a single interior zone.
- The air pressure difference between the indoor and outdoor air is measured. A measured flow of air is blown into or drawn out of the building, which changes the indoor-outdoor pressure difference. When the indoor-outdoor pressure difference changes by a specified amount, the airflow required to induce that change is recorded. The tighter the building, the smaller airflow is needed to induce the specified pressure. So two things are measured—a change in indoor-outdoor air pressure difference and the flow rate associated with that change. Historically, building enclosure airtightness has been reported in two ways:
    - The flow rate in cubic feet per minute required to induce 75 Pa change in indoor-outdoor air pressure difference divided by the building enclosure surface area (cfm 75/ft^2 enclosure)
    - The flow rate in air changes per hour (ACH) required to induce 50 Pa change in indoor-outdoor air pressure difference (ach 50).

The flow rates are normalized to surface area or volume so that buildings of different sizes can be compared.

Currently there are three target airtightness levels used by several building programs in the United States:

- 0.40 cfm 75/ft^2 enclosure used by the Government Services Administration (*Facilities Standards for the Public Buildings Service* [P-100]), the 2012 International Energy Conservation Code.
- 0.25 cfm 75/ft^2 enclosure used by the Army Corps of Engineers
- 0.60 ach 50 used by the Passive House Institute U.S. (PHIUS) (Note: 0.60 ach 50 roughly translates to 0.09 cfm 75/ft^2 enclosure for a three story building with 30,000 ft^2 (2800 m^2) of gross floor area. This makes the PHIUS requirement the most stringent (and most difficult to achieve) in the United States.

What Does It Take to Achieve these Airtightness Targets?

- Perhaps half of the newly constructed commercial buildings in the United States meet or beat the target of 0.40 cfm 75/ft^2 enclosure. This level of airtightness can be achieved by sealing the largest holes in a building. To achieve the airtightness target of 0.25 cfm 75/ft^2, a continuous air barrier system must be installed. To get a building to this level of airtightness follow these steps:

1. Identify a target airtightness level in the specifications.
2. Identify air barrier materials and connecting accessories in the specifications, submittals, sections, plans, and details.
3. Require training, in-house and third party inspection, QA testing, and documentation in the specifications.
4. Specify intermediate qualitative fan pressure testing of full-scale mock-ups that include a corner and roof/wall connection. A top floor, corner room enclosed early in the construction process is a practical way of achieving this requirement. Use fan pressurization and a fog machine or infrared imaging to identify air leaks in the mock-up.
5. Specify a final, whole-building fan-pressurization test to determine whether or not the building meets the airtightness specification. Conduct the test in accordance with the Army Corps of Engineers test protocol.

These same steps, conducted with great rigor, must be followed to achieve Passive House levels of airtightness in commercial buildings. The chances of being successful are improved if people who have been involved in the design and construction of buildings this airtight before are part of the design and construction team for the current project under consideration.

Investigate Regional and Local Outdoor Air Quality

The outdoor atmosphere contains particles and gases that can adversely affect IAQ. A primary resource for information on outdoor air pollution is in the Green Book on the EPA website (see References section). EPA illustrates on maps areas that are not in compliance (nonattainment) with the National Ambient Air Quality Standards (NAAQS). Following the requirements in ASHRAE Standard 62.1-2013 (ASHRAE 2013), the first step in ventilation design for IAQ is to determine compliance with outdoor air quality standards in the region where the building will be located. The next step is to determine if there are any local sources of outdoor air pollution that may affect the building. Filtration, air cleaning, and location of outdoor air intakes can then be considered as a means of reducing the entry of these outdoor contaminants into the indoor environment. Operating scenarios can also be developed to reduce entry of pollutants into the building if the pollutant levels vary over time. For instance, carbon monoxide (CO) from cars will vary with traffic volumes and patterns. Ozone also varies by time of day, with higher concentrations usually occurring in the afternoon.

Local sources of pollution can include industrial or commercial operations in the area surrounding the building site. Visual inspection and consultation with local planning and air pollution control agencies can help identify sources of concern.

The NAAQS addresses particles, ozone, lead, CO, nitrogen dioxide (NO_2), and sulfur dioxide (SO_2). Particulate matter is designated PM10 for particles smaller than 10 µm in diameter and PM2.5 designates particles that are smaller than 2.5 µm in diameter. Filters tested by ANSI/ASHRAE Standard 52.2, *Method of Testing General Ventilation Air-Cleaning Devices for Removal Efficiency by Particle Size* (ASHRAE 2012) are assigned minimum efficiency reporting value (MERV) ratings. Filters need to be at least MERV 8 to have any effective removal efficiency on these smaller particles and filters with MERV ≥ 11 are much more effective at reducing PM2.5. Lead is a solid that will be a particle or attached to other particles and MERV 11 or higher will also be effective at removing lead. Ozone air treatment is provided by carbon or other sorbent filters that cause the ozone to react on the surface. SO_2 can be cleaned by gas-phase air cleaners. Certain filter materials (for example, activated alumina/$KMnO_4$) adsorb SO_2. Odors are often a localized concern. Activated carbon may be effective at reducing odors. For more information, see Strategy 3.3 in the *Indoor Air Quality Guide*.

Locate Outdoor Air Intakes to Minimize Introduction of Contaminants

Outdoor air enters a building through its air intakes when mechanically ventilated and through openings in the building envelope when naturally ventilated.

Whenever possible, the outdoor air entry locations should be separated from known external pollutant sources such as loading docks, exhaust stacks from the building itself and others in its vicinity, and regions of heavy vehicular traffic. Industrial processes in the vicinity can also be important pollutant sources that should be shielded from building air intakes.

Table 5.1 of ASHRAE Standard 62.1 specifies that outdoor air intakes must be located at least 25 ft (7.6 m) from plume discharges and upwind of cooling towers, evaporative condensers, and fluid coolers.

All nearby potential odor sources and prevailing wind conditions need to be evaluated. (See Strategy 3.2, *Indoor Air Quality Guide*)

Use Air Filtration/Cleaning to Remove Contaminants in Outdoor Supply Air

Although epidemiological studies linking ambient fine particle and ozone levels with morbidity and premature mortality have generated increased concern, indoor pollutant levels of outdoor origin in a building have a very large effect on a person's total exposure. The use of air filtration or air cleaning to remove contaminants in outdoor air supply in the HVAC system can reduce outdoor contaminants ingress. Incorporating air-cleaning technologies for outdoor contaminant removal involves both particulate and

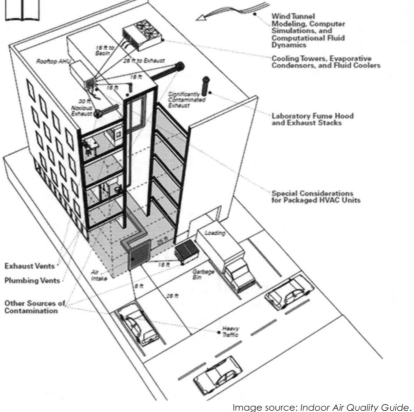

Image source: *Indoor Air Quality Guide*.

Figure 9-2 Examples of problem pollutants and air intake locations for good IAQ.

gas-phase cleaning. Typically, filtration of outdoor particles include the use media filters and electrostatic precipitators while air cleaning of outdoor ozone, nitrogen dioxide, and volatile organic compounds uses adsorption and chemisorption filtration.

Control Entry of Radon

Radon is a radioactive gas formed from the decay of uranium in rock, soil, and groundwater. Exposure to radon is the second leading cause of lung cancer in the United States after cigarette smoking and is responsible for about 10% to 14% of lung cancer deaths.

Radon most commonly enters buildings in soil gas that is drawn though joints, cracks, penetrations, or porous regions of concrete masonry units when the building is at a negative pressure relative to the ground. The potential for high radon levels varies regionally, with additional variation from building to building in the same region and even from room to room in the same building. Radon levels are also highly subject to seasonal fluctuations as well as variations due to discrete weather events.

Design for control of radon entry has three major elements:

1. Active soil depressurization (ASD) uses one or more suction fans to extract radon from the area below the building slab and exhaust it above the building where it is diluted into the surrounding air. Keeping the sub-slab pressure below the pressure in the building reduces the soil gas entering the building and radon entry in this path. A permeable sub-slab layer enhances the area affected by sub-slab depressurization.
2. Sealing of radon entry routes reduces the volume of sub-slab air entering the building.
3. Use of HVAC systems to maintain a positive pressure in ground contact rooms and other spaces enhances the suppression of soil gas entry into these spaces.

An active soil depressurization system reduces entry of other soil gas constituents such as vapors from brownfield sites and water vapor. This design feature can be added to an existing building but is much more effective when added to the design in new construction (ASHRAE 2009a).

Control Vapor Intrusion from Subsurface Contaminants

Vapor Intrusion. The physics of radon entry into buildings from subsurface contamination of soil gas is mirrored in another important group of contaminants, organic vapors from subsurface soils or groundwater. This process, called *vapor intrusion,* occurs when contaminants exist below ground due to accidental spills, improper disposal, leaking landfills, or leaking storage tanks (both below grade and/or above grade).

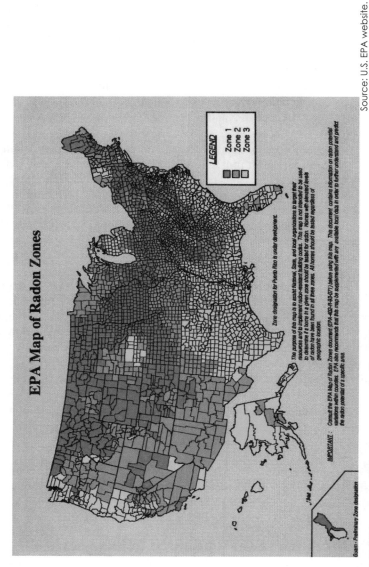

Figure 9-3 U.S. EPA map of Radon zones in the U.S.

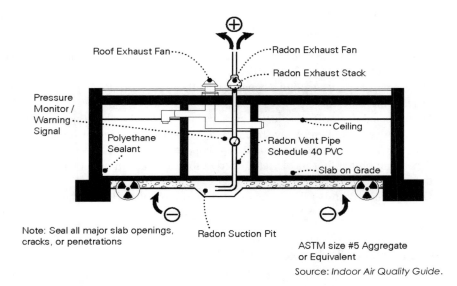

Figure 9-4 Schematic of an active soil-depressurization scheme for control of radon entry into buildings.

These phenomena are commonly associated with construction sites near abandoned gas stations, dry cleaners, and industrial sites using vapor degreasers, etc. It is important to note that vapor intrusion can occur in regions separated from sites where the contaminants were originally released since these contaminants can travel in groundwater plumes for several miles (ASHRAE 2009a).

Vapor intrusion is a concern because of chronic health effects due to long-term exposure to low-level contaminants. ASTM International recently developed a national standard for the assessment of vapor intrusion on properties involved in real estate transactions (ASTM 2012a).

Provide Effective Track-Off System at Entryways

Tracked-in dirt contains contaminants that can result in health effects for the occupants, contains abrasives that contribute physical damage to floor finishes, and is reported to be the source of the majority of dirt that must be cleaned from floors in buildings.

The simplest and most effective way to reduce this impact is to design the landscape and entryways to prevent dirt entry, because around two thirds of the dust in low-rise buildings appears to be tracked in from the outside. Suggestions for the design of entryways include the following:

- Hold back 10% of the sidewalk budget until after the first year of operation; pave the dirt paths created by foot traffic.
- Use fine mesh grates with pits beneath to collect the grit (outside the building or in vestibule is best).
- Use carpeted track-off mats inside entries to dry feet and collect finer particles.
- The final section of entryway should be a hard surface flooring that resists abrasion, is slip resistant, and can be damp mopped, with the length of this section depending on the foot traffic rates. (Strive for 15 ft or 5 m on either side of the door as a goal.)
- Dirt abrasion destroys floors; this system will pay for itself in longer floor life (unless the occupants wear caulked logging boots, hob-nailed boots, or tap dance shoes).

Design and Build to Exclude Pests

There are several different types of pests with a diverse set of entry pathways and supportive conditions. Starting with the ground, subterranean termites are a common problem. Keeping the ground dry under and immediately around the building and eliminating cellulosic debris are both essential and effective ways to eliminate the potential for subterranean termites. (A bonus for dry ground under a building is superior resistance to movement in a seismic event through the foundation's stability and performance.)

Flies and mosquitoes can be prevented from entering by screens sized for the species typically found in the area.

Rodents can enter through amazingly small openings, so extreme care to create a solid barrier around all horizontal and vertical envelope penetrations and all vertical conduits, chases, shafts, and pipes is essential. This latter measure is also important to control pressure gradients between and among spaces that often serve as a driving force for pollutant migration.

Wood, metal, and concrete posts can be pathways for vertical migration of various types of pests. Rat guards are often used to prevent mice and rats from climbing. These are typically sheet metal attachments to the post that project outward, surrounding the post completely with their surfaces slanted downward at approximately a 30% angle.

Limit Penetration of Liquid Water into the Building Envelope

The building envelope consists of the roof/ceiling assemblies, walls, windows, doors, and foundations that intervene between the indoor climate we desire and the outdoor climate most locations in the U.S. provide. One of the primary functions of the building envelope is to protect building materials, finishes, and occupants and their possessions from rain and snow. Here, liquid water is meant to include rain and meltwater falling from the sky, blown by the wind, coursing as runoff or migrating through subgrade soil, fill, or permeable bedrock, that comes into to contact with the building enclosure. This section does not address liquid water that results from condensation on cool surfaces.

The work falls to the geometry and nature of the materials that form the outer layers of the building enclosure. The first line of defense is drainage. Use gravity to drain roofs, roof penetrations, walls, window and door openings, surrounding landscapes, and subgrade foundations away from the building. Sometimes sumps and pumps are needed to drain subgrade foundations, but they are no help if the power is out for three weeks after a hurricane. Use flashing, overhangs, and drip edges to direct water away from vulnerable joints, intersections, and penetrations.

The second line of defense is to specify materials that form the outer skin of a building that are able to tolerate getting wet and, in some climates, wet and frozen. Stone, concrete, brick, decay-resistant woods, paints and coatings, synthetic membranes and boards, glass, corrosion resistant metals, composite wood, and fiber- and cement-based materials have all been successfully used as exterior materials on buildings.

Using only moisture-tolerant materials on the outside of the capillary break is a third line of defense. If materials that must be kept dry are used in the enclosure, put them all inside of a capillary break. Capillary breaks can be made using materials that are impermeable to liquid water; that is, they will not wick water. Low-slope roofing membrane and metal roof systems provide both moisture tolerance and a capillary break. Traditional and synthetic building papers and self-adhering membranes provide a capillary break between moisture-sensitive materials (such as sheathings, insulation, framing and finishes) and the outer barriers, such as shingles on roofs, outer veneer materials like brick, stone or concrete masonry units (CMU), wood or fiber-cement siding, or metal claddings.

An air gap between the capillary break and the outer layers of the envelope, combined with flashing at large penetrations and intersection and sealant at fasteners, greatly improves drainage of any water that seeps past the roofing or cladding systems. The air gap does not have to be large to drain water. A gap of 1/8 in. (3 mm) should be sufficient. If an air gap is intended to reduce drying time, then the gap must be much larger, 2 in. (50 mm) or so minimum. A ventilated attic is an example of a large air gap.

Limit Condensation of Water Vapor within the Building Envelope and on Interior Surfaces

It is critical that building envelopes be properly designed to minimize entry of bulk water and control water vapor. Specifically, envelope systems need to include a drainage plane so that water can exit the envelope system and is kept away from materials that might absorb water or otherwise be damaged. Similarly, placement of vapor barriers in the envelope assembly must be determined so as to manage condensation and minimize impacts, including both vapor diffusion and bulk airflows. These measures are primarily aimed at avoiding water damage and long term wetting that may result in microbial growth. These measures should also help in avoiding thermal bridges and uncontrolled air leakage. The interactions that lead to the necessary amount and duration of moisture accumulation are complex. One example of interactions between different building elements that combine to result in moisture accumulation is vinyl wallpaper on the indoor surfaces of exterior walls in combination with an air-conditioned space in a hot, humid climate. Outdoor air with a high dew point infiltrates the wall and, because the vinyl wallpaper prevents the moisture from passing into the interior of the home, moisture levels increase within the cool interior gypsum wallboard. The problem is aggravated if the HVAC system is designed and/or installed such that it overcools wall surfaces, which, in combination with humid air infiltration (which may be promoted by the HVAC system), can lead to high surface relative humidity (RH) inside the wall for extended periods.

It is critical to keep the indoor dew point low enough to ensure that there is no condensation on the exposed surfaces of cool HVAC components or persistent, excessive moisture accumulation on sensitive building materials or furnishings (ASHRAE 2009b). It is also important to have an effective, continuously sealed air barrier covering all six sides of the building envelope to address infiltration/exfiltration issues and to avoid interstitial moisture accumulation in building materials as well as on cold interior surfaces (ASHRAE 2011a, 2011d, and 2010e). The long-term average indoor air pressure should be kept positive with respect to the outdoors when the outdoor dew point is higher than indoor surface temperatures. In cold weather, one should not humidify the indoor air to dew points high enough to create conditions where there are entire days or weeks of surface RH above 80% inside cooled walls and attics. Architectural features and HVAC system design and operation observed to have contributed to the reduction of the risks of condensation, moisture accumulation, mold, and microbial growth are documented in an ASHRAE Position Document, "Limiting Indoor Mold and Dampness in Buildings" (ASHRAE 2012).

Maintain Proper Building Pressurization

Control of building pressures is required because air will flow from high pressure to low. Maintaining proper relationships ensures that air flows in desirable directions. There are three primary reasons for maintaining differential pressures between different spaces or between the indoors and outdoors. The most critical is to ensure that air in a space with high contaminant levels does not escape into adjacent spaces that are expected or need to be cleaner. Pressurization is also used to minimize the flow of warm, humid air through the building envelope toward cooler surfaces, which may result in condensation inside the envelope and eventual microbial growth. Relative pressurization is also used to minimize the introduction of contaminants and moisture from the soil into the building. ASHRAE Guideline 10-2011 provides additional guidance on pressure relationships.

Pressurization is achieved by maintaining supply airflow higher than exhaust and return airflow in positively pressurized spaces and higher exhaust flows than supply flow in negatively pressurized spaces. Buildings with higher outdoor airflow than exhaust or relief airflow will be positively pressurized relative to the outdoors. Stack effects and wind pressures, however, must also be considered. The effectiveness of a building pressurization strategy will be greatly enhanced by decreasing the leakage paths available for airflow, such as by designing a building with a continuous air barrier system as described in the *Indoor Air Quality Guide*. In fact, decreasing leakage pathways is an important consideration in all buildings.

Positive pressure of 3–7 Pa (0.012–0.028 in. of water) is recommended for large commercial buildings in most climate zones. This may not be appropriate for very cold conditions in order to minimize moisture condensation in wall cavities. Although desirable, pressure control is more difficult to maintain in residential and small commercial buildings.

Control Indoor Humidity

Indoor humidity must be controlled for two primary reasons. Most importantly, if humidity is too high, condensation on cool, indoor surfaces is more likely. If such condensation occurs frequently or for extended periods of time, microbial growth will occur, which can then cause indoor air quality problems and damage to building materials. The other reason to control indoor humidity is that low humidity can result in occupant discomfort. There is also some evidence that airborne pathogens may survive longer in low-humidity indoor environments, possibly increasing disease transmissions.

Humidity control is less of an issue in cold climates where humidification is used, although it must be borne in mind that extremely low humidity levels in the indoor environment can lead to adverse health impacts, such as dry skin, dry eyes, and static electricity. In comparison, hot, humid climates pose a greater challenge where high humidity can lead to potential risks of mold growth. The challenge in such climates is not so much in being able to dehumidify but to do so without having to overcool. Based on considerations provided in ASHRAE Standards 55, 62.1, and 90.1, it is imperative that the efficiency dimension associated with the cooling and dehumidification process involve an absolute improvement in component/system efficiency as well as some means of energy recovery (ASHRAE 2010b, 2013a, 2013b). The following list, by no means exhaustive, contains some key considerations towards the objective of controlling indoor humidity:

1. Oversizing of A/C systems reduces run times and therefore also reduces dehumidification potential.
2. Tight envelopes promote substantially better control over indoor humidity.
3. Buildings in hot, humid climates should be designed with the control of relative humidity as a priority. (Current design practices often lead to overcooling in these regions.)
4. Design systems that are sufficient to control humidity at the design as well as at part loads.
5. Thermal comfort acceptability of tropically acclimatized subjects may well be different from their temperate climate counterparts. Warmer and drier thermal environments or a cooler microenvironment (immediate breathing zone) coupled with warmer ambient conditions may be more acceptable.
6. Consider the use of desiccant dehumidification systems, dedicated outdoor air systems (DOAS), and personalized ventilation systems when designing the air-conditioning and air distribution systems, as appropriate.
7. Take into account the humidity control in energy recovery systems, such as a runaround coil or heat pipe system.

Select Suitable Materials, Equipment, and Assemblies for Unavoidably Wet Areas

Some places in buildings are going to regularly get wet. Bathrooms, showers, spas, indoor pools, kitchens, entryways, custodial closets, and conditioned garages are examples of spaces that are likely to get wet during ordinary operation.

Use materials that can tolerate regular wetting without damage and that can dry quickly enough to avoid moisture-related problems. Among these materials are ceramic tile, glass, plastic resins, metals, and cement-based products.

Moisture-resistant forms of oriented strand board and gypsum board are available. Specify mold-resistance testing criteria that are appropriate for materials (e.g., a score of 10 when tested using ASTM D3273, *Standard Test Method for Resistance to Growth of Mold on the Surface of Interior Coatings in an Environmental Chamber*)(2012b).

Control Impacts of Landscaping and Indoor Plants on Moisture and Contaminant Levels

There are potential advantages and disadvantages associated with the presence of plants as a component of the building envelope (e.g., green roofs or roof gardens, living facades, or vertical gardens) or on walls or other locations in the interior space (e.g., atrium gardens, living walls, vertical gardens, and biowalls).

As part of their physiology, plants emit water molecules into the air through the process of transpiration. In an outdoor environment like a building roof, this provides evaporative cooling. Plants also provide shading to the microenvironment. Inside buildings, an average sized houseplant emits up to 0.22 lb (100 g) of water per day into the indoor air. An increased amount of water vapor in the air will raise the relative humidity. In a building, this is an advantage during the dry season, depending on the source of the water, but can be a disadvantage in a warm, humid condition if not well managed. Benefits of green roofs are thought to include reduction in stress (i.e., thermal stress) on the waterproofing membrane, reduction in heat island effects in urban areas, and reduction in storm water runoff. It is widely recognized that both the integrity and protection of the waterproofing membrane beneath the roof garden need to be of very high quality if leakage into the building is to be avoided. In this regard, the building architect and the landscape (garden) architect need to work together to ensure that the waterproofing membrane is installed with excellent workmanship and that penetrations through the membrane are avoided. The presence of indoor flora (potted plants, atrium gardens, etc.) is generally perceived as beneficial to occupants. However, this assumes that the water transpired does not exceed the capacity of the HVAC system to manage the increased water in the room, that the potted plants are not overwatered, and that atrium gardens are well maintained.

Control of Moisture and Dirt in Air-Handling Systems

Moisture and dirt in air handlers can lead to several problems. Moisture can build up at several locations within the air handlers including at the outdoor air intake, in the condensate pans, and downstream from the air humidification unit. The moisture can lead to rust on equipment and possible microbial growth.

Dirt and debris can reduce the airflow below the design levels, particularly when dirt buildup on the fan blades is sufficient to alter the fan curve. Also, dirt buildup on cooling coils can restrict airflow. Air resistance provided by dirt buildup can increase the fan power needed, and, coupled with the insulating properties of the dirt, can lead to energy waste. Finally, moisture and dirt can facilitate microbial growth in the air-handling unit. The microbial growth in turn can lead to the release of potentially irritating or disease-causing biological and chemical agents into the occupied spaces.

ASHRAE/ACCA/ANSI Standard 180 (2008) (section 5.2) identifies ten inspection/maintenance tasks that relate to control of dirt, microbial growth, and/or excess moisture in air handlers. Control of moisture and dirt in these systems is facilitated by designing and implementing a maintenance program in accordance with ASHRAE Standard 180.

Control Moisture Associated with Piping, Plumbing Fixtures, and Ductwork

Mold growth can occur on cold water pipes or cold air supply ducts with inadequate thermal insulation or a failed vapor retarder and result in material damage or significant IAQ problems, leading to potential adverse health impacts on occupants. Liquid water from condensation can damage materials nearby such as ceiling tiles, wood materials, and paper-faced wallboard located below or adjacent to the piping or ducts. Leaks from poorly designed plumbing within walls or risers may go unnoticed until damage, including mold growth, becomes evident in occupied spaces. Implementation of design strategies that limit condensation on cold-water piping and ducts and that reduce the likelihood of piping leaks hidden in building infrastructure will lessen the likelihood of these potential problems.

Facilitate Access to HVAC Systems for Inspection, Cleaning, and Maintenance

Access to the HVAC systems at critical locations such as after supply air silencers, at junctions, and near duct ends should be made available for regular inspection, cleaning and maintenance. To access the duct interiors, access panels should be considered at the design stage. Specific locations, size, and type of access panels have been described in the HVCA and NAIMA (North American Insulation Manufacturers Association) guidelines (HVCA 2002; NAIMA 2007). In existing ductwork systems without access panels, it is possible to cut through duct walls or use existing openings such as grilles, registers, and diffusers (NAIMA 2007; NADCA 2006). When cutting openings, detailed procedures that take into account the main insulated duct systems (fibrous glass duct board, lined sheet metal ducts, wrapped sheet metal ducts, metal ducts insulated with rigid fibrous glass board, flexible ducts) have been described (NAIMA 2007). Prior to cutting, considerations should be given to the potential presence of asbestos-containing wrapping and lead paint on the ductwork.

Regardless of the access means used, the openings should be closed airtight after inspection, cleaning and maintenance. There should be no obstruction or alteration to the airflows and no degradation to the thermal or functional integrity of the HVAC system.

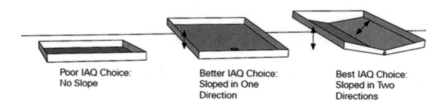

Source: *Indoor Air Quality Guide.*

Figure 9-5 Drain pain examples with and without slope.

Control *Legionella* in Water Systems

Legionnaires' disease can develop when a susceptible individual inhales aerosolized water droplets containing *Legionella* bacteria. *Legionellae* grow in both natural and man-made aquatic environments such as lakes, river, streams, as well as cooling towers, whirlpool spas, humidifiers, water fountains, and potable water distribution lines.

Legionnaires' disease is a bacterial disease commonly associated with water-based aerosols that have originated from warm water sources. It is often associated with poorly maintained cooling towers and potable water systems. The Center for Disease Control and Prevention (CDC) estimates that between 8,000 and 18,000 cases of Legionnaires' disease result in hospitalization in the United States annually (CDC 2013). This disease typically is fatal in 5% to 30% of the cases diagnosed.

Preventative Strategies

Preventive maintenance should be carried out for effectively operating cooling towers and other evaporative equipment. These maintenance activities include using effective drift eliminators, periodical cleaning, and employing a water treatment program with a proper biocide. Regular monitoring of *Legionella* levels in cooling-tower water is recommended by many experts. The European Working Group for *Legionella* Infections (EWGLI) shares knowledge about monitoring potential sources of *Legionella* as well as providing technical guidelines for investigation, control and prevention of Legionnaires' disease.

Consider Ultraviolet Germicidal Irradiation

Ultraviolet germicidal irradiation (UVGI) has been used successfully for many years to control airborne infective microorganisms such as *Mycobacterium tuberculosis*, the bacterium that causes tuberculosis. Ultraviolet light at all wavelengths but especially at around 265 nm modern ultraviolet lamps have an optimal discharge at 254 nm) damages the DNA of irradiated microorganisms. Ultraviolet light emitted and localized in the upper portion of room air (referred to as upper-air UVGI) has been used for controlling tuberculosis, especially in poorly ventilated or crowded indoor spaces where other interventions, such as increasing HVAC outdoor air ventilation rates or raising filtration efficiency, are not practical. Droplet nuclei from infected occupants in a room can migrate on air currents into the upper room air to be inactivated by UVGI from lamps along upper portions of walls. Inactivation of airborne microorganisms depends on both the intensity of the UVGI and the length of time that the particle containing the microbe is irradiated.

In addition to inactivating airborne microorganisms, UVGI directed at environmental surfaces can damage culturable microorganisms present or growing on the surface. Lower-intensity UVGI is effective for surface inactivation because irradiation is applied continuously. UVGI from lamps in air-handling unit (AHU) plenums has been used successfully to inactivate microorganisms present on airstream surfaces such as cooling coils and drain pans. One study reported decreased sick-building syndrome (SBS) symptom prevalence associated with the use of UVGI in an office building AHU.

The designer must be aware that the use of UVGI lamps in AHUs, ductwork, and upper air requires careful attention to safety considerations to prevent inadvertent exposure of people to ultraviolet light. For example, lockout/tagout procedures are necessary to prevent accidental turning on of UVGI lamps when facility maintenance personnel are working in AHUs. Refer to Chapter 17 of the *ASHRAE Handbook—HVAC Systems and Equipment* (ASHRAE 2012) and Chapter 60 of the *ASHRAE Handbook—HVAC Applications* (ASHRAE 2011) for a comprehensive review of safety considerations associated with the use of UVGI in buildings. Well-designed upper-air UVGI systems have been used safely for many years. For more complete information, refer to *Indoor Air Quality Guide* (ASHRAE 2009a) Section 4.5 and references cited therein.

Limit Contaminants from Indoor Sources

Understanding the emission characteristics of various indoor sources is an important step in devising strategies to limit the contaminants emissions. In this section, we first classify the typical indoor contaminant sources in terms of their

Photo courtesy of Hal Levin.

Figure 9-6 Bellevue Stratford Hotel, Philadelphia: Site of 1976 Legionnaires' disease outbreak resulting in which 29 people died from exposure in the hotel and 5 people died from exposure in the vicinity.

emission behavior; discuss the major influencing factors and general control strategies. We then provide more specific guides for selecting low-emitting materials, and methods for limiting the impact of the emissions. See *Indoor Air Quality Guide*, Strategy 5.1, for extensive information on this complicated and important subject.

Indoor Contaminant Source Classification and their Emission Characteristics

Indoor contaminant sources can be classified into several groups based on their emission characteristics:

Passive Sources. These are the building materials, finishes, and furnishings that do not involve occupants and exist in buildings regardless of the activities of the building occupants or operation of the ventilation system. Major contaminants of concern from these sources include VOCs and semivolatile organic compounds (SVOC). The emission rates are generally highest following the installation or application and then decrease over time throughout a building's service life. However, the actual emission pattern depends strongly on the material and its position in a material assembly, the chemical compound of interest, and the environmental conditions (temperature, relative humidity, and surrounding air velocity and concentration level). For example, architectural coatings such as paint, wood stain, and varnishes that are applied wet can initially have emission rates that are several orders of magnitude higher than dry materials such as engineered wood products, but then decrease quickly; they may drop well below long-term emission rates of dry materials (such as medium density fiberboard [MDF]). After several weeks, the emission rate of the dry material may be an order of magnitude higher than the wet material. The emission rates of dry materials start at a relatively low level, but can persist over a long period of time (months to years). The types of emitted compounds also vary dramatically among different material types (ASHRAE 2009a). Oil-based wet coating materials using petroleum products as raw materials emit significantly higher number of VOCs, while water-based coating materials generally have fewer compounds emitted as well as a lower emission rate. The importance of the compound to IAQ depends on its irritancy or toxicity, established indoor exposure limit, and the amount of its emitting materials used in the building, as well as the emission rate. Unless otherwise indicated, specific emission rate here refers to an emission rate per unit of the material (surface area, length, weight, or number). In addition, the following discussion is intended as an example rather than an extensive list, though the compounds listed are of importance in various IAQ guides or emission testing protocols.

Active Sources. These source types also generate heat, such as computers, printers, copiers, other office equipment, and room air cleaners. These sources may be associated with particle emissions, especially ultrafine particulate matter, as well as with emissions of VOCs and SVOCs. These types of source are associated with occupant activities or use of the building, and hence their emission rates are highly dependent on the timing of their respective activities or operation. The emission rates of the sources are typically dramatically higher during the operation or activity than during idle or power-off conditions. While proper selection of materials or equipment can reduce emissions from the sources, using local exhaust for these sources or increased dilution flow during the process can further limit the contamination from these sources.

Occupants. Contaminants from occupants are mainly bioeffluents (including exhaled breath, sweat, and shed skin cells) and depend on personal hygiene, health conditions, and the immediate environment they were in before entering the space or building of interest. The can include VOCs, SVOCs, and both viable and nonviable particles. There are also emissions from clothes such as dry cleaning chemicals and chemicals adsorbed from air during previous exposure situations.

Building Operation and Maintenance Related. These emission sources include items such as cleaning of the floor, furniture, and windows, and involve cleaning agents or emissions from filters, HVAC condensates, ductwork, etc. The type of compounds emitted and their emission rates are more difficult to characterize, depending on the cleaning agents used and the cleaning procedure or the nature of the environments the building system components are exposed to.

Secondary Sources. This type of emission source results from emissions from sink materials that had previously adsorbed contaminants from the space or from chemical reactions such as ozone-initiated gas-phase and surface reactions. Their emissions result from interaction among multiple materials or gases coexisting in the space or building, and hence their emission profiles depend greatly on such interaction. Contaminants resulting from ozone-initiated chemistry include ultrafine particles, formaldehyde, and other aldehydes that are of IAQ concern. The most effective approach is to limit the entry of ozone from the outside or emissions from indoor devices. Compounds reemitted from the sinks are those adsorbed previously when the air-phase concentrations were high. Properly staging the installation and application of construction materials can help reduce such emissions. For example, carpet should be installed after the emissions from the paint have reduced to significantly low levels to avoid excessive adsorption of the pollutants emitted from the paint.

Selection of Low-Emitting Products in Building Design and Operation

This strategy is applicable to both passive and active sources. Over the last 20 years, standard test methods and guideline emission limits have been established for many building materials, furnishings, furniture, copiers and printers, and computers (e.g., California Department of Public Health, the Collaborative for High Performance Schools, GreenGuard, ASTM, Carpet and Rug Institute, and Business and Institutional Furniture Manufacturers Association [BIFMA]).

The emission test methods are used to measure the emission rate per unit of materials or products tested (called emission factors). Green-building certification programs such as LEED have assigned credits for using materials and products that have measured emission factors less than the established limits to encourage the use of low-emitting materials and products.

Designers can also require such data from material or product manufacturers or suppliers. The emission factor data can be used to estimate the total emission rates based on the amount of materials/products used in the building. In cases where the data are not available but the materials or products have been labeled as low-emitting ones because their emission factors are less than the established guideline emission limits, the emission limits can be used as worst-case scenarios to estimate the maximum total amount of emissions for the contaminants of concern. Such estimations can be carried out for all materials that are potential emitters to determine the total internal pollutant load and help to determine if further reduction of the emissions are possible with alternative materials or products or if the use of local exhaust, air cleaning, or increasing ventilation rate strategies are more practical. Readers should refer to the indoor air quality procedure in ASHRAE Standard 62.1 to perform such analysis.

Besides material emission tests-based labeling or certification programs, there are also green-product labeling programs that are based on the chemical contents in the materials or products (e.g., GreenSeal, EcoLogo). At best, this type of labeling program can be used for relative comparison among the same type of materials or products, because the actual chemical emission rates can differ significantly from the chemical content. Materials safety data sheet (MSDS) from manufacturers may only be useful for initial screening since only chemicals with more than 0.1% of the material mass are listed, and requirements for listing are limited to chemicals identified by OSHA as hazardous. Chemicals of IAQ concern often occur in much smaller mass amount in the materials. For those compounds that are listed, listing the actual amount in the product is not required.

Odor acceptability is also determined in some emission test standards. Odor acceptability alone or in combination with emission rate data when available can be used to screen low-emitting materials/products, because strong odor intensity is highly correlated with high emission rates of organic compounds.

Selecting low-emitting materials/products in building design can result in reduction of indoor contaminant concentrations by a factor of two or more (ASHRAE 2009a). Careful material selection is considered the most economical and effective approach to IAQ improvement. In 1858, a pioneer in indoor air quality, von Pettenkofer wrote: "If there is a pile of manure in a space, do not try to remove the odor by ventilation. Remove the pile of manure."

Strategies to Limit Impact of Emissions

Emissions from indoor sources can be minimized through proper selection of materials and products, but are often inevitable due to functional requirements, practicality, and economic considerations. Strategies to limit the impact of emissions vary among the type of indoor emission sources.

For passive sources such as building materials, finishes, and furniture, allowing a period for the emission rates to decline immediately following the production of materials is an effective strategy. This approach can also be applied before the installation. Actively conditioning products after production and/or before installation has been shown to be an effective measure. The installation schedule should also be sequenced to the extent possible in such a way that materials or products with high peak emissions are installed first and allowed a certain period to air out with sufficient ventilation before lower-emitting and high-surface-area ("fleecy") materials are installed. This can reduce the significant adsorption of the contaminants from high-emitting materials by the low-emitting materials and reemit later (the so called *sink effect*). Airing out the whole building with the highest practical outdoor air ventilation rate for a period of one to two weeks before occupancy (flush-out) will also help reduce the exposure of occupants to the indoor source contaminants. The so-called *bake out* procedure, that is, airing the building with elevated temperature, is however, generally not recommended due to the potential of increasing the emissions from inner layer materials and possible damage to the material's structural and other performance properties.

For active sources such as copiers, printers, and computers, apply local exhaust when possible. When a dedicated room equipped with separate exhaust system is not available, consider placing these sources in such a way that their emissions would have minimal transport to the breathing zone of the occupant. This needs to be achieved in conjunction with the room air distribution design or flow management. For example, personal ventilation with fresh air supply to the breathing zone and employing local air cleaning can reduce the impact of the emission sources.

For emissions from occupants, adequate personal hygiene, proper selection of personal deodorants, cosmetics, fragrances, and laundry detergents are perhaps a good starting point. Emissions from smokers even when they are not smoking can still be significant.

For building HVAC system operation- and maintenance-related emission sources, the key is to regularly check the filters and replace them in a timely manner to avoid emissions from old filters. For the cleaning of floors, furniture, and windows with cleaning agents, the activity should be scheduled during the unoccupied period and with increased ventilation to minimize adsorption of the cleaning chemicals by the building materials and furnishings. Refer to ASHRAE's *Indoor Air Quality Guide: Best Practices for Design, Construction and Commissioning* for more specific guidance.

For secondary emissions that result from the interactions between different contaminants or sources, the key is to prevent the introduction of reactive compounds and isolate the interactive sources. In an area where the outdoor ozone concentration level is high, properly selected filters (including activated carbon) should be applied in the HVAC air intake to minimize the entry of the ozone into the building. The building should be operated under slightly positive pressure to minimize the entry of ozone via infiltration. Indoor ozone-emission sources should be controlled. For example, ionization-based air-cleaning devices may be effective for removing particulates, but are often accompanied by ozone generation. Ozone emissions from copiers and printers should be minimized first by careful selection of low-emitting devices and secondly with carbon filters or by removal with an adequate local exhaust system.

Finally, it should be emphasized that the understanding of indoor emissions, especially secondary emissions, are still very limited. A minimum level of ventilation rate is always required to dilute and remove currently unknown emissions as well as for the bioeffluent intrinsic to building occupants.

Properly Vent Combustion Equipment

Combustion appliances are divided into two primary categories—vented and unvented. Vented equipment is designed to be attached to a venting system that carries the products of combustion to the outdoors. As discussed in Objective 6.1 of the *Indoor Air Quality Guide*, it is important that the venting system be designed, installed, and maintained to properly vent the appliance; otherwise, combustion gases, including potentially dangerous pollutants, may be released into occupied indoor spaces.

All vented combustion appliances will be supplied with information describing the requirements of the venting system that must be used with that appliance. These specifications include the size of the vent system piping, required materials, requirements for pipe pitch, and required connections to the appliance. For example, natural draft equipment usually requires a draft hood that allows room air to flow up the venting system in addition to the combustion products from the appliance. The venting system must also be pitched such that the buoyancy of the hot flue gases results in flow to the outdoors. Condensing equipment will require smaller pipe sizes and has requirements that the system be pitched to drain the liquid condensate.

When natural- or induced-draft vented combustion appliances are located within the building envelope and installed with venting systems that comply with manufacturer instructions, they will be able to operate in the presences of typically occurring negative pressures. However, excessive negative pressures, such as those that can be created in a home with large, unbalanced exhaust fans, may cause back drafting of the appliances. This can best be avoided by providing a makeup air fan that is interlocked with the exhaust or otherwise ensuring that extreme negative pressures are not created by the ventilation systems in the building.

Venting systems must be maintained to ensure that they remain intact with no leakage and that no blockage of the system has developed. Blockages can develop due to dirt and debris, water traps, and animal or bird nesting.

Unvented combustion appliances are designed such that the combustion products can be released into the occupied space. The combustion products of properly operating equipment will be primarily carbon dioxide and water, but small amounts of carbon monoxide and oxides of nitrogen, particularly nitrogen dioxide, will also be present. It is important to follow the manufacturer's installation instructions, which will include information on acceptable room size and provision of adequate ventilation air. It is also important that the appliance not be operated for extended periods of time that might allow excessive contaminant concentrations to build up. Finally, unvented combustion appliances need regularly scheduled maintenance to ensure that emission rates for contaminants remain within allowable tolerances. See ASHRAE's "Position Document on Unvented Combustion Devices and Indoor Air Quality" (ASHRAE 2012) for additional information on these appliances.

Provide Local Capture and Exhaust for Point Sources of Contaminants

It is much more efficient to capture contaminants at the source and directly exhaust to the outdoors than it is to allow them to be released into the space and then remove them with dilution ventilation. ASHRAE Standards 62.1 and 62.2 contain requirements for local exhaust for some contaminant sources, such as residential kitchens and bathrooms (ASHRAE 2013a, 2010a). When equipment or a process in a space releases pollutants, an exhaust system may be installed to perform capture and removal. Ideally, the equipment will be designed to provide necessary pollutant capture and will have exhaust system connections. Otherwise, a capture hood may need to be designed and installed as part of the exhaust system. Objective 6.2 of the *Indoor Air Quality Guide* provides additional guidance on local capture and exhaust, including the necessity to discharge exhaust in a manner to prevent reentrainment of contaminants and the optional strategy of maintaining contaminant generation areas at a negative pressure to prevent migration of contaminants to other spaces.

Design Exhaust Systems to Prevent Leakage of Exhaust Air into Occupied Spaces or Air Distribution Systems

Objective 6.3 of the *Indoor Air Quality Guide* discusses the importance of ensuring that pollutants that have been captured and are being transported through an exhaust system do not leak back into occupied space or into air distribution systems. Exhaust systems should be at a negative pressure relative to surrounding space such that any leaks in the system result in airflow into the exhaust. This is done by placing exhaust fans such that they are sucking air out of the exhaust rather than blowing it in. Alternatively, exhaust systems must have special measures applied to ensure that they have no leaks. Standard 62.1 requires exhaust systems that are not negatively pressurized relative to their surroundings be sealed in accordance with Sheet Metal and Air Conditioning Contractors' National Association (SMACNA) Seal Class A. Standard 62.2 contains requirements for multibranch exhaust ducting, such as a continuously operating downstream fan or individual back draft dampers at exhaust inlets to prevent cross-contamination between dwelling units of a multifamily building.

Maintain Proper Pressure Relationships between Spaces

In many buildings, certain spaces have significantly higher contaminant levels than adjacent spaces. When this is the case, it is important that the dirtier spaces have lower air pressure (i.e., be negatively pressurized) relative to nearby clean spaces. This pressurization minimizes transport of contaminants into the cleaner spaces. Facilities built for health care, laboratory, food service, swimming (natatoriums), and many industrial uses have spaces that require negative pressurization. Negative pressure is obtained by having exhaust airflow rates that exceed supply airflow, while positive pressurization is obtained by higher supply flows.

Careful planning of airflows from supplies to exhausts and returns must be performed to ensure that proper pressurization relationships are maintained. The pressure differences achieved by a given difference between supply and exhaust flow rates are a function of how open the space is to adjacent spaces. Rooms that are containment rooms, designed to minimize leakage, require lower airflow differences than more open rooms. The HVAC system design must include consideration of changing airflows as thermal or ventilation conditions change. In some buildings, there may be spaces that need to be positive relative to some spaces but negative relative to others. Obviously, particular care is required to maintain these relationships.

A number of steps can be taken to ensure that desired space pressurization is attained. These include the following:

- Verifying proper construction of space envelopes, including design information on contract documents
- Performing testing and balancing to verify all airflows and performing pressure differential mapping (ASHRAE 2009a)

Provide Appropriate Outdoor Air Quantities for Each Room or Zone

The ventilation rate procedure in ASHRAE Standard 62.1-2013 specifies minimum ventilation rates for the United States. Local building codes usually reference or include these rates but may differ in various ways from the standard. If local codes require more ventilation air than Standard 62.1, these values must be used.

After the designer determines the outdoor air required for each zone, the quantity of air for the ventilation system must be adjusted to account for air distribution effectiveness and air-handling system ventilation efficiency.

ASHRAE Standard 62.1-2013 specifies two distinct ventilation rate requirements. The first is a per-person requirement to dilute pollutant sources associated with human activity that are considered to be proportional to the number of occupants. The second is the additional per-unit-area requirement designed to dilute pollutants generated by building materials, furnishings, and other sources not associated with the number of occupants.

The ventilation rates are specific to the type of occupant activity. Thus, the outdoor air ventilation rates for different parts of an office building may vary depending on the office activity in the zones.

During short-term episodes of poor outdoor air quality, ventilation can be temporarily decreased using a short-term conditions procedure from Standard 62.1-2013. Similarly, during unusual indoor conditions, increasing outdoor air ventilation rates can be done when the quality of the outdoor air is high and the energy penalty is not excessive. Air-cleaning devices or filters should also be considered in the design to provide sufficient clean ventilation air while conserving energy (see Strategy 7.1, *Indoor Air Quality Guide*).

Continuously Monitor and Control Outdoor Air Delivery

Accurate monitoring and control of outdoor air intake at the air handler is important for providing the correct amount of outdoor airflow to a building. In particular, it has been common practice for designers to use fixed minimum outdoor air damp-

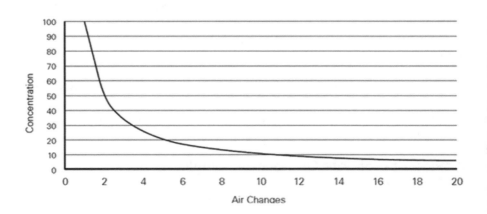

Figure 9-7 Relationship between ventilation rate and pollutant concentration from an indoor source with no outdoor source.

ers. However, the approach does not necessarily provide good control of outdoor air intake rates, particularly in VAV systems. (ASHRAE 2009a).

In most systems, it is difficult to measure accurately outdoor airflows at the outdoor air dampers during balancing, commissioning, or operation. As a result, both overventilation and underventilation can commonly occur. Furthermore, in occupied buildings, overventilation is common because occupancy rates per unit floor area are often less than design values. It is estimated that the current amount of energy for ventilating U.S. buildings could be reduced by as much as 30% (first order estimate of savings potential) if the average minimum outdoor air rate were reduced to meet current standards.

Accurate measurement of airflows in ducts also requires careful design, proper commissioning, and ongoing verification. Under carefully controlled laboratory conditions, commercially available airflow sensors are very accurate. However, in most cases, laboratory conditions and accuracies cannot be replicated in the field. Therefore, appropriate correction factors in the programming of the controls may be required.

Continuous monitoring of the outdoor air rates at the air handler does not guarantee that the proper amount of ventilation is delivered locally within the building. Poor air mixing both in the ductwork and in the occupied space, especially in larger and more complex air distribution systems, can result in parts of a building receiving less than the design minimum amount of ventilation.

Effectively Distribute Ventilation Air to the Breathing Zone

Ventilation only works when the air is effectively distributed to the breathing zone. Different methods of air distribution have different efficiencies. For an inefficient system, the quantity of outdoor air at the air handler needs to be increased in order to provide the minimum quantities in the breathing zone that are required by code and by ASHRAE Standard 62.1 (ASHRAE 2013).

The airflow rate that needs to be distributed to a zone varies by the effectiveness of the distribution within the room. Correspondingly, the ventilation airflow rate provided to the zone needs to be sufficient to provide the required ventilation air to the breathing zone. Zone air distribution effectiveness is an important criterion in ensuring that adequate ventilation air is delivered to the breathing zone and is included in the following expression to compute the zone outdoor airflow (Equation 6.2 in ASHRAE Standard 62.1):

$$V_{oz} = V_{bz}/E_z$$

where:
V_{oz} = quantity of ventilation air delivered to the occupied zone, cfm (L/s)
V_{bz} = quantity of ventilation air delivered to the breathing zone, cfm (L/s)
E_z = zone air distribution effectiveness

Thus, the less inefficient an air distribution system is within a zone, the greater will be the required outdoor airflow to the zone. Different air distribution strategies can result in varying levels of zone air distribution effectiveness, which can be typically categorized as follows (ASHRAE 2011c).

Superior Distribution. In a truly stratified ventilation air distribution system, such as low-velocity displacement ventilation (DV) that achieves unidirectional flow and thermal stratification, E_z can be 1.2 or even greater.

Effective Distribution. In a mixing ventilation (MV) air distribution system, $E_z = 1.0$ if the zone air is perfectly mixed. Ceiling supply of cool air and ceiling return, ceiling supply of warm air and floor return, floor supply of cool air and ceiling return (provided that the 150 fpm [0.8 m/s] supply air jet reaches 4.5 ft [1.4 m]) or more above the floor [such as UFAD system]), floor supply of warm air, and floor return are some of the examples of effective distribution.

Adequate Distribution. Ceiling supply of warm air 15°F (8°C) or more above space temperature and ceiling return is an example of adequate distribution.

Ineffective Distribution. Floor supply of warm air and ceiling return is an example of ineffective distribution.

Effectively Distribute Ventilation Air to Multiple Spaces

As discussed in Strategy 7.3 of the *Indoor Air Quality Guide*, ventilation only effectively dilutes and removes contaminants when the air is effectively distributed to the breathing zone. Different methods of air distribution have different efficiencies and these methods (systems) include constant air volume (CAV), variable air volume (VAV), dedicated outdoor air systems (DOASs), and secondary recirculation systems, such as parallel fan-powered boxes, series fan-powered boxes, ducted/plenum returns, and transfer fans. For an inefficient system, the quantity of outdoor air at the air handler must be increased to provide the required minimum quantities in the breathing zone that are required by code and by ASHRAE Standard 62.1 (ASHRAE 2013). For multiple-zone recirculating systems, the system will have efficiency (E_v) that must be calculated to determine the outdoor airflow rate at the air handler. This efficiency is for the system, and must be used in addition to the corrections for effectiveness of distributing the air within the zone (E_z). The values for system efficiency can range from 1.0 or lower, with higher values being more efficient.

For a single-zone system or a DOAS, the airflow rate at the outdoor air intake (V_{ot}) is equal to the air required for the zone(s). For systems that supply multiple zones, the calculations are more complex. The airflow rate required at the outdoor air intake is driven by the critical zone. Systems that serve multiple zones and recirculate air from one or more of these zones have an inherent inefficiency if the percentage of outdoor air required is not the same for each zone (ASHRAE 2010b). This is because the percentage of outdoor air in the supply air is the same for all zones, so zones that require a high ratio of outdoor air to supply air will be under-ventilated if outdoor air rates at the air-handling unit are not increased. Table 6-3 in ASHRAE Standard 62.1 is a simplified method of accounting for this phenomenon and provides the values for system ventilation efficiency to be used in design on the basis of an average outdoor air fraction of 0.15 for the system (the ratio of uncorrected outdoor air intake to the total zone primary airflow for all the zones served by the air handler). For systems with higher values of outdoor air fraction, Table 6-3 may result in unrealistically low values of E_v and necessitate a very high outdoor airflow rate. Appendix A in ASHRAE Standard 62.1 offers an alternative method for calculating the system ventilation efficiency (E_v) for multiple zone systems. In this alternative method, the percent outdoor air required at the system intake can be less than the percent outdoor air required at the critical zone, as some credit is given to the unused outdoor air from other zones that can be reused in the recirculation system.

Provide Particle Filtration and Gas-Phase Air Cleaning

Canadian houses built to the R-2000 Standard comply with one of the few energy efficiency rating systems in the world to include enhanced particle filtration to improve indoor air quality. In the United States, the LEED rating system doesn't focus much on air treatment systems (other than for air filters used during and after construction). However, it is likely the matter will receive increasing attention in the future.

The incorporation of air-cleaning systems for particle and gas contaminant removal in any building requires investment in design, labor, equipment, and maintenance, and, as for all building systems, must be evaluated in economic terms. Because airborne particle and ozone exposures have been established to cause detrimental health effects, air cleaning will provide noticeable improvement in air quality, especially for susceptible individuals. At present, there are a few published studies that have reported the cost/benefits of indoor air cleaning with regards to airborne particle removal (e.g., Fisk et al. 2011). These studies have reported insignificant filtration costs relative to salaries, rent, health insurance costs, loss of productivity, morbidity, and mortality. An ongoing project is looking into the cost/benefits of indoor air cleaning regarding ozone removal. Despite little information on the recommended filtration to be provided in green buildings, it is recommended that individuals who require higher levels of air cleanliness should consider designing and installing a higher level of air filtration (a minimum efficiency reporting value [MERV] 10 or 11 filter) and/or an activated carbon filter.

Because using of ventilation to remove indoor contaminants in buildings is associated with increased energy use, air cleaning has the potential to be a sustainable alternative. However, few studies have demonstrated energy benefits of using air cleaning compared to ventilation, especially in large commercial office buildings. Additional research is needed.

Various technologies exist in the marketplace for cleaning air being circulated in buildings. These technologies include both particulate and gas-phase cleaning. Types of technologies and products can range from the well-established particle media filtration to newer concepts, such as electrostatic precipitators, electret filtration, negative, positive, or bipolar air ionization, photocatalytic oxidation, adsorption and chemisorption filtration for gaseous pollutants, ultraviolet treatment, and even ozone generators. Some of these technologies are known to generate byproducts such as ultrafine particles and ozone. Air cleaning can be in the form of in-duct systems, integrated in the HVAC unit removing pollutants from outdoor air or supply air, or portable units for removing pollutants in a single room or specific areas for residential application. Portable units generally contain a fan to circulate the air, may be moved from room to room, and can be used when continuous and localized air cleaning is needed. Both in-duct system and portable systems may use one or more of the air-cleaning technologies described below.

Particle Filtration Technologies. Mechanical air filters use a fan to force air through a medium that traps particles such as dusts, pollens, smoke, and other airborne allergens. The high-efficiency particulate air (HEPA) filter is the best-known air filter. To qualify as a true HEPA filter, it must be able to capture at least 99.9997% of all particles 0.3 μm or larger in diameter that enter it. Electrostatic

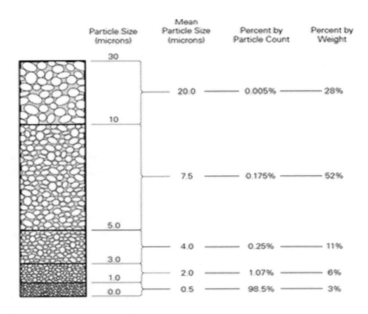

Figure 9-8 Particle distribution by size, count, and weight in typical outdoor air samples.

precipitators use electrical charges to attract and deposit particles on their collection plates. Ion generators use corona discharge to generate ions with the intention of charging and clumping particles until they are settled to the ground due to gravity. Other charged ions may deposit to room surfaces and have to be cleared away. A disadvantage of electrostatic precipitators and ion generators is that almost all of them create small amounts of ozone or ultrafine particles.

Gas-Phase Air Cleaning. Gas-phase air-cleaning technologies include adsorption filters (e.g., activated carbon), chemisorption filters, and photocatalytic oxidation. The main purpose of these technologies is to remove VOCs, formaldehyde, odors, and other nonparticulate gases such as ozone and nitrogen dioxide in the indoor air. Disadvantages of adsorption filters are that they quickly become ineffective over time and need regular replacement. A major disadvantage of photocatalytic oxidation technology includes the generation of byproduct such as ozone.

Air Cleaning Using Plants. Living vegetation or plants have been advocated as a natural biofilter to reduce levels of gaseous pollutants and particulates. In addition, vegetation for air cleaning has been advocated for the improved sense of well-being and psyche of the building occupants. A number of houseplants have been identified that may contribute to passive cleaning of indoor air pollutants, but these are

Table 9-1: Comparison of MERV Level, Prior Rating, and Efficiency by Particle Size

MERV Level	Dust Spot %	Typical Particulate Filter Type	% 0.3–1 μm	% 1–3 μm	% 3–10 μm
1	N/A	Low-efficiency fiberglass and synthetic media disposable panels, cleanable filters, and electrostatic charged media panels	Too low efficiency to be applicable to ASHRAE Standard 52.2 (ASHRAE 2007) determination		
2	N/A				
3	N/A				
4	N/A				
5	N/A	Pleated filters, cartridge/cube filters, and disposable multi-density synthetic link panels			20–35
6*	N/A				36–50
7	25%–30%				50–70
8	30%–35%				>70
9	35%–40%	Enhanced media pleated filters, bag filters of either fiberglass or synthetic media, rigid box filters using lofted or paper media		>50	>85
10	50%–55%			50–65	>85
11	60%–65%			65–85	>85
12	70%–75%			>80	>90
13	80%–85%	Bag filters, rigid box filters, minipleat cartridge filters	>75	>90	>90
14	90%–95%		75–85	>90	>90
15	>95%		85–95	>90	>90
16	98%		>95	>95	>95
The following classes are determined by a different methodology than that of ASHRAE Standard 52.2 (ASHRAE 2007)					
NA	N/A	HEPA/ULPA filters evaluated using IEST Recommended Practice CC001.3 (IEST 1993). Types A through D yield efficiencies at 0.3 mm and Type F at 0.1 mm			99.97% IEST Type A
NA	N/A				99.99% IEST Type C
NA	N/A				99.999% IEST Type D
NA	N/A				>99.999% IEST Type F

* MERV 6 is prescribed by ASHRAE Standard 62-2001 (ASHRAE 2001) for minimum protection of HVAC systems.

Source: *Indoor Air Quality Guide*

very ineffective. A critical review of results of both laboratory chamber studies and field studies leads to the conclusion that indoor plants have little, if any, benefit for removing VOCs in residential and commercial buildings. Furthermore, their effects as a source for indoor airborne microbes are not known.

Provide Comfort Conditions that Enhance Occupant Satisfaction

Thermal conditions indoors, combined with occupant activity and clothing, determine occupant thermal comfort, which in turn impacts occupant productivity and perceptions of air quality. Dry-bulb temperature is only one physical parameter out of many that interact in a complex manner to produce occupant satisfaction. (See strategies discussed in the *Indoor Air Quality Guide*.)

Thermal conditions affect chemical and biological contaminant levels and/or the intensity of occupants' reactions to these contaminants, but our knowledge of these effects and their mechanisms is limited. Despite this limited knowledge, achieving high performance in thermal comfort is likely to result in lower contamination levels and better occupant perceptions of IAQ.

In traditional designs, the HVAC designer's role in achieving comfort conditions often begins and ends at the selection of an indoor design condition and the sizing of the HVAC system to provide these conditions at peak load. The selection of a design dry-bulb condition involves comfort, cost, and energy considerations and can dictate critical design features of the system. For example, some designers may pick a relatively high design cooling condition such as 78°F (26°C) to conserve energy, while others may select one such as 72°F (22°C) to improve thermal comfort perceptions. Selecting systems and controls that perform efficiently at part load can mitigate the energy downside of the latter option. Use of displacement ventilation for space air diffusion combined with radiant cooling can also reduce the cooling energy consumption while achieving comfort conditions.

Each person having control over his or her own environment, referred to as personalized ventilation and conditioning (as provided in many automobiles and airplanes, for example) is the ideal situation but is not easily attained in buildings. It is wise, therefore, to select zones carefully and consider using as many as needed to create sufficient capacity within each zone to improve the ability to satisfy the comfort needs of occupants in the zones. Rooms and areas having loads that vary over time in patterns that are significantly different from areas that surround them benefit from having their own conditioning control loops and thermostats. Figure 9-9 illustrates many of the basic relationships important to effective maintenances of thermal comfort (ASHRAE 2009a).

The control of humidity at part load is a comfort goal that needs to be considered in the design of systems and their control sequences. Controlling moisture is also important to limit condensation and mold, as discussed in Strategy 2.4, "Control Indoor Humidity," of the *Indoor Air Quality Guide*.

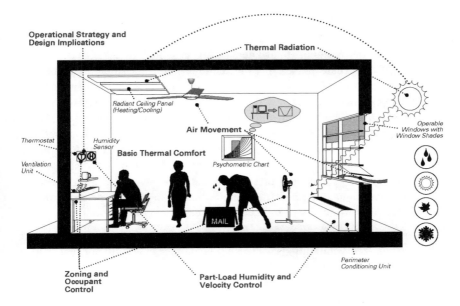

Source: *Indoor Air Quality Guide*, based on a sketch by Larry Schoen.

Figure 9-9 Thermal comfort is maintained by correctly understanding all of the relationships in the figure and applying effective tools, as illustrated in Strategy 7.6 in the *Indoor Air Quality Guide*.

Air diffusion devices need to be selected so that the required air velocity conditions in the occupied zones are maintained at low airflow, as would occur in a VAV system. It is also important to choose thermostat locations that best represent the conditions that occupants will experience and are not exposed to solar radiation and other heat sources.

Taking thermal radiation into account in system capacity can mitigate the negative effects of strong radiant sources, while the comfort benefits, for example, radiant heating, can be realized.

Use Dedicated Outdoor Air Systems and Personalized Ventilation Where Appropriate

All DOASs are 100% outdoor air systems. The DOAS approach makes calculating the required outdoor ventilation airflow more straightforward than for multiple-space systems. Having the ventilation system decoupled from the heating and air-conditioning system can provide many advantages for HVAC system design despite the disadvantage of an additional item of equipment, the DOAS unit itself (ASHRAE 2011a). DOASs must address latent loads, primarily from the outdoor air. In some cases, DOASs may also be designed to remove latent load from both outdoor air and the building (total latent load), in which case there are multiple advantages. If the exhaust airstream is located near the ventilation airstream, both sensible and latent energy can be recovered, which makes DOASs more energy efficient. DOASs are customized with specific selection of components and can be built up or manufactured. Cooling coils are typically required for cooling and dehumidifying the air, and in some cases heating coils may be required. Integration of energy recovery technology is easily achieved with DOASs and can reduce the load on the cooling and heating coils. Such energy recovery components can be either total (enthalpy) energy recovery or sensible energy recovery. In high-latent-load outdoor air conditions, an active desiccant wheel or a passive dehumidification component, both of which assist in managing humidity in the building, may be cost effective.

The personalized ventilation (PV) system is another emerging concept that can be readily integrated with DOASs. It is fundamentally aimed at improving ventilation in the immediate breathing zones of occupants in the built environment. PV

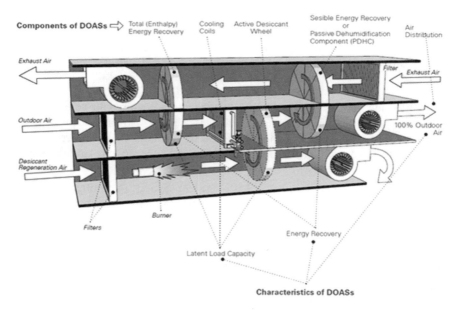

Source: *Indoor Air Quality Guide*.

Figure 9-10 Characteristics of dedicated outdoor air systems.

systems can greatly improve ventilation effectiveness, leading to improved exposure control, which then offers the possibility of reducing the absolute quantity of outdoor air to be provided to the breathing zone. In addition, PV systems also allow the immediate breathing zone to be maintained at the required thermal comfort conditions and offer the possibility of operating the background secondary system in the space at warmer temperatures (heat gain situations) or cooler temperatures (heat loss situations). Both the reduction in the absolute quantity of outdoor air and the different thermal conditions of the background system can lead to significant energy savings. The issue of cold feet with underfloor air distribution (UFAD) systems can also be overcome with the integration of such systems operating at a warmer supply air temperature with a cooler PV system so that the whole-body thermal comfort is achieved without the adverse effect of stratification.

Use Energy Recovery Ventilation Where Appropriate

Energy recovery ventilation is required for certain applications by energy standards (ASHRAE 2010a). In other cases, energy recovery systems provide such sufficient payback in overall system sizing and reduced operating costs over the life of the system that they are installed voluntarily. They are also attractive because they could yield credit towards green rating systems, such as LEED.

In general, there are two types of energy recovery ventilation devices: (1) total energy recovery ventilators (ERVs) that transfer heat and moisture between incoming and exhaust air, and (2) heat recovery ventilators (HRVs) that do not transfer moisture (ASHRAE 2011a). Types of energy recovery systems include an energy recovery wheel and fixed plate with or without latent transfer, heat pipe, and runaround loop systems. In the case of total ERVs, improved humidity control is an added benefit, critical for controlling condensation and mold growth and for thermal comfort. Along with generic IAQ concerns about keeping mechanical air delivery equipment clean, proper application of equipment needs to address correct selection, sizing, application, and maintenance. Energy recovery systems also increase system fan power due to increased flow resistance. General design considerations include those associated with filtration, controls, sizing, condensation, and sensible heat ratio. Control systems, possibly including bypass dampers, must be designed to minimize fan power, particularly during conditions of low heat recovery. Proper commissioning of ERVs is also important, because they must to perform correctly under all outdoor weather conditions.

Demand-Controlled Ventilation (DCV) in Appropriate Applications

To avoid energy penalties associated with overventilation, standards and codes allow or require DCV to reduce outdoor airflow rates. Reasonable DCV approaches can reduce the annual heating and cooling loads up to 20% or up to $1/ft^2 ($11/m^2) (ASHRAE 2009a).

Chapter 9: Indoor Environmental Quality

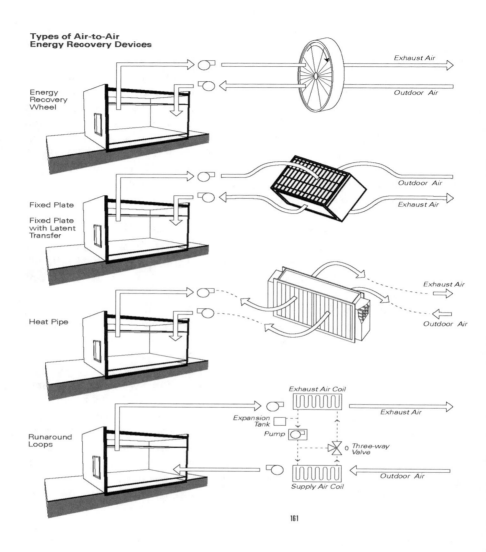

Source: *Indoor Air Quality Guide.*

Figure 9-11 Types of heat recovery ventilation systems.

Demand-controlled ventilation, defined as "a ventilation system capability that provides for the automatic reduction of outdoor air intake below design rates when the actual occupancy of spaces served by the system is less than design occupancy" (ASHRAE 2010a), offers a means to ensure adequate outdoor airflow (OA) for acceptable IAQ as described in Standard 62.1 (ASHRAE 2013) while reducing energy consumption at part-load conditions. HVAC design engineers consider using DCV for densely occupied zones and zones with large occupancy fluctuations. Standards 90.1 and 189.1 set requirements for when DCV is required, such as defining the minimum occupancy density for requiring DCV.

Different Approaches for DCV. Many different control approaches have emerged, including direct count of people, presence of people, and time-of-day schedule, none of which require the use of carbon dioxide (CO_2) sensors. But people produce CO_2 and bioeffluents at a related rate, so CO_2-based DCV approaches have also emerged. Relatively inexpensive CO_2-sensing technology attracts many designers to this approach, and standards and codes, including California's Title 24, ASHRAE Standard 62.1, and the International Mechanical Code, allow it.

The *62.1 User's Manual* (ASHRAE 2010c) presented one DCV approach for single-zone systems. However, the manual states that DCV approaches for multiple-zone recirculating systems have not been adequately developed, implying a need for more research.

Currently underway, ASHRAE research project RP-1547 is studying single-path VAV system control logic that effectively reduces energy use without underventilation.

Caution is urged when using CO_2 in DCV systems. CO_2-monitoring devices must be installed in appropriate locations, usually not on a wall (where thermostats are typically located for convenient access). Devices must be properly calibrated after installation and periodically during their useful lives. It is also essential to measure accurately the outdoor CO_2 concentration when it is subtracted from the indoor concentration to estimate outdoor air ventilation per person. Outdoor CO_2 can vary from 400 ppm to 650 ppm in urban and suburban areas. It can be less than 400 ppm in forests and other heavily vegetated areas. Concentrations do vary over the course of the day, so the timing of the outdoor measurements is important. Furthermore, temperature can impact performance of most commonly used measurement instruments, so appropriate calibration and/or corrections should be made.

Use Natural or Mixed-Mode Ventilation where Appropriate

Natural ventilation has been used for thousands of years to ventilate and cool spaces. Naturally ventilated buildings do not aim to achieve constant environmental conditions but do take advantage of, and adapt to, dynamic ambient conditions to provide a controllable, comfortable indoor environment for the occupants. Large portions of the temperate climate zones of the world can take advantage of natural ventilation strategies for a significant portion of the year. When considering the use of natural ventilation, the prevailing climate must be evaluated in some detail. Climatic issues such as the ambient air temperatures, humidity, cleanliness of the outdoor air, and wind airflow patterns must be considered. Ambient noise levels and security issues should also be considered.

Natural ventilation and cooling generally works well with other sustainable strategies; for instance, energy-efficient design typically requires minimizing thermal gains and losses, which in turn is an essential design component for natural ventilation. Daylit buildings with narrow floor plates and high floor-to-ceiling areas work well for natural ventilation. ASHRAE Standard 62.1-2013 requires that spaces using the natural ventilation procedure also be provided with a mechanical ventilation system for use when natural ventilation is not appropriate. With mixed-mode systems, natural ventilation is commonly used for ventilating/cooling for most of the year and mechanical HVAC systems are used for peak cooling or when natural ventilation is not available or practical. The air-based heating and cooling systems should be provided with controls to avoid unnecessary operation when natural ventilation systems are operating. Radiant heating and cooling methods uncoupled from the delivery of ventilation air are becoming more popular. Sophisticated systems using pressure sensors and motor-driven dampers can be used to control pressures in various parts of buildings and to take advantage of stack effect or wind pressure to deliver ventilation where and when it is needed. These ventilation control systems need considerable care in design and operation as well as end-user education. The high potential return in energy performance and indoor air quality can often justify the extra effort.

The *Indoor Air Quality Guide* discusses other key design issues, such as ensuring a satisfactory acoustic environment and smoke control.

Use the ASHRAE Standard 62.1 IAQ Procedure where Appropriate

Standard 62.1's IAQ Procedure (IAQP) provides designers with an important option or adjunct to the prescriptive Ventilation Rate Procedure (VRP) in ASHRAE Standard 62.1, thereby increasing the potential for good IAQ control.

In general, the attainment of acceptable IAQ can be achieved through the removal or control of irritating, harmful, and unpleasant constituents in the indoor environment. The established methods of contaminant control are source control, ventilation, and filtration and air cleaning (FAC). Source control approaches should always be explored and applied first, because they are usually more cost-effective than either ventilation (dilution) or FAC (extraction). Other strategies in this guide discuss various aspects of ventilation for attainment of acceptable IAQ as presented by Standard 62.1. However, the main focus of Standard 62.1 is ventilation control (i.e., the VRP) that specifies the minimum outdoor ventilation rates to dilute indoor contaminants.

Application of the IAQP typically uses a combination of source control and enhanced extraction through filtration and/or gas-phase chemical air cleaning, in some cases resulting in a reduction in the minimum outdoor air intake required, compared to the more commonly used prescriptive VRP. (See Strategy 7.5 of ASHRAE's *Indoor Air Quality Guide*, "Provide Particle Filtration and Gas-Phase Air Cleaning Consistent with Project IAQ Objectives.")

The IAQP's use of all three control methods—source control, ventilation and FAC—takes advantage of the combined strengths of each to yield the following potential benefits:

- It provides a methodology for documenting and predicting the outcome of source control approaches and rewarding source reduction tactics by potentially lowering ventilation requirements, which can lower the heat, moisture, and pollutant burden of outdoor air.
- The use of enhanced FAC lowers the constituent concentration of contaminants of concern contained in the outdoor air.
- The use of enhanced FAC can lower the concentrations of contaminants of concern created and recirculated within the constituent space.
- Enhanced FAC can result in cleaner heat exchange surfaces and more energy-efficient HVAC system operation.
- Lower outdoor air intake rates can allow lower system capacity and reduce operating costs.

Applying the IAQP. Standard 62.1 allows several alternative approaches of applying the IAQP. They include a mass balance approach using steady-state calculations of the contaminants of concern, a comparison with similar buildings approach to document successful usage elsewhere, and a contaminant monitoring approach where actual contaminant levels are monitored. By following a set of predefined steps, it is possible to both enhance the IAQ and substantially reduce the outdoor airflow requirements in some buildings. The steps include evaluation of the contaminants of concern, target levels of acceptability, methods of determining acceptability, examination of ventilation requirements, material selection, FAC options, and implementation and documentation of the IAQP.

THERMAL COMFORT AND CONTROL

The Importance of Thermal Comfort

Achieving and maintaining thermal comfort for the occupants of buildings drives much of the energy consumption by buildings for human occupancy. Thermal comfort is an essential requirement for building users, is a key metric of a building's overall IEQ, and has an influence on the health and productive performance of building occupants. The challenge facing HVAC&R engineers and designers of green buildings is to provide healthy, thermally comfortable, and productive environments as energy efficiently as possible, using active and, where possible and appropriate, passive means.

What Is Thermal Comfort, and What Factors Are Involved?

ANSI/ASHRAE Standard 55-2010, *Thermal Environmental Conditions for Human Occupancy* (ASHRAE 2010b), the primary reference standard for designing a thermally comfortable space, defines thermal comfort as "that condition of mind that expresses satisfaction with the thermal environment." The standard acknowledges that the sensation of thermal comfort experienced is specific to the individual and that it is difficult to satisfy all occupants. Therefore, the standard specifies the combination of environmental conditions, in relation to a given mix of metabolic rates and clothing ensembles, that will satisfy a large majority of occupants within the space.

There are six primary factors that affect the conditions for thermal comfort of occupants:

Environmental factors

- Air temperature
- Radiant temperature
- Air speed
- Humidity

Personal factors

- Metabolic rate
- Thermal insulation of clothing

Other, secondary factors can affect thermal comfort in some circumstances.

Evaluating Thermal Comfort

These factors can be used to evaluate the thermal sensation of occupants within an environment controlled by mechanical means. However, in environments where thermal conditions are regulated primarily by the occupants themselves through opening and closing of windows, thermal comfort is found in practice to be achieved over a wider band of environmental conditions. This is the result of an individual's expectations and prior experience of thermal environments (including the environment outdoors), availability of control, as well as other adaptive effects where such adaptive opportunities are available to individuals. This suggests that occupants take a more active role in achieving thermal comfort, rather than simply being passive recipients of their thermal environment.

Role of HVAC and Control

The above factors imply the need for adequate control of the following aspects of an indoor environment:

- Air movement (e.g., from diffusers, avoidance of drafts)
- Air temperature and humidity conditions
- Vertical air temperature gradients (floor to ceiling air temperatures)
- Radiant surface temperatures (e.g., warm or cold floors, ceilings, or other surfaces)

An indoor thermal environment created by an HVAC system or by occupant-controlled natural conditioning and designed in accordance with ASHRAE Standard 55-2010 will provide a thermal environment that will not adversely affect the occupants' perception of comfort and that will support the productive performance of the building occupants. Where possible for green-building design, systems should be as energy efficient as possible, and advantage should be taken of opportunities to use passive design (of the building itself) to minimize the need for mechanical systems to achieve comfort conditions.

Monitoring and control of air temperature and humidity will assist in maintaining acceptable indoor thermal environments. Allowance for localized delivery of appropriately tempered air movement under the individual's control can also help meet individual preferences, as well as potentially reduce some of the energy demand in heating or cooling the overall space. Shutdown of localized delivery when individual locations are unoccupied can also help energy efficiency, as can reduction of overcooling, especially during unoccupied hours.

LIGHT AND ILLUMINATION

Energy efficiency in illumination is an increasingly important part of the effort to reduce buildings' environmental impacts, including, but not limited to, those associated with energy use. As buildings become less energy intensive overall through increased use of more efficient ventilation, heating, and cooling approaches (both passive and active), many of the further energy savings will be achieved through more effective use of daylight and more efficient use of electricity for illumination.

An efficient illumination systems design can be combined with a natural ventilation scheme to maximize economic and environmental benefits.

The major environmental health issues associated with illumination in green buildings relate to the use of more energy-efficient electric illumination sources and the increased use of daylight. Setting adequate but proper illumination levels has the potential both to improve, and possibly to worsen, the well-being of people in buildings. Light reveals the world to us through vision; it also influences other physiological and psychological processes through eye-brain connections that are the target of active research programs.

Features of Illumination in Low-Energy Buildings that Can Impact Environmental Health

- **Illumination Sources.** There is some evidence that higher daily light doses might benefit well-being even for healthy individuals. Increasing use of daylighting together with changing to more energy-efficient light sources (light-emitting diode, LED, systems, for example) might provide the opportunity to increase light levels during daytime, without increasing energy use. It is known that outcomes depend on the spectrum, intensity, duration, pattern, and timing of light exposure; however, guidelines on the best combinations of these factors do not yet exist.
- **Digital Devices.** As personal computers, smartphones, personal data devices, and tablets gain more of our daily time and visual attention, the illumination from the displays of these devices themselves will become increasingly important sources of illumination. The typically close proximity of the screen to our eyes and faces further affects the intensity of the incident illumination. The timing, spectrum, and duration of light exposure from these devices (as well as from televisions) are known to affect circadian rhythms.
- **Television Screens.** are becoming larger at the same time as they become more energy efficient and ubiquitous. So, while their energy impact may be small over the long run, it will modify and be modified by the levels of room illumination we maintain while watching television screens. Use will increase with "smart" TVs that are enabling people to use their televisions as displays for normal internet and other computer-based activities.

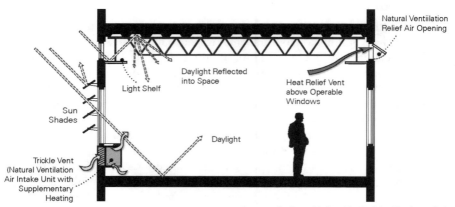

Source: *Indoor Air Quality Guide*, Strategy 8.4.

Figure 9-12 Daylight used in combination with shading and mixed-mode ventilation.

- **Automatic Control.** Illumination will increasingly be controlled by occupancy sensors, timers, and, in some cases, by timer-limited switches. This is common in Europe and Japan already and will become increasingly common in green buildings in the United States. A failed sensor or control system can cause inconvenience and even accidental injury.
- **Individual Control.** There is strong evidence that individuals differ in their preferences for illumination levels. Having the opportunity to choose one's own light levels creates a positive mood that carries over into other dimensions of well-being. On average, individual (personal) control also reduces energy use by 10%–20% over typical fixed light levels. There is some research interest in the possibility that individual control over light source spectrum, which is possible with LED lighting, might add to these benefits as the technology matures.
- **Sophisticated Glazings.** A well designed window glazing system should consider how to reduce indoor and/or solar heat gain or loss. Included in various types of "smart" windows are those that use electrochromic devices, suspended particle devices, microblinds, and liquid crystal devices to adjust the amount of incoming light according to the designer or occupant's intent.
- **Exterior Shading.** Increased shading of windows subject to direct insolation is important, as well as using more sophisticated light guides and interior shade to increase useful daylight while preventing both visual and thermal discomfort.
- **Light Tubes.** These roof-mounted tubes can carry daylight into the core of a large building and supply a large fraction of a building's lighting requirements.

The Electromagnetic Radiation (EMR) Spectrum and Light (Visible Portion or EMR)

The electromagnetic spectrum spans a broad range of wavelengths from $<10^{-12}$ m to $>10^{10}$ m. The shortest wavelengths are from cosmic and x-rays; the longest are from microwaves and electrical power generation. Ultraviolet, visible, and infrared radiations occupy the middle of spectrum. The effect on humans (or other living things) is a function of the wavelength, frequency, and energy (power density). Throughout the spectrum, the product of frequency and wavelength is the speed of light, 0.3×10^9 m/s. Frequencies are measured in hertz (Hz) and wavelengths are measured in meters. The electromagnetic spectrum is shown with its various components in Figure 9-13.

The Harmfulness or Helpfulness of Light in Human Affairs Depends on the Situation

The health effects of light exposure are known to depend on the spectrum, intensity, timing, duration, and pattern of exposure. Moreover, individual susceptibility differs. A healthful pattern of light exposure includes a period of darkness each day to keep physiological systems appropriately entrained to rest and activity (circadian) cycles.

Risks from Exposure to Electrical Illumination

In general, the probability that electric lighting for visibility purposes induces any acute pathologic conditions is low; international standards are developed to reduce the risks of such problems when light sources are used as intended.

Flicker

Modern compact fluorescent lamps (CFLs) are basically flicker-free due to their electronic high-frequency ballasts. However, perceivable flicker can occur during certain conditions with CFLs (e.g., if a nondimmable lamp is used on a dimmer). The obvious biological effects that occur from flicker that is visible include seizures (in a small percentage of the population) and less specific neurological symptoms including malaise and headache. The less obvious (and less common) biological effects occur from flicker that is invisible (above ~65 Hz) and after exposure of several minutes. Invisible flicker health effects have been reported to include headaches and eye strain (Wilkins et al. 2010). LED light sources have widely varying levels of imperceptible flicker and there is no way to predict by looking at a device whether a given LED replacement lamp might

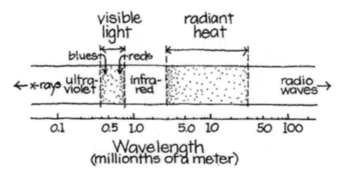

Image courtesy of Dean, 1981.

Figure 9-13 The electromagnetic spectrum including visible light.

flicker in this way. For this reason, there are international standards committees working to develop guidelines for lighting products that will prevent unintended problems.

Circadian Rhythms

Entrainment of human circadian rhythms depends on timing, the length of exposure, intensity, and spectrum. Retinal cells that influence circadian rhythms are more sensitive to short-wavelength (violet-blue) light than to longer wavelengths, particularly at night.

Exposure to light at night (independent of lighting technology) while awake (e.g., shift work) is associated with an increased risk of breast cancer and can also cause sleep, gastrointestinal, mood, and cardiovascular disorders. Usually discussed as disruption of circadian rhythms, this subject has received a lot of attention from the lighting research community in recent years.

A desynchronization of circadian rhythms plays a role in various tumoral diseases, reduced function of the immune system, diabetes, obesity, and depression. Shift workers are known to experience increased morbidity and mortality for diseases including cardiovascular disorders and cancer. In fact, in 2007, the International Agency for Research on Cancer (IARC) of the World Health Organization decreed that shift work is a risk factor for breast cancer, and on that basis, in 2009, the Danish government began compensating some female shift workers who had developed breast cancer.

Keeping circadian rhythms synchronized requires a regular daily pattern of light and dark, preferably at the same times each day. Avoiding light exposure at night, particularly short-wavelength light, contributes to better sleep quality and supports parallel processes in the immune and digestive systems.

Mood

There is also evidence that increased light exposure by day has beneficial effects on well-being. Timed exposure to bright white light is an established nonpharmacological treatment for seasonal mood disorders. Higher light doses can also improve reported well-being among healthy individuals.

User Control

As with other characteristics of the indoor environment, occupant or user control has been shown to be beneficial both in terms of energy consumption and occupant satisfaction with the indoor environment. Self-reported productivity is higher and SBS symptom prevalence is lower where occupants control aspects of their indoor environment important to them. Engineering approaches that delegate a substantial amount of control to occupants can benefit the individual occupants as well as the larger environment.

Design Considerations

Design considerations for lighting include the following:

- Electrical illumination should be accomplished with user-controlled task lighting wherever practical.
- Lighting controls in buildings should be localized, permit variable light levels, and be accessible to users wherever practical.
- A variety of lighting conditions through diverse illumination sources should be provided to allow maximum user choice in control and regulation of interior lighting.
- Design of lighting and performance standards for lighting should be based on light delivered to actual task stations in buildings, not on illumination of general spaces. Standards should discourage delivery of over-illumination to nontask areas.
- Public health authorities and building operators and occupants should consider the possible connection between occupant discomfort or illness and environmental lighting, including spectral characteristics, illumination levels, electromagnetic and chemical emissions from fixtures, and user access to effective lighting control.
- The use of newer light sources should be applied cautiously until the health effects research indicates that the impacts are within acceptable limits.

ACOUSTICS

Poor acoustic conditions are cited as the top source of dissatisfaction in many postoccupancy evaluations of building occupants, particularly in green buildings. It is incumbent that architects and designers consider the resulting acoustic quality during green-building design to ensure high indoor environmental quality and occupant satisfaction. The primary acoustic problems are related to insufficient sound isolation between spaces; poor speech privacy, common in open plan building designs; and overly reverberant spaces, making speech nearly unintelligible and noise levels high. Many of the design trends for green buildings exacerbate these acoustic problems:

- To increase daylighting and reduce energy usage, larger amounts of glass and more open-plan layouts are often used in green buildings. These choices lead to more sound transmission into and between spaces, as glass is not as effective acoustically in blocking sound as other building envelope or opaque wall constructions, and the open-plan furniture often has low height partitions. Another consequence is that interior spaces become more reverberant because there are fewer surfaces on which absorptive materials may be applied.

- To improve indoor air quality and reduce mechanical system ventilation costs, green buildings may use natural ventilation methods. However, natural ventilation via operable windows also leads to greater noise levels coming in from the outside and carrying between spaces. Thus, background noise levels may be higher, reducing speech intelligibility and overall comfort.
- Mechanical systems favored for green buildings (e.g., radiant heating/cooling, thermal mass, underfloor air distribution) often result in less available ceiling surface for the placement of acoustical materials to control reverberation and lower HVAC background noise levels that reduce speech privacy even further.
- In addition to the diminishing surface area on which acoustically absorptive materials may be applied in green buildings, traditional fibrous absorptive materials are less used because of maintenance and indoor air quality concerns. Hard reflecting materials like wood or stone are favored for their natural, sustainable character. The lack of acoustic absorption, though, creates highly reverberant ("echoey") interior spaces that can exacerbate sound isolation and speech privacy problems.

Good acoustic environments are possible in green buildings, but designers should be careful during the design phase in balancing the choices made to satisfy other green-building goals (daylighting, energy efficiency, indoor air quality, etc.) with knowledge of the acoustic consequences. Some suggestions include the following:

- Limit the amount of glass in a building to the amount just needed for daylighting requirements.
- For natural ventilation designs, place ventilation inlets on the building envelope so that they do not face external noise sources (e.g., traffic), and use louvered openings designed for acoustic attenuation purposes on the exterior and interior of the building.
- Consider using a sound-masking system in low-background-noise conditions to improve speech privacy.
- Place acoustical absorption on allowable surfaces within spaces and/or consider using hanging acoustical elements such as banners, drapes, etc. Also explore acoustically absorptive materials that may be more green (i.e., made from recycled content and/or recyclable).

More information to help with designing appropriate acoustics in green buildings may be found in Field (2008) and Chapter 48 of *ASHRAE Handbook—HVAC Applications* (ASHRAE 2011b).

REFERENCES AND RESOURCES

Note: The primary reference for this chapter is *Indoor Air Quality Guide*. The guide is a 718-page document available for free download from the ASHRAE website.

Published

ASHRAE 2005. Guideline 0-2005, *The Commissioning Process*. Atlanta: ASHRAE.

ASHRAE 2009a. *Indoor Air Quality Guide*. Atlanta: ASHRAE. Available for free download at www.ashrae.org/FreeIAQGuidance.

ASHRAE. 2009b. ASHRAE Standard 160, *Criteria for Moisture-Control Design Analysis in Buildings*. Atlanta: ASHRAE.

ASHRAE. 2010a. ANSI/ASHRAE Standard 62.2-2010, *Ventilation and Acceptable Indoor Air Quality in Low-Rise Residential Buildings,* ASHRAE, Atlanta.

ASHRAE. 2010b. ANSI/ASHRAE Standard 55-2010, *Thermal Environmental Conditions for Human Occupancy*. Atlanta: ASHRAE.

ASHRAE, 2010c. *Standard 62.1-2010 User's Manual*. Atlanta: ASHRAE.

ASHRAE. 2011a. *ASHRAE Handbook—HVAC Applications*, Chapter 44: Building Envelopes. Atlanta: ASHRAE.

ASHRAE. 2011b. *ASHRAE Handbook—HVAC Applications*, Chapter 48: Building Envelopes. Atlanta: ASHRAE.

ASHRAE. 2011c. Guideline 10-2011, *Interactions Affecting the Achievement of Acceptable Indoor Environments*. Atlanta: ASHRAE.

ASHRAE. 2011d. ANSI/ASHRAE/USGBC/IES Standard 189.1-2011, *Standard for the Design of High-Performance Green Buildings Except Low-Rise Residential Buildings*. Atlanta: ASHRAE.

ASHRAE, 2012. ASHRAE Position Document. Limiting Indoor Mold and Dampness in Buildings. Atlanta: ASHRAE. www.ashrae.org/about-ashrae/position-documents.

ASHRAE 2012. *ASHRAE Handbook—Fundamentals,* Chapter 10. Environmental Health. Atlanta: ASHRAE.

ASHRAE. 2012. ASHRAE Position Document. Unvented Combustion Devices and Indoor Air Quality. Atlanta: ASHRAE.

ASHRAE. 2013a. ANSI/ASHRAE Standard 62.1-2013, Ventilation for Acceptable Indoor Air Quality. Atlanta: ASHRAE.

ASHRAE. 2013b. ASHRAE/ANSI/IES Standard 90.1-2013, *Energy Standard for Buildings except Low-Rise Residential Buildings*. Atlanta: ASHRAE.

ASTM 2012a. ASTM D7663 - 12, *Standard Practice for Active Soil Gas Sampling in the Vadose Zone for Vapor Intrusion Evaluations*. West Conshohocken: ASTM International.

ASTM 2012b. ASTM D3273 - 12, *Standard Test Method for Resistance to Growth of Mold on the Surface of Interior Coatings in an Environmental Chamber*. West Conshohocken: ASTM International.

Dean, E., 1981. *Energy Principles in Architectural Design*. Sacramento: California Energy Commission, (out of print, but photocopy versions are available by request from Publications, California Energy Commission).

Fisk, W.J., H. Destaillats, and M.A. Sidheswaran, 2011. Saving Energy and Improving IAQ through Application of Advanced Air Cleaning Technologies, *REVA Journal*, May 2011 27:29.

Field, C. 2008. Acoustic Design in Green Buildings. *ASHRAE Journal* 50(9): 60–70.

GSA. 2005. P-100, *Facilities Standards for the Public Buildings Service*. Washington: U.S. General Services Administration.

HVCA. 2002. TR/19, *Internal Cleanliness of Ventilation Systems. Guide to Good Practice*. Heating and Ventilating Constructors' Association, London UK.

NADCA ACR. 2006. ACR, *The NADCA Standard for Assessment, Cleaning & Restoration of HVAC Systems*. The Industry Standard for HVAC Cleaning Professionals. National Air Duct Cleaners Association. Mt. Laurel, NJ.

NAIMA. 2007. *Cleaning Fibrous Glass Insulated Duct (Ductboard/Duct Board) Systems—Recommended Practices*. Recommendations for Opening and Closing Insulated Ducts Before and After Cleaning. North American Insulation Manufacturers Association. Alexandria, VA.

OSHA. 2002. 3096 - 2002 (Revised) Asbestos Standard for the Construction Industry. Washington, DC: Occupational Safety and Health Administration.

Online

Business and Institutional Furniture Manufacturers Association (BIFMA)
www.bifma.org/.

California Department of Public Health, Indoor Air Quality Program
www.cal-iaq.org/.

Carpet and Rug Institute,
www.carpet-rug.org/.

CDC, Top 10 Things Every Clinician Needs to Know About Legionellosis
www.cdc.gov/legionella/clinicians.html.

Collaborative for High Performance Schools
www.chps.net/dev/Drupal/node.

The European Working Group for *Legionella* Infections (EWGLI)
http://ecdc.europa.eu/en/activities/surveillance/ELDSNet/Documents/EWGLI-Technical-Guidelines.pdf.

GreenGuard Environmental Institute
www.greenguard.org/en/index.aspx.

OSHA Fact Sheet No. 93-47. Lead Exposure in Construction (#1 in a Series of 6) Natural Resources Canada's (NRCan), Office of Energy Efficiency, R-2000 Standard
http://oee.nrcan.gc.ca/residential/builders-renovators-trades/12148.

U.S. Environmental Protection Agency, The Green Book Nonattainment Areas for Criteria Pollutants
www.epa.gov/air/oaqps/greenbk/.

U.S. Green Building Council, Leadership in Energy and Environmental Design, Green Building Rating System
www.usgbc.org/DisplayPage.aspx?CategoryID=19.

Wilkins et al, 2010. LED Lighting Flicker and Potential Health Concerns: IEEE Standard PAR1789 Update.
www.ece.neu.edu/groups/power/lehman/Publications/Pub2010/2010_9_Wilkins.pdf.

CHAPTER TEN

ENERGY DISTRIBUTION SYSTEMS

For there to be heating, cooling, lighting, and electric power throughout a building, the energy required by these functions is usually distributed from one or more central points.

This is usually accomplished through the flow of steam or a hydronic fluid, air, electrons (electricity), and sometimes a refrigerant. Air and hydronic flows in particular also serve the function of disposition of used air (exhaust) or liquid waste. Because the air and hydronic media being moved are often at different thermal levels (i.e., warm or cold), opportunities are offered to incorporate green design techniques. There are several practical techniques whereby energy from one flow stream can be transferred usefully to another. Several GreenTips on energy recovery systems are included in this chapter.

ENERGY EXCHANGE

Cooling and Dehumidifying Media

Chilled Water (CHW). Circulating CHW is generally the least energy-efficient process for refrigerated air conditioning. However, in many cases it is the most cost-effective, particularly for large buildings. The relatively low thermal efficiency results from it being a two-step process. The most common process uses a refrigerant circulated at a low temperature through an electrically powered recycling vapor compression system. The refrigerant then chills a secondary refrigerant (water) to the temperature needed for cooling or both cooling and dehumidifying the air supplied to the occupied zones.

CHW can also be produced by a heat-driven absorption process that uses a mixture of water with lithium bromide or ammonia.

Some installations can benefit from energy reduction by using thermal storage using either CHW or ice, if further reduced in temperature and melted for later use. The CHW is pumped through air-conditioning coils that may be in a central plant room, in several building zones, or in individual rooms. Further energy losses result from heat gained through insulation on the CHW piping.

Liquid Desiccant (LD). Circulating LD is a single-step process that uses heat (i.e., that from natural gas or from liquid petroleum gas [LP], solar, or waste heat sources) when available. LD generally does not need piping to be insulated.

Synthetic Refrigerant (SR). Circulating SR from an electrically powered recycling vapor-compression system can be directly used in air-conditioning coils that are in the same housing or plant room. Alternatively, the SR can be distributed to air-conditioning coils in several building zones or in individual rooms (rooms with variable-refrigerant-flow [VRF] systems).

Natural Refrigerant (NR). Ammonia, hydrocarbons, and CO_2 should also be considered.

Leakage

There are differing hazards and costs associated with media leaks, and all should be considered serious faults.

- *CHW.* Leaks mainly cause building damage. Pressure testing procedures must be specified.
- *LD.* Some can have minor toxicity and can also cause building damage. Piping, being uninsulated, allows ready inspection.
- *SR.* Different types of SR damage the ozone layer and increase global warming potential. These leaks can be hard to detect, especially in systems with extended piping, and may develop long after installation. As Pearson (2003) noted, high losses have been reported, leading to concern that SR should perhaps be confined to plant room or factory-certified, leak-free systems.
- *NR.* Hydrocarbons are highly flammable, and ammonia is toxic and has limited flammability. These refrigerants are required by relevant codes to be confined to plant rooms having appropriate ventilation and spark protection.

ENERGY DELIVERY METHODS

Media Movers (Fans/Pumps)

Basics: Power, Flow, and Pressure. If air conditioning (heating/cooling) could be produced exactly where it is needed throughout a building, overall system efficiency would increase, because there would be no additional energy used to move (distribute) conditioned water or air. For acoustic, aesthetic, logistic, and a variety of other reasons, this ideal seldom is realized. Therefore, fans and pumps are used to move energy in the form of water and air. Throughout this process, the goal is to minimize system energy consumption.

Minimization of a media mover's power at full-load and part-load conditions is the goal. Understanding how fan and pump power change with flow and pressure is imperative. (Refer also to Chapter 2, "Background and Fundamentals.")

$$Power \sim Flow \times \Delta Pressure$$

As flow drops, so does the pressure differential through pipes, chillers, and coils. Pressure drop through these devices varies approximately with $flow^{1.85}$. This, in turn, means that the power required to cause flow through these devices varies approximately as $flow^{2.85}$. In an ideal world, power changes with the cube of the flow. While the 2.0 and 3.0 exponent is not fully achieved in practical application, it is nevertheless clear that reducing flow can drastically reduce energy use. For example, reducing the flow by 20% will result in a 50% reduction in energy use.

However, there are some pressure drops in typical systems that do not change as flow decreases. These include

- the pressure differential setpoint that many system controls use,
- cooling tower static lift, and
- pressure drop across balancing devices.

Therefore, for flow through these system components, power varies directly with flow.

To reduce pressure drop, the pipe or duct size should be maximized and valve and coil resistance minimized. (Duct and pipe sizes are discussed later in this chapter.) Coil sizes should be maximized within the space allowed to reduce pressure drop on both the water side and the air side. The ideal selection requires striking an economic balance between first cost and projected energy savings (i.e., operating cost).

Chilled-Water Pumps. Historically, a design chilled-water temperature differential (ΔT) across air-handling unit cooling coils of 10°F (5.5°C) was used, which resulted in a flow rate of 2.4 gpm/ton (2.6 L/min per kW). In recent years, the 60% increase in required minimum chiller efficiency from a 3.8 coefficient of performance (ASHRAE Standard 90-75, *Energy Conservation in New Building Design* [ASHRAE 1975]) to a 6.1 coefficient of performance (ANSI/ASHRAE/IES Standard 90.1-2013 *Energy Standard for Buildings Except Low-Rise Residential Buildings* [ASHRAE 2013]) has led to a reexamination of the assumptions used in designing hydronic media flow paths and in selecting movers (pumps) with an eye to reducing energy consumption.

The CoolTools team came to the following conclusion:

> ...the trend for most applications is that higher chilled-water delta-Ts result in lower energy costs, and they will always result in the same or lower first costs. (Taylor et al. 1999)

Simply stated, increase the temperature difference in the chilled-water system to reduce the chilled-water pump flow rate and increase chiller efficiency with warmer return water. This reduces installed cost and operating costs. *The CoolTools Chilled*

Water Plant Design and Performance Specification Guide (Taylor et al. 1999) recommends starting with a chilled-water temperature difference of 12°F to 20°F (7°C to 11°C). It is important to understand that for the chiller plant to use a higher chilled-water ΔT, you must start at the load coils (i.e., the air-handling units).

Condenser Water Pumps. In the same manner, design for condenser water flow has traditionally been based on a 10°F (5.6°C) ΔT, which equates to 3 gpm/ton (3.2 L/min/kW). Today's chillers will give approximately a 9.4°F (5.2°C) ΔT with that flow rate. The CoolTools guide states, "Higher delta-Ts will reduce first cost (because pipes, pumps, and cooling towers are smaller), but the net energy-cost impact may be higher or lower depending on the specific design of the chillers and tower."

The CoolTools team, in their summary, state:

> In conclusion, there are times you can "have your cake and eat it too." In most cases larger DT's and the associated lower flow rates will not only save installation cost but will usually save energy over the course of the year. This is especially true if a portion of the first cost savings is reinvested in more efficient chillers. With the same cost chillers, at worst, the annual operating cost with the lower flows will be about equal to "standard" flows but still at a lower first cost (Taylor et al. 1999).

They further recommend a design method that starts with a condenser water temperature difference of 12°F to 18°F (7°C to 10°C).

Thus, reducing chilled- and condenser-water flow rates (conversely, increasing the ΔTs) can not only reduce operating cost but, more importantly, can free funds from being applied to the less-efficient infrastructure and allow them to be applied toward increasing overall efficiency elsewhere.

Variable-Flow Systems. The above discussion suggests that variable-flow air and water systems are an excellent way to reduce system energy consumption. Variable flow (either air or water) is required by Standard 90.1 (ASHRAE 2013) for many applications, but it may be beneficial in even more applications than the standard requires. Today's most used technology for reducing flow is the variable-frequency drive.

Chilled-Water Plants Overall Design

From July 2011 to June 2012, *ASHRAE Journal* has published a series of five articles (authored by S. Taylor) describing a systematic approach to the design of chilled-water plants.

This series of articles also summarized the ASHRAE Self-Directed Learning (SDL) course called "Fundamentals of Design and Control of Central Chilled-Water Plants" and the research that was performed to support its development. The series included five segments, as follows

"**Chilled-water distribution system selection**" (*ASHRAE Journal*, July 2011). This article discussed distribution system options, such as primary-secondary and primary-only pumping, and provided a simple application matrix to assist in selecting the best system for the most common applications.

"**Condenser water distribution system selection**" (*ASHRAE Journal*, September 2011). This article discussed piping arrangements for chiller-condensers and cooling towers, including the use of variable-speed condenser water pumps and water-side economizers.

"**Pipe sizing and optimizing ΔT**" (*ASHRAE Journal*, December 2011). This article discussed how to size piping using life-cycle costs, then how to use pipe sizing to drive the selection of chilled-water and condenser water temperature differences (ΔTs).

"**Chillers and cooling tower selection**" (*ASHRAE Journal*, March 2012). This article addressed how to select chillers using performance bids and how to select cooling tower type, control devices, tower efficiency, and wet-bulb approach.

"**Optimized control sequences**" (*ASHRAE Journal*, June 2012). This article included a discussion of how to optimally control chilled-water plants, focusing on all variable-speed plants.

The intent of the SDL (and these articles) is to provide simple yet accurate advice to help designers and operators of chilled-water plants to optimize life-cycle costs without having to perform rigorous and expensive life-cycle cost analyses for every plant. In preparing the SDL, a significant amount of simulation, cost estimating, and life-cycle cost analysis was performed on the most common water-cooled plant configurations to determine how best to design and control them. The result is a set of improved design parameters and techniques that will provide much higher-performing chilled-water plants than common rules-of-thumb and standard practice.

Distribution Paths (Ducts/Pipes/Wires)

Ducts, pipes, and wires are used to move media. Proper sizing is a balancing act between energy use and cost and between material use and first cost. In terms of space consumed in running these carriers, and in terms of the energy carried for the cross section of the carrier involved, wires (through electricity) have the capacity for carrying the most energy, followed by pipes (hydronics), and followed, in turn, by ducts (air). Another consideration is that the different energy-carrying media have different characteristics and capabilities in terms of meeting the requirements of the spaces served. The above factors will have some influence on determining the type of HVAC system to be used.

Sizing Considerations. The previous section, "Media Movers," stated that reducing flow rates may reduce both installed cost (by reducing duct, pipe, fan, and pump sizes) and operating cost (by reducing pump and fan energy use). Decreased duct and pipe sizes also lead to less insulation. However, if the incremental installed cost savings would be relatively small, the design professional may want to leave pipe and duct sizes larger to minimize energy cost. The best designs begin with generalized ranges (as were stated above for chilled and condenser water) that are fine-tuned for the specific application. This fine-tuning may be done with commonly available analysis and design software.

To reduce pressure drop, pipes and ductwork should be laid out prior to locating pumps, chillers, and air handlers.

STEAM

Advantages

- Steam flows to the terminal usage point without aid of external pumping.
- Steam systems are not greatly affected by the height of the distribution system, which has a significant impact on a water system.
- Steam distribution can readily accommodate changes in the system terminal equipment.
- Major steam distribution repair does not require piping draindown.
- The thermodynamics of steam use are effective and efficient.

Disadvantages

- Steam traps, condensate pumps, etc., require frequent maintenance and replacement.
- Steam piping systems often have dynamic pressure differentials that are not easily controlled by typical steam control appurtenances.
- Returning condensate by gravity is not always possible. Lifting the steam back to the source can be a challenge due to pressure differential dynamics, space constraints, wear and tear on equipment, etc.
- Venting of steam systems (through pressure relief valves, flash tanks, boiler feed tanks, condensate receivers, safety relief valves, etc.) must be done properly and must be brought to the outside of the building.
- High-pressure steam systems require a full-time boiler plant operator, which adds to operating costs.

With the advantages listed, steam is often a logical choice for commercial and industrial processes, and for large-scale distribution systems, such as on campuses and in large or tall buildings. However, steam traps require periodic maintenance and can become a source of significant energy waste. Condensate return venting issues can add to operation and design challenges. Venting of flash steam can also cause significant energy to be wasted. (See subsequent section on steam traps, condensate return, and pressure differential.)

Classification

Steam distribution is either one-pipe or two-pipe. One-pipe distribution is defined to be where both the steam supply and the condensate travel through a single pipe connecting the steam source and the terminal heating units. Two-pipe distribution is defined to be where the steam supply and the steam condensate travel through separate pipes. Two-pipe steam systems are further classified as gravity return or vacuum return.

Steam systems are also classified according to system operating pressure:

- Low pressure is defined to be 15 psig (103 kPa) or less.
- Medium-pressure steam is defined to be between 15 and 50 psig (103 and 345 kPa).
- High pressure is defined to be over 50 psig (345 kPa).

Selection of steam pressure is based on the constraints of the process served. The level of system energy rises with the system pressure. Higher steam pressure may allow smaller supply distribution pipe sizes, but it also increases the temperature difference across the pipe insulation and may result in more heat loss. Higher steam pressure also dictates the use of pipes, valves, and equipment that can withstand the higher pressure. This translates to higher installation costs.

Piping

Supply and return piping must be installed to recognize the thermodynamics of steam and to allow unencumbered steam and condensate flow. Piping that does not slope correctly—that is, it is installed with unintended water traps or has leaks—will not function properly and will increase system energy use. Careful installation will result in efficient and effective operation. If steam or condensate leaks from the piping system, additional water must be added to make up for the losses. Makeup water is chemically treated and is an operating cost.

Control

Control of steam flow at the terminal equipment is very important. Steam control valves are selected to match the controlled process. If steam flow varies over a wide range, it may be necessary to have multiple control valves. On large terminal heat exchange units, a single control valve usually cannot provide effective control. In such cases, multiple control valves of various sizes operated in a sequential manner are used. Consult sizing data available from manufacturers.

Condensate Return and Flash Steam

The energy conservation and operational problems that arise when using steam often occur because the system design and/or installation does not properly address the issue of returning condensate to the boiler or cogeneration plant. The simplest and most efficient way to return condensate is via gravity. When this is not possible, the designer needs to clearly understand issues of lift, condensate rate, and pressure differential to ensure that operational problems (e.g., water hammer and reduction in capacity) do not occur.

In addition, the proper sizing and routing of vents for both the condensate receiver and the flash tank need to be clearly understood by the designer and the operator.

When considering recovery of flash steam, it is important to accurately calculate the expected operating pressure of the flash steam to ensure that the system or equipment served by the flash steam will be able to operate under all system conditions. Alternatively, if the flash tank is to be vented to atmosphere, the designer must fully understand the amount of energy that will be relieved and wasted by venting the flash tank to the atmosphere.

Steam Traps

Selection of steam traps is related to the function of the terminal device or pipe distribution served. Steam traps have the function of draining condensate from the supply side of the system to the condensate return side of the system without allowing steam to flow into the return piping. The flow of steam into return piping unnecessarily wastes energy, and significant energy waste can occur if periodic maintenance is not performed. Properly sized and installed steam traps allow terminal heat exchange equipment to function effectively. If condensate is not fully drained from the terminal heat exchange equipment, the heat transfer area is reduced, resulting in a loss of capacity.

Efficient Steam System Design and Operation Tips

- Preheat boiler plant makeup water with waste heat.
- Use ecofriendly chemical treatment.
- Recover flash steam, making sure to understand pressure differential of flash steam compared to system pressure serving equipment where flash steam recovery is used.
- Minimize use of pumped condensate return.
- Do not use steam pressure for lifting condensate return.
- Consider clean steam generators where steam serves humidification or sterilization equipment.
- Consider the use of blowdown heat recovery when economically feasible

Sources of Further Information

Manufacturers of equipment—control valves, steam traps, and other devices—are valuable resources. There are multiple websites that contain system design information. Perform a search on the Internet for *steam piping design*. "Steam Systems" is Chapter 11 of the 2012 *ASHRAE Handbook—HVAC Systems and Equipment* (ASHRAE 2012). See the "References and Resources" section at the end of this chapter for additional resources.

HYDRONICS

Pumping heated water and chilled water is common system design practice in many buildings. Water is often diluted with an antifreeze fluid to avoid freezing in extremely cold conditions (thus, referring to these systems as *hydronic*.) There are many approaches in using these systems.

Classification

Hot-water heating systems are classified as low-temperature water, medium-temperature water, or high-temperature water.

- Low-temperature water systems operate at temperatures of 250°F (121°C) and below.
- Medium-temperature water systems operate at temperatures between 250°F and 350°F (121°C and 177°C).
- High-temperature water systems operate at temperatures above 350°F (177°C).

Chilled-water systems distribute cold water to terminal cooling coils to provide dehumidification and cooling of conditioned air or cooling of a process. They can also serve cooling panels in occupied spaces. Chilled-water panels, which serve as a heat sink for heat radiated from occupants and other warm surfaces to the radiant panel, can be used to reduce the sensible load normally handled solely by mechanical air cooling. The percentage by which the load can be reduced depends upon the panel surface area and dew-point limitations, which are necessary to avoid any possibility of condensation.

Condenser water systems connect mechanical refrigeration equipment to outdoor heat dissipation devices (e.g., cooling towers or water- or air-cooled condensers). These, in turn, reduce the temperature of the condenser water by rejecting heat to the outdoor environment.

Piping, Flow Rates, and Pumping

Each of these systems uses two pipes—a supply and a return—to make up the piping circuit, and each uses one or more pumps to move the water through the circuit. Information on the design and characteristics of these various systems can be found in Chapters 13, 14, and 15 of the 2012 *ASHRAE Handbook—HVAC Systems and Equipment* (ASHRAE 2012).

Cost-effective design depends on consideration of the system constraints. Piping must be sized to provide the required load capacities and arranged to provide necessary flow at full- and part-load conditions. Design is determined by several system characteristics and selections including

- supply and return water temperatures,
- flow rates at individual heat transfer units,
- system flow rate at design condition,
- piping distribution arrangement,
- system water volume,
- equipment selections for pumps, boilers, chillers, and coils, and
- temperature control strategy.

Pumping energy can be a significant portion of the energy used in a building. In fact, pumping energy is roughly equal to the inverse fifth power of pipe diameter, so a small increase in pipe size has a dramatic effect on lower pumping energy (horsepower or kW). Traditionally, it was common to select heating water flow for coils based on a temperature difference of 20°F (11.1 K) between the supply and return. Flow rate in gal/min (gpm) was calculated by dividing the heating load in Btu/h by 10,000 (1 Btu/lb· °F×8.33 lb/gal×60 min/h×20°F temperature difference) (L/s was calculated by dividing the kW heating load by 4.187 [specific heat capacity of water, kJ/kg· K] × 11.1 K). As long as the cost of energy was cheap, this method was widely used.

Flow rate can easily be reduced by one-half of the 20°F (11.1 K) value by using a 40°F (22.2 K) temperature difference. The impact on pump flow rate is significant. The temperature difference selected depends upon the ability of the system to function with lower return water temperatures. Certain types of boilers can function with the low return water temperatures, while others cannot. Care must be taken in selecting the boiler type, coupled with supply and return water design temperatures. In specific instances, a low return water temperature could damage the boiler due to the condensation of combustion gases.

Lower flow rates could allow smaller pipe sizes, and pipe size, along with flow, affects pumping energy. A goal should be established for the pump power to be selected. A small increase in some or all of the pipe distribution sizes could reduce the pump energy (horsepower or kilowatts) needed for the system. When this goal is established and attained in the finished design, the concept and energy usage will be achieved. A reasonable goal can be expressed using the water transport factor equation adjusted to reflect kilowatts (multiply horsepower by 0.746). Measurements of efficient designs indicate a performance of 0.026 kW/ton (0.007 kW/kWR [with kWR being refrigeration cooling capacity]) being served as a reasonable goal for 10°F (5.6 K) ΔT systems. Adjusting the flow rate and ΔP variables in this formula will quickly show the benefits of larger pipes or lower flow rates (greater ΔT).

For instance, calculate the energy (kW) required with the modified water transport factor equation:

$$\text{pump Hp} = \frac{(Q)(\Delta P)(sg)}{3968(\eta_P)(\eta_M)}$$

$$\text{pump kW} = \frac{(Q)(\Delta P)(sg)(9.81)}{(1000)(\eta_P)(\eta_M)}$$

where
Q = flow rate, gpm (L/s)
ΔP = pump head, ft (m)
sg = specific gravity (—)
η_P = pump efficiency
η_M = electric motor efficiency

Create a performance index by solving the equation for 1 ton of cooling. That produces an answer in kilowatts per ton (kilowatts per kilowatts of refrigeration). A typical condenser system might use 3 gpm/ton (0.054 L/s/kWR) of cooling and a ΔP of 100 ft (300 kPa) with an 82% efficient pump and a 92% efficient motor. Inserting these variables into the above equation, the derived answer for 1 ton (3.5 kWR) of cooling would be 0.075 kW or 0.075 kW/ton (0.021 kW/kWR).

Compare this index to the performance of an efficient design that increases piping and fitting size, and therefore reduces ΔP to, say, 30 ft (90 kPa) total dynamic head (TDH), keeps the water flow (gpm or L/s) the same, and uses an 85% efficient pump with a 92% efficient motor. (Because the most efficient motors cost the same as less efficient motors, almost everyone is now using premium efficiency motors.) If we solve again for kilowatts, we get 0.022 kW/ton (0.006 kW/kWR).

This shows us we can reduce pumping energy by 71% by lowering the TDH from 100 to 30 ft (300 to 90 kPa) and selecting a more efficient pump. One author of this guide has personally measured systems with the characteristics indicated above. If the average cost of electricity per kWh is $0.08, and we are pumping for a 1000 ton (3517 kWR) chiller, the annual operating cost difference would be more than $37,000/yr.

If the system lasts 20 years, the improved system would save $740,000 in electricity costs. Of course, the obvious question is: "How much does it cost to increase pipe and fitting size and pump efficiency?" The answer will vary by project, but keep in mind that efficient pumps cost no more than inefficient ones. Larger pipes and fittings cost more than smaller pipes and fittings and, in one author's experience, the cost of increasing the pipe size one size (going from a 10 in. [250 mm] pipe to a 12 in. [300 mm] pipe) is recovered in electrical cost in less than 18 months.

Using the formula above and carefully measuring existing projects, we can establish performance goals for our designs such as the ones in Table 10-1. The reader is cautioned that the percentages savings shown in Table 10-1 are individual savings from each component and should not be mistaken with the overall building energy performance, which is a more holistic number involving all components.

At times, reductions in pipe sizes to reduce first cost are suggested as value engineering. However, energy usage of the building may be greatly impacted: pump size and horsepower could well be increased. In order to be truly valid, value engineering should also include refiguring the life-cycle cost of owning and operating the building. These factors can also be applied to chilled-water systems, except that chilled-water systems have a smaller range of temperatures within which to work.

System Volume

In small buildings, water system volume may relate closely to boiler or water-chiller operation. When pipe distribution systems are short and of small water volume, both boilers and water chillers may experience detrimental operating effects. Manufacturers of water chillers state that system water volume should be a minimum of 3 to 10 gal/installed ton of cooling (0.054 to 0.179 L/s/·kWR). In a system less than this and under light cooling load conditions, thermal inertia coupled with the reaction time of chiller controls may cause the units to short-cycle or shut down on low-temperature safety control.

Similar detrimental effects may occur with small modular boilers in small systems. Under light load conditions, boilers may experience frequent short cycles of operation. An increase in system volume may eliminate this condition.

Sources of Further Information

Manufacturers of equipment—pumps, boilers, water chillers, control valves, and other devices—are valuable sources. There are also multiple websites that contain system design information. (Perform a search on the Internet for *hydronic piping design*.)

Antifreeze Additions

Fluid properties can greatly affect system performance. Antifreeze generally increases pressure drop and decreases heat transfer effectiveness. This leads to reduced system efficiency and perhaps increased cost due to the need for larger components. Therefore, the design professional should first examine the system to determine if antifreeze is an absolute necessity or whether water, with proper antifreeze safeguards, could be used.

If antifreeze is truly necessary, remember the following:

Table 10-1: Example Performance Goals

Component	Typical kW/ton (kW/kWR)	Efficient kW/ton (kW/kWR)	Delta	% Savings
Chiller	0.62 (0.1763)	0.485 (0.1379)	0.135 (0.0384)	22%
Cooling tower	0.045 (0.0128)	0.012 (0.0034)	0.033 (0.0094)	73%
Condenser water pump	0.0589 (0.0167)	0.022 (0.0063)	0.0369 (0.0104)	63%
Chilled-water pump	0.0765 (0.0218)	0.026 (0.0074)	0.0505 (0.0144)	66%
Total water-side system	0.8004 (0.2276)	0.545 (0.155)	0.2554 (0.0726)	32%

- Determine whether it is freeze or burst protection that is being sought. If an affected component (such as a chiller) does not need to be operated during freezing conditions, perhaps only burst protection is necessary. The amount of antifreeze needed can be greatly reduced, although slush may form in the pipes.
- Use only the minimum antifreeze necessary to provide protection; higher concentrations will simply reduce performance.
- Balance all environmental aspects of the antifreeze. Understand that while some antifreeze solutions are viewed as less toxic, they can significantly increase system installation and operating costs. Often the greatest environmental cost of antifreeze is the increased energy consumption.
- Consult the manufacturer for more information on burst and freeze protection.

AIR

Using air as a means of energy distribution is almost universal in buildings, especially as a means of providing distributed cooling to spaces that need it. A key characteristic that makes air so widely used, however, is its importance in maintaining good indoor air quality. Thus, air distribution systems are not only a means for energy distribution, they serve the essential role of providing fresh or uncontaminated air to occupied spaces.

That said, it is important to understand that properly distributing air to heat and cool spaces in a building is less efficient from an energy usage perspective than using hydronic or steam distribution. Air distribution systems are often challenging to design because, for the energy carried per cross-sectional area, they take up the most space in a ceiling cavity and are frequent causes of space conflicts among disciplines (i.e., structural members, plumbing lines, heating/cooling pipes, light fixtures, etc.). Many approaches have been tried to better coordinate duct runs with other services—or even to integrate them in some cases.

Another tricky aspect of air system design is that there are temperature limitations on supply air (due to the fact that air is the energy medium that directly impacts space occupants). While care must always be taken in how air is introduced into an occupied space, it is especially critical the colder the supply temperature gets. Low-temperature air supply systems offer many advantages in terms of green design, but an especially critical design aspect is avoiding occupant discomfort at the supply air/occupant interface.

Most of the same principles that were discussed under hydronic energy distribution systems apply to air systems, with respect to temperature differences, carrier size, and driver power and energy. Thus, there are plenty of opportunities for applying green design techniques to such systems and for seeking innovative solutions.

Energy Usage

The "Advanced Variable Air Volume (VAV) Systems Design Guide" (California Energy Commission, 2005) recommendations for duct design include the following:

- Run ducts as straight as possible to reduce pressure drop, noise, and first cost.
- Use standard length straight ducts and minimize both the number of transitions and of joints.
- Use round spiral duct wherever it can fit within space constraints.
- Use radius elbows rather than square elbows with turning vanes whenever space allows.
- Use either conical or 45° taps at VAV box connections to medium pressure duct mains.
- Specify sheet metal inlets to VAV boxes; do not use flex duct.
- Avoid consecutive fittings because they can dramatically increase pressure drop.
- For VAV system supply air duct mains, use a starting friction rate of 0.25 in. to 0.30 in. per 100 feet (2 to 2.5 Pa/m). Gradually reduce the friction rate at each major juncture or transition down to a minimum friction rate of 0.10 in. to 0.15 in. per 100 feet (0.8 to 1.2 Pa/m) at the end of the duct system.

- For return air shaft sizing, maximum velocities should be in the 800 to 1200 fpm (4 to 6 m/s) range through the free area at the top of the shaft (highest airflow rate).
- To avoid system effect, fans should discharge into duct sections that remain straight for as long as possible, up to 10 duct diameters from the fan discharge to allow flow to fully develop.
- Use duct liner only as much as required for adequate sound attenuation. Avoid the use of sound traps.

In addition, below are additional ways to reduce energy usage when designing air distribution systems:

- Require SMACNA's (Sheet Metal and Air Conditioning Contractors' National Association) Seal Class A for all duct systems.
- Use variable-air-volume concepts for constant-volume systems.
- Use static pressure reset logic for all variable-air-volume applications.
- Minimize pressure drops of air-handling unit components and duct-mounted accessories (e.g., sound attenuators, dampers, diffusers, filter, energy wheels, etc.)

Sources of Further Information

Manufacturers of equipment—fans, ductwork, air-handling units, dampers, and other air devices—are valuable sources. There are also multiple websites that contain system design information. Air systems are also discussed in Chapters 20 and 21 of the 2013 *ASHRAE Handbook—Fundamentals* (ASHRAE 2013), and in Chapters 19 through 28 of the 2012 *ASHRAE Handbook—HVAC Systems and Equipment* (ASHRAE 2012). Another source for information includes SMACNA, www.smacna.org. Refer also to the "Advanced Variable Air Volume (VAV) System Design Guide" published by the California Energy Commission (Publication No. 500-03-082-A-11) for a detailed discussion of variable-flow air systems.

ELECTRIC

From the standpoint of space consumed, distribution of energy by electric means (e.g., wire, cables, etc.) is the most efficient. This advantage has often overcome the usual higher cost of electricity (per energy unit) as an energy form and has been one of the major reasons for electing to design all-electric buildings. The relative inefficiencies of electric resistance heating no longer make it a viable choice from an environmental or life-cycle cost perspective. However, when used in conjunction with a heat pump system, electrical energy as a source is efficient and less costly from an operating perspective.

REFERENCES AND RESOURCES

California Energy Commission. 2007. Advanced Variable Air Volume (VAV) System Design Guide, Public Document.

Taylor, S., P. Dupont, M. Hydeman, B. Jones, and T. Hartman. 1999. *The CoolTools Chilled Water Plant Design and Performance Specification Guide.* San Francisco, CA: PG&E Pacific Energy Center.

Taylor, S. 2011–2012. Optimizing Design and Control of Chilled-Water Plants. *ASHRAE Journal.* 53: (7),(9), and (12) and 54 (3) and (6).

ASHRAE. 2012. Chilled-water distribution system selection. *ASHRAE Journal* 54(7).

ASHRAE. 2011. Condenser water distribution system selection. *ASHRAE Journal* 53(9).

ASHRAE. 2012. Pipe sizing and optimizing DT. *ASHRAE Journal* 53(12).

ASHRAE. 2012. Chilled-water distribution system selection. *ASHRAE Journal* 54(3).

ASHRAE. 2012. Optimized control sequences. *ASHRAE Journal* 54(6).

ASHRAE GreenTip #10-1

Variable-Flow/Variable-Speed Pumping Systems

GENERAL DESCRIPTION

In most hydronic systems, variable flow with variable-speed pumping can be a significant source of energy savings. Variable flow is produced in chilled- and hot-water systems by using two-way control valves and in condenser-water cooling systems by using automatic two-position isolation valves interlocked with the chiller machinery's compressors. In most cases, variable flow alone can provide energy savings at a reduced first cost, since two-way control valves cost significantly less to purchase and install than three-way valves. In condenser-water systems, even though two-way control valves may be an added first cost, they are still typically cost-effective, even for small (1 to 2 ton [3.5 to 7 kWR]) heat pump and air-conditioning units. (Standard 90.1 [ASHRAE 2013] requires isolation valves on water-loop heat pumps and some amount of variable flow on all hot-water and chilled-water systems.)

Variable-speed pumping can dramatically increase energy savings, particularly when it is combined with demand-based pressure reset controls. Variable-speed pumps are typically controlled to maintain the system pressure required to keep the most hydraulically remote valve completely open at design conditions. The key to getting the most savings is placement of the differential pressure transducer as close to that remote load as possible. If the system serves multiple hydraulic loops, multiple transducers can be placed at the end of each loop, with a high-signal selector used to transmit the signal to the pumps. With direct digital control (DDC) systems, the pressure signal can be reset by demand and controlled to keep at least one valve at or near 100% open. If valve position is not available from the control system, a trim-and-respond algorithm can be employed.

Even with constant-speed pumping, variable-flow designs save some energy, as the fixed-speed pumps ride up on their impeller curves, using less energy at reduced flows. For hot-water systems, this is often the best life-cycle cost alternative, as the added pump heat will provide some beneficial value. For chilled-water systems, it is typically cost-effective to control pumps with variable-speed drives. It is very important to right size the pump and motor before applying a variable-speed drive, as a means of keeping drive cost down and performance up.

WHEN/WHERE IT'S APPLICABLE

Variable-flow design is applicable to chilled-water, hot-water, and condenser-water loops that serve water-cooled, air-conditioning, and heat-pump units. The limitations on each of these loop types are as follows:

- Chillers require a minimum flow through the evaporators. (Chiller manufacturers can specify flow ranges if requested.) Flow minimums on the evaporator side can be achieved via hydronic distribution system design using either a primary/secondary arrangement or primary-only variable flow with a bypass line and valve for minimum flow.
- Some boilers require minimum flows to protect the tubes. These vary greatly by boiler type. Flexible bent-water-tube and straight-water-tube boilers can take huge ranges of turn-down (close to zero flow). Fire-tube and copper-tube boilers, on the other hand, require a constant-flow primary pump.

Variable-speed drives on pumps can be used on any variable-flow system. As described above, they should be controlled to maintain a minimum system pressure. That system pressure can be reset by valve demand on hot-water and chilled-water systems that have DDC control of the hydronic valves.

PRO

- Both variable-flow and variable-speed control save significant energy.

- Variable-speed drives on pumps provide a "soft" start, extending equipment life.
- Variable-speed drives and two-way valves are self-balancing.
- Application of demand-based pressure reset significantly reduces pump energy and decreases the occurrence of system overpressurization, causing valves near the pumps to lift.
- Variable-speed systems are quieter than constant-speed systems.

CON

- Variable-speed drives add cost to the system. (They may not be cost-effective on hot-water systems.)
- Demand-based supply pressure reset can only be achieved with DDC of the heating/cooling valves.
- Variable flow on condenser-water systems with open towers requires that supplementary measures be taken to keep the fill wet on the cooling towers. Cooling towers with rotating spray heads or wands can accept a wide variation in flow rates without causing dry spots in fill. Fitting the cooling tower with variable-speed fans can take advantage of lower flow rates (there's more free area) to reduce fan energy while providing the same temperature of condenser water.

KEY ELEMENTS OF COST

The following provides a possible breakdown of the various cost elements that might differentiate a variable-flow/variable-speed system from a conventional one and an indication of whether the net cost for the hybrid option is likely to be lower (L), higher (H), or the same (S). This assessment is only a perception of what might be likely, but it obviously may not be correct in all situations. There is no substitute for a detailed cost analysis as part of the design process. The listings below may also provide some assistance in identifying the cost elements involved.

First Cost

- Hydronic system terminal valves: two-way vs. three-way
  (applicable to hot-water and chilled-water systems) — L
- Bypass line with two-way valve or alternative means
  (if minimum chiller flow is required) — H
- Hydronic system isolation valves: two-position vs. none
  (applicable to condenser-water systems) — H
- Cooling tower wet-fill modifications (condenser-water systems) — H
- Variable-speed drives and associated controls — H
- DDC system (may need to allow demand-based reset)
  or pressure transducers — H
- Design fees — H

Recurring Costs

- Pumping energy — L
- Testing and balancing of hydronic system — L
- Maintenance — H
- Commissioning — H

SOURCES OF FURTHER INFORMATION

ASHRAE. 2013. ANSI/ASHRAE/IES Standard 90.1-2013, *Energy Standard for Buildings Except Low-Rise Residential Buildings*. Atlanta: ASHRAE.

CEC. 2002. *Part II: Measure Analysis and Life-Cycle Cost 2005, California Building Energy Standards*, P400-02-012. Sacramento, CA: California Energy Commission.

Taylor, S.T. 2002. Primary-only vs. primary-secondary variable-flow chilled-water systems. *ASHRAE Journal* 44(2):25–29.

Taylor, S.T., and J. Stein. 2002. Balancing variable flow hydronic systems. *ASHRAE Journal* 44(10):17–24.

ASHRAE GreenTip #10-2

Variable-Refrigerant Flow (VRF) Systems

GENERAL DESCRIPTION

VRF systems have been used in Asia and Europe for almost 25 years. The main advantage of a VRF system is its ability to respond to fluctuations in space load conditions. By comparison, conventional direct expansion (DX) systems offer limited or no modulation in response to changes in the space load conditions. The problem worsens when conventional DX units are oversized, or during part-load operation (because the compressors cycle frequently). A simple VRF system is composed of an outdoor condensing unit and several indoor evaporators. The systems are interconnected by refrigerant piping with integrated oil and refrigerant management controls, which allows each individual thermostat to modulate its corresponding electronic expansion valve to maintain its space temperature setpoint.

There are two basic types of VRF multisplit systems: heat pump and heat recovery (Figure 10-1). Both heat pump and heat recovery VRF systems are available in air-to-air and water-source (water-to-refrigerant) configurations. Heat pumps can operate in heating or cooling mode. A heat recovery system, by managing the refrigerant through a gas flow device, can simultaneously heat and cool—some indoor fan-coil units in heating and some in cooling, depending on the requirements of each building zone. The majority of VRF systems are equipped with variable-speed compressors. Often called variable-frequency drives or inverter compressors, this component responds to indoor temperature changes, varying the speed to operate only at the levels necessary to maintain a constant and comfortable indoor environment. Heat recovery systems increase VRF efficiency because, when operating in simultaneous heating and cooling, energy from one zone can be transferred to meet the needs of another.

WHEN/WHERE IT'S APPLICABLE

VRF systems offer controls that match the space heating/cooling loads to that of the indoor coil over a range of operation. Variable-speed compressors and fans in the outdoor units modulate their speed, saving energy at part-load conditions.

VRF systems are best suited to buildings with diverse zoning especially buildings requiring individual control, such as office buildings, residential, schools, or hotels.

PRO

- A single condensing unit can serve multiple indoor units.
- VRF systems are generally modular and can easily be modified, which makes it easy to adapt the system to expansion and/or reconfiguration.

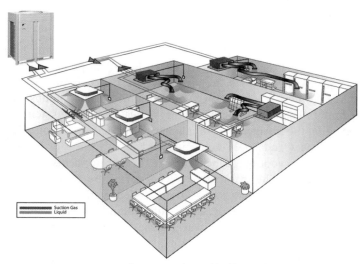

Image courtesy of Daikin AC (www.daikinac.com)

Figure 10-1 Heat-pump system in cooling mode.

- VRF indoor unit capacities are generally smaller, allowing more individual zones and individual zone controls. Systems can be designed with individual space zoning and controls.
- Variable-speed compressors enable capacity modulation, which translates to higher part load efficiencies in VRF systems.

CON

- Indoor units on VRF systems generally do not have high latent capacities and are not suitable for applications requiring a high percentage of outdoor air.
- The external static pressure available for ducted indoor sections is limited. For ducted indoor sections, the permissible ductwork lengths and fittings must be kept to a minimum.
- There is the potential for refrigerant leaks in the building.

KEY ELEMENTS OF COST

The following provides a possible breakdown of the various cost elements that might differentiate a VRF system from a conventional one and an indication of whether the net cost for the hybrid option is likely to be lower (L), higher (H), or the same (S). This assessment is only a perception of what might be likely, but it obviously may not be correct in all situations. There is no substitute for a detailed cost analysis as part of the design process. The listings below may also provide some assistance in identifying the cost elements involved.

First Cost

- Conventional heat pumps/DX systems   H
- Ground-source heat pumps   L
- Refrigerant piping   L
- Installation costs   H
- Controls   S
- Design fees   H

Recurring Costs

- Overall energy cost   L
- Maintenance   L

SOURCES OF FURTHER INFORMATION

Afify, R. 2008. Designing VRF systems. *ASHRAE Journal* 52–55.

Aynur, T., Y. Hwang, and R. Radermacher. 2008. Experimental evaluation of the ventilation effect on the performance of a VRV system in cooling mode—Part I: Experimental evaluation. *HVAC&R Research* 14(4):615–30.

Aynur, T., Y. Hwang, and R. Radermacher. 2008. Simulation evaluation of the ventilation effect on the performance of a VRV system in cooling mode—Part II: Simulation evaluation. *HVAC&R Research* 14(5):783–95.

Cendón, S.F. 2009. *New and Cool: Variable Refrigerant Flow Systems*. AIArchitect.

Goetzler, W. 2007. Variable refrigerant flow systems. *ASHRAE Journal* 24–31.

REFERENCES AND RESOURCES

Published

ASHRAE. 1975. ASHRAE Standard 90-75, *Energy Conservation in New Building Design*. Atlanta: ASHRAE.

ASHRAE. 2012. *ASHRAE Handbook—HVAC Systems and Equipment*. Atlanta: ASHRAE.

ASHRAE. 2013a. *ASHRAE Handbook—Fundamentals*. Atlanta: ASHRAE.

ASHRAE. 2013b. ANSI/ASHRAE/IES Standard 90.1-2013, *Energy Standard for Buildings Except Low-Rise Residential Buildings*. Atlanta: ASHRAE.

California Energy Commission. 2005. *Advanced Variable Air Volume System Design Guide*.

NADCA. *General Specifications for the Cleaning of Commercial Heating, Ventilating and Air Conditioning Systems*. Washington, DC: National Air Duct Cleaners Association.

Pearson, F. 2003. ICR 06 Plenary Washington. *ASHRAE Journal* 45(10).

SMACNA. 2005. *HVAC Duct Construction Standards,* 3d ed., Chapter 1. Chantilly, VA: Sheet Metal and Air Conditioning Contractors National Association, Inc.

Taylor, S., P. Dupont, M. Hydeman, B. Jones, and T. Hartman. 1999. *The CoolTools Chilled Water Plant Design and Performance Specification Guide.* San Francisco, CA: PG&E Pacific Energy Center.

Taylor, S. 2011–2012. Optimizing Design and Control of Chilled-Water Plants. *ASHRAE Journal.* 53 (7), (9), and (12) and 54 (3) and (6).

Online

Armstrong Intelligent Systems Solutions
www.armstrong-intl.com.

Spirx Sarco Design of Fluid Systems, Steam Learning Module
www.spiraxsarco.com/learn/modules.asp.

ASHRAE
GreenTip # 10-3

Displacement Ventilation

GENERAL DESCRIPTION

With a ceiling supply and return air system, the ventilation effectiveness may be compromised if sufficient mixing does not take place. While there are no data suggesting that cold air supplied at the ceiling will short circuit, it is possible that a fraction of the supply air may bypass directly to the return inlet without mixing at the occupied level when heating from the ceiling. For example, when heating with a typical overhead system with supply temperatures exceeding 15°F (8.3°C) above room temperature, ventilation effectiveness will approach 80% or less. In compliance with Table 6.2, ASHRAE Standard 62.1-2013, zone air distribution effectiveness is only 0.8, so ventilation rates must be multiplied by 1/0.8 or 1.25. While proper system design and diffuser selection can alleviate this problem, another potential solution is displacement ventilation.

In displacement ventilation, conditioned air with a temperature slightly lower than the desired room temperature is supplied horizontally at low velocities at or near the floor. Returns are located at or near the ceiling. The supply air is spread over the floor and then rises by convection as it picks up the load in the room. Displacement ventilation does not depend on mixing. Instead, you are literally displacing the stale polluted air and forcing it up and out the return or exhaust grille. Ventilation effectiveness may actually exceed 100%, and Table 6.2 of ASHRAE Standard 62.1-2013 indicates a zone air distribution effectiveness of 1.2 must be used.

Displacement ventilation is a fairly common practice in Europe, but its acceptance in North America has been slow primarily because of the conventional placement of ductwork at the ceiling level and more extreme climatic conditions.

WHEN/WHERE IT'S APPLICABLE

Displacement ventilation is typically used now in industrial plants and data centers, but it can be applied in almost any application where a conventional overhead forced-air distribution system could be used and the load permits.

Because the range of supply air temperatures and discharge velocities is limited to avoid discomfort to occupants, displacement ventilation has a limited ability to handle high heating or cooling loads if the space served is occupied. Some designs use chilled ceilings or heated floors to overcome this limitation. When chilled ceilings are used, it is critical that building relative humidity be controlled to avoid condensation. Another means of increasing cooling capacity is to recirculate some of the room air.

Some associate displacement ventilation solely with underfloor air distribution and the perceived higher costs associated with it. In fact, most underfloor pressurized plenum, air distribution systems do not produce true displacement ventilation but, rather, well-mixed airflow in the lower section of the conditioned space. It can, however, be a viable alternative when considering systems for modern office environments where data cabling and flexibility concerns may merit a raised floor.

PROS AND CONS

Pro

- Displacement ventilation offers the potential for improved thermal comfort and IAQ due to increased ventilation effectiveness.
- There is reduced energy use due to extended economizer availability associated with higher supply temperatures.

Con

- It may add complexity to the supply air ducting.
- It is more difficult to address high heating or cooling loads.
- There are perceived higher costs.

KEY ELEMENTS OF COST

The following provides a possible breakdown of the various cost elements that might differentiate a system utilizing displacement ventilation from one that does not and an indication of whether the net cost is likely to be lower (L), higher (H), or the same (S). This assessment is only a perception of what might be likely, but it obviously may not be correct in all situations. There is no substitute for a detailed cost analysis as part of the design process. The listings below may also provide some assistance in identifying the cost elements involved.

First Cost

- Controls — S
- Equipment — S
- Distribution ductwork — S/H
- Design fees — S

Recurring Cost

- Energy cost — L
- Maintenance of system — S
- Training of building operators — S/H
- Orientation of building occupants — S/H
- Commissioning cost — S

SOURCES OF FURTHER INFORMATION

Advanced Buildings Technology and Practice, www.advancedbuildings.org.

ASHRAE. 2013. *2013 ASHRAE Handbook—Fundamentals*. Atlanta: ASHRAE.

ASHRAE. 2013. ASHRAE Standard 62.1-2013, *Ventilation for Acceptable Indoor Air Quality*. Atlanta: ASHRAE.

Bauman, F., and T. Webster. 2001. Outlook for underfloor air distribution. *ASHRAE Journal* 43(6). Atlanta: ASHRAE.

Interpretation IC-62-1999-30 of ANSI/ASHRAE Standard 62-1999, *Ventilation for Acceptable Indoor Air Quality*, August 2000. Atlanta: ASHRAE.

Public Technology Inc., U.S. Department of Energy and the U.S. Green Building Council. 1996. *Sustainable Building Technical Manual—Green Building Design, Construction and Operations*.

ASHRAE GreenTip #10-4

Dedicated Outdoor Air Systems

GENERAL DESCRIPTION

ASHRAE Standard 62 describes in detail the ventilation required to provide a healthy indoor environment as it pertains to IAQ. Traditionally, designers have attempted to address both thermal comfort and IAQ with a single mixed-air system. But ventilation becomes less efficient when the mixed-air system serves multiple spaces with differing ventilation needs. If the percentage of outdoor air is simply based on the critical space's need, then all other spaces are overventilated. In turn, providing a separate dedicated outdoor air system (DOAS) may be the only reliable way to meet the true intent of ASHRAE Standard 62.

A DOAS uses a separate air handler to condition outdoor ventilation air before delivering it directly to the occupied spaces. The air delivered to the space from the DOAS should not adversely affect thermal comfort (i.e., too cold, too warm, too humid); therefore, many designers call for systems that deliver neutral air. However, there is a strong argument for supplying cool dry air and decoupling the latent conditioning as well as the IAQ components from the thermal comfort (sensible only) system.

The only absolute in a DOAS is that the ventilation air must be delivered directly to the space from a separate system. Control strategy, energy recovery, and leaving air conditions are all variables that can be fixed by the designer.

WHEN/WHERE IT'S APPLICABLE

While a DOAS can be applied in any design, it is most beneficial in a facility with multiple spaces with differing ventilation needs. A DOAS can be combined with any thermal comfort conditioning system, including, but not limited to, all-air systems, fan-coil units, and hydronic radiant cooling. Note, however, that a design incorporating a separate 100% outdoor air unit delivering air to the mixed-air intakes of other units is not a DOAS as defined here. While this type of system may have benefits, such as using less energy or providing more accurate humidity control, it still suffers from the multiple space dilemma described above.

PROS AND CONS

Pro

- A DOAS ensures compliance with ASHRAE 62.1 for proper multiple space ventilation and adequate IAQ.
- It reduces a building's energy use when compared to mixed-air systems that require overventilation of some spaces in order to ensure adequate ventilation.
- It allows the designer to decouple the latent load from the sensible load, hence providing more accurate space humidity control.
- It allows easy airflow measurement and balance and keeps ventilation loads off main HVAC units.

Con

- Depending on overall design (thermal comfort and IAQ), it may add additional first cost associated with providing parallel systems.
- Depending on overall design, it may require additional materials with their associated embodied energy costs.
- Depending on overall design, there may be more systems to maintain.
- With two airstreams, proper mixing may not occur when distributed to the occupied space.
- The total airflow of two airstreams may exceed airflow of a single system.

KEY ELEMENTS OF COST

The following provides a possible breakdown of the various cost elements that might differentiate a building with a DOAS from one with another system and an indication of whether the net cost is likely to be lower (L), higher (H), or the same (S). This assessment is only a perception of what might be likely, but it obviously may not be correct in all situations. There is no substitute for a detailed cost analysis as part of the design process. The listings below may also provide some assistance in identifying the cost elements involved.

First Cost

- Controls   H
- Equipment   S/H
- Distribution ductwork   S/H
- Design fees   S/H

Recurring Cost

- Energy cost   S/L
- Maintenance of system   S/H
- Training of building operators   S/H
- Orientation of building occupants S
- Commissioning cost   S/H

SOURCES OF FURTHER INFORMATION

ASHRAE. 2013. ASHRAE Standard 62.1-2013, *Ventilation for Acceptable Indoor Air Quality*. Atlanta: ASHRAE.

Coad, W.J. 1999. Conditioning ventilation air for improved performance and air quality. *Heating/Piping/Air Conditioning*, September.

Morris, W. 2003. The ABCs of DOAS. *ASHRAE Journal* 45(5).

Mumma, S.A. 2001. Designing dedicated outdoor air systems. *ASHRAE Journal* 43(5).

ASHRAE GreenTip #10-5

Ventilation Demand Control Using CO_2

GENERAL DESCRIPTION

A significant component of indoor environmental quality (IEQ) is the indoor air quality (IAQ). ASHRAE Standard 62.1 describes in detail the ventilation required to provide a healthy environment. However, providing ventilation based strictly on the peak occupancy using the ventilation rate procedure (Section 6.1) will result in overventilation during some periods. Any positive impact on IAQ brought on by overventilation could be outweighed by the costs associated with the energy required to condition the ventilation air. Demand-controlled ventilation can be done based on a number of different methods to determine the room occupancy, but by far the most common method is through the use of CO_2 measurement.

CO_2 can be used to measure or control the per-person ventilation rate in a given space and, in turn, allow the designer to introduce a ventilation demand control strategy. Simply put, the amount of CO_2 present in the air is an indicator of the number of people in the space relative to the ventilation being provided. In turn, this level can help determine if an adjustment in the amount of ventilation air that is being provided. CO_2-based ventilation control does not affect the peak design ventilation capacity required to serve the space as defined in the ventilation rate procedure, but it does allow the ventilation system to modulate in sync with the building's occupancy.

The key components of a CO_2 demand-based ventilation system are CO_2 sensors and a means by which to control the outdoor fresh air intake (i.e., a damper with a modulating actuator). There are many types of sensors, and the technology is evolving while, at the same time, costs have dropped over the past decade or so. Sensors can be wall-mounted or mounted in the return duct, but it is recommended that the sensor be installed within the occupied space whenever possible.

WHEN/WHERE IT'S APPLICABLE

CO_2 demand control is best suited for buildings with a variable occupancy. The savings will be greatest in spaces that have a wide variance, such as gymnasiums, large meeting rooms, and auditoriums. For buildings with a constant occupancy rate, such as an office building or school, a simple nighttime setback scenario may be more appropriate for ventilation demand control, but CO_2 monitoring may still be used for verification that high IAQ is achieved.

PROS AND CONS

Pro

- CO_2 demand control reduces a building's energy use as it relates to providing ventilation above that needed for adequate IAQ.
- It assists in maintaining adequate ventilation levels regardless of occupancy.

Con

- There is an added first cost associated with the sensors and additional controls.
- There are additional materials and their associated embodied energy costs.
- Evolving sensor technology may not be developed to full maturity.

KEY ELEMENTS OF COST

The following provides a possible breakdown of the various cost elements that might differentiate a building using a CO_2 ventilation demand control strategy from one that does not and an indication of whether the net cost is likely to be lower (L), higher (H), or the same (S). This assessment is only a perception of what might be likely, but it obviously may not be correct in all situations. There is no substitute for a detailed cost analysis as part of the design process. The listings below may also provide some assistance in identifying the cost elements involved.

First Cost

- Controls   H
- Design fees   S/H

Recurring Cost

- Energy cost   L
- Maintenance of system   S/H
- Training of building operators   S/H
- Orientation of building occupants   S/H
- Commissioning cost   S/H

SOURCES OF FURTHER INFORMATION

Advanced Buildings Technologies and Practices, www.advancedbuildings.org.

ASHRAE. 2013. ANSI/ASHRAE Standard 62.1-2013, *Ventilation for Acceptable Indoor Air Quality*. Atlanta: ASHRAE.

ASTM. 1998. ASTM D 6245-1998: *Standard Guide for Using Indoor Carbon Dioxide Concentrations to Evaluate Indoor Air Quality and Ventilation*. ASHRAE.

Lawrence, T.M. 2004. Demand-controlled ventilation and sustainability. *ASHRAE Journal* 46(12):117–21.

Schell, M., and D. Int-Hout. 2001. Demand control ventilation using CO_2. *ASHRAE Journal* 43(2). Atlanta: ASHRAE.

Trane Company. *A Guide to Understanding ASHRAE Standard 62-2001*. Lacrosse, WI: Trane Company. http://trane.com/commercial/issues/iaq/ashrae2001.asp.

ASHRAE GreenTip #10-6

Hybrid Ventilation

GENERAL DESCRIPTION

A hybrid ventilation system allows the controlled introduction of outdoor air ventilation into a building by both mechanical and passive means; thus, it is sometimes called mixed-mode ventilation. It has built-in strategies to allow the mechanical and passive portions to work in conjunction with one another so as to not cause additional ventilation loads compared to what would occur using mechanical ventilation alone. It thus differs from a purely passive ventilation system, consisting of operable windows alone, which has no automatic way of controlling the amount of outdoor air load.

Two variants of hybrid ventilation are the *changeover* (or *complementary*) type and the *concurrent* (or *zoned*) type. With the former, spaces are ventilated either mechanically or passively, but not both simultaneously. With the latter variant, both methods provide ventilation simultaneously, though usually to zones discrete from one another.

Control of hybrid ventilation is obviously an important feature. With the changeover variant, controls could switch between mechanical and passive ventilation seasonally, diurnally, or based on a measured parameter. In the case of the concurrent variant, appropriate controls are needed to prevent "fighting" between the two ventilation methods.

WHEN/WHERE IT'S APPLICABLE

A hybrid ventilation system may be applicable in the following circumstances:

- When the owner and design team are willing to explore employing a nonconventional building ventilation technique that has the promise of reducing ongoing operating costs as well as providing a healthier, stimulating environment.
- When it is determined that the building occupants would accept the concept of using the outdoor environment to determine (at least, in part) the indoor environment, which may mean greater variation in conditions than with a strictly controlled environment.

- When the design team has the expertise and willingness—and has the charge from the owner—to spend the extra effort to create the integrated design needed to make such a technique work successfully.
- Where extreme outside conditions—or a specialized type of building use—do not preclude the likelihood of the successful application of such a technique.

Buildings with atriums are particularly good candidates.

PROS AND CONS

Pro

- Hybrid ventilation is an innovative and potentially energy-efficient way to provide outdoor air ventilation to buildings and, in some conditions, to cool them, thus reducing energy otherwise required from conventional sources (power plant).
- Corollary to the above, it could lead to a lower building life-cycle cost as the operational costs are lower.
- It would generally be expected to create a healthier environment for building occupants.
- It offers a greater sense of occupant satisfaction due to the increased ability to exercise some control over the ventilation provided.
- There is more flexibility in the means of providing ventilation; the passive variant can act as backup to the mechanical system and vice versa.
- It could extend the life of the equipment involved in providing mechanical ventilation since it would be expected to run less.

Con

- Failure to integrate the mechanical aspects of a hybrid ventilation system with the architectural design could result in a poorly functioning system. Some architectural design aspects could be constrained in providing a hybrid ventilation system, such as building orientation, depth of occupied zones, or grouping of spaces.

- Additional first costs could be incurred since two systems are being provided where only a single one would be provided otherwise, and controls for the passive system could be a major portion of the added cost.
- If automatic operable window openers are used, these could result in security breaches if appropriate safeguards and overrides are not provided.
- If integral building openings are used in lieu of, or in addition to, operable windows, pathways for the entrance of outside pollutants and noise or of unwanted insects, birds, and small animals would exist. If filters are used to prevent this, they could become clogged or could be an additional maintenance item to keep clear.
- Building operators may have to have special training to understand and learn how best to operate the system. Future turnovers in building ownership or operating personnel could negatively affect how successfully the system performs.
- Occupants would probably need at least some orientation so that they would understand and be tolerant of the differences in conditions that may prevail with such a system. Future occupants may not have the benefit of such orientation.
- Special attention would need to be given to certain safety issues, such as fire and smoke propagation in case of a fire.
- Although computer programs (such as computational fluid dynamics) exist to simulate, predict, and understand airflow within the building from passive ventilation systems, it would be difficult to predict conditions under all possible circumstances.
- Few codes and standards in the U.S. recognize and address the requirements for hybrid ventilation systems. This would likely result in local code enforcement authorities having increased discretion over what is acceptable or not.

KEY ELEMENTS OF COST

The following provides a possible breakdown of the various cost elements that might differentiate a hybrid ventilation system from a conventional one and an indication of whether the net cost for the hybrid option is likely to be lower (L), higher (H), or the same (S). This assessment is only a perception of what might be likely, but it obviously may not be correct in all situations. There is no substitute for a detailed cost analysis as part of the design process. The listings below may also provide some assistance in identifying the cost elements involved.

First Cost

- Mechanical ventilation system elements — S
- Architectural design features — H/L
- Operable window operators — H
- Integral opening operators/dampers — H
- Filters for additional openings — H
- Controls for passive system/coordination with mechanical system — H
- Design fees — H

Recurring Cost

- Energy for mechanical portion of system — L
- Maintenance of above — L
- Energy used by controls, mechanical operators — H
- Maintenance of passive system — H
- Training of building operators — H
- Orientation of building occupants — H
- Commissioning cost — H
- Occupant productivity — H

SOURCE OF FURTHER INFORMATION

Kosik, W.J. 2001. Design strategies for hybrid ventilation. *ASHRAE Journal* 43(10).

CHAPTER ELEVEN

ENERGY CONVERSION SYSTEMS

HEAT GENERATORS (HEATING PLANTS)

Considerable improvements in the seasonal efficiency of conventional heating plant equipment (e.g., boilers and furnaces) were made over the last several decades. Designers should verify claims of equipment manufacturers by reviewing documented data of this equipment to prove the efficiency ratings are accurate.

Some unconventional equipment and techniques to achieve greater efficiency or other possible green-building design goals are described in the GreenTips at the end of this chapter.

COOLING GENERATORS (CHILLED-WATER PLANTS)

Chilled-water plants are most often used in medium to large facilities. Their benefits include higher efficiency, reduced maintenance costs, and redundant capacity (in comparison to decentralized plants).

Because chilled-water temperature can be closely controlled, chilled-water plants have an advantage over direct-expansion systems, because they allow air temperatures to be closely controlled.

Generally, a chilled-water plant consists of the following elements:

- Chillers
- Chilled-water pumps
- Condenser-water pumps (for water-cooled systems)
- Cooling towers (for water-cooled systems) or air-cooled condensers
- Associated piping, connections, and valves

Often, chilled-water plant equipment will be in a single, central location, allowing system control, maintenance, and problem diagnostics to be performed efficiently. Chilled-water plants also allow redundancy to be easily designed into the system by designing for firm capacity. Firm capacity is calculated with the largest piece of equipment not operating. With firm capacity, one piece of equipment can be maintained or repaired and the system still has the ability to meet peak loads.

Chiller Types

Electric chillers used within chiller plants employ either a scroll-, reciprocating-, screw-, or centrifugal-type compressor (in order of increasing size). Models can be offered with either air-cooled or water-cooled condensers, with the exception that today's centrifugal compressors are water-cooled only.

Steam-driven turbines or absorption chillers, powered by steam, hot water, natural gas, or other hot gases, are used in many central plants to balance steam and electric demands in combined heat and power plants, and to offset high electric demand or consumption charges.

The various chiller types are used most often, though not exclusively, in the following situations.

Electric Motor-Driven Chillers:

- Low to moderate electric consumption and demand charges prevail
- Air-cooled heat dissipation is preferred
- Condenser heat recovery is desired

Steam-Driven and Absorption Chillers:

- Part of a combined heat and power plant
- Low fuel (e.g., natural gas) costs prevail
- High electric demand charges prevail
- Plentiful source of heat available (its main use usually for other functions)

Heat Pumps. A heat pump is another means of generating cooling as well as heating using the same piece of equipment. There is usually an array of them used for a project, and they are generally distributed throughout the building (i.e., they're not part of a central plant). Buildings that have consistent demands for both chilled and hot water, such as hospitals, are good candidates to use a larger heat pump as a central plant. (See the GreenTips at the end of this chapter on various heat pump system types.)

Thermal Energy Storage (TES). TES is a technique that has been encouraged by electricity pricing schedules where the off-peak rate is considerably lower than the on-peak rate. Cooling, in the form of chilled water or ice, is generated during off-peak hours and stored for use during on-peak hours. Although not refrigeration equipment per se, the technique can usually reduce the size of refrigeration equipment—or obviate the need for adding a chiller to an existing plant.

The characteristics, merits, and cost factors of TES for cooling, as well as numerous reference sources, are presented in GreenTip #11-8.

System Design Considerations

When a designer puts together a chilled-water plant, there are many design parameters to optimize. They include fluid flow rates and temperatures, pumping options, plant configuration, and control methods. For each specific application, the design professional should understand the client's needs and desires and implement the chiller plant options that best satisfy him or her.

Fluid Flow Rates. Flow rates were discussed in the "Media Movers" section of Chapter 10.

Fluid Temperatures. To allow lower flow rates, the chiller must be able to supply colder chilled-water temperatures and to tolerate higher condenser leaving temperatures.

Pumping Options. Pumps may be selected to operate with a specific chiller or they may be manifolded.

Pump-per-chiller arrangement advantages include the following:

- Hydronic simplicity
- Chiller and pump are controlled together
- Pumps and chillers may be sized for one another

Manifolded-pump arrangement advantages include the following:
- Simpler redundancy
- Pumps may be centrally located

Plant Configuration (Multiple Chillers). The most prevalent chiller plant configuration is the primary-secondary (decoupled) system. This system allows the flow rate through each chiller to remain constant, yet accommodates a reduction in pumping energy, since the system water flow rate varies with the load.

Becoming more common are variable-primary-flow systems that also vary the flow through the chiller evaporators. New chiller controls allow this. Often these systems can be installed at a reduced cost when compared to primary-secondary systems since fewer pumps (and their attendant piping, connections, valves, fittings, and electrical draws) are required. These systems can also save energy in comparison to the primary-secondary configuration, due to more efficient chiller operation and reduced pumping energy.

(The subjects of plant configuration, pumping options, and control methods are discussed in detail in ASHRAE's *Fundamentals of Water System Design* [ASHRAE 1998], Trane's *Multiple-Chiller-System Design and Control Manual* [Trane 2000], and the *Chilled Water Plant Design Guide* from Energy Design Resources [EDR 2009].)

The designer should always review the overall use of energy within a facility and employ systems (including heat recovery systems) that interact with one another so as to minimize the overall energy consumption of the entire chilled-water plant.

COOLING SYSTEM HEAT SINKS

The function of building HVAC systems is to exchange a significant amount of thermal energy between the building and the surrounding environment, which could be air, ground, or a water body. Historically, air has been the primary source of heat rejection from buildings because of its availability. A good example of that process is a cooling tower, which is covered in the next section. In the last decade, the ground has emerged as a good and green choice for heat rejection (refer to GreenTip #11-6, Ground-Source Heat Pump).

Designers should pay attention when selecting unconventional heat rejection sources or sinks, as some may have indirect environmental impacts. For example, systems have been installed that use nearby deep water in a lake or ocean as a heat sink, which results in significant energy savings. This technique has been used in Scandinavian systems for approximately 20 years and more recently in colder regions in North America. However, this heat rejection technique, if not studied properly, may increase the water temperature in the source and may affect the maritime wildlife. The possible net energy savings are impressive, but these must be balanced against potential adverse impacts on local aquatic environment.

COOLING TOWER SYSTEMS

Cooling towers, which are generally the heat rejection equipment in a water cooled system, are a very efficient method of heat rejection to air, especially in dry climates. Cooling towers remove heat by evaporation and can cool close to the ambient wet-bulb temperature (the difference between the leaving water and ambient wet-bulb being defined as the *approach temperature*). Unless the air is totally saturated (i.e., 100% relative humidity), the wet-bulb temperature is always lower than the dry-bulb. Thus, water cooling allows more efficient condenser operation than air cooling. In a typical cooling tower operation, about 1% of the recirculated water flow is evaporated. This evaporation will cool the remaining 99% of the water for reuse. In addition to evaporation, some recirculated water must be bled from the system to prevent soluble and semisoluble minerals from reaching too high a concentration. This bleed or blowdown is usually sent to a publicly owned treatment works (POTW). The HVAC designer should always consult with the plumbing engineer on the local regulations governing the discharge of cooling tower blowdown.

Water Treatment

The water in the evaporative cooling loop must be treated to minimize biological growth, scaling, and corrosion. Typically, a combination of biocides, corrosion inhibitors, and scale inhibitors are added to the system.

Corrosion inhibitors are usually phosphate or nitrogen based (e.g., fertilizers) or molybdenum- or zinc-based (e.g., heavy metals). These inhibitors are more effective when added in combinations. These materials have low vapor pressures and thus do not evaporate from the system. The inhibitors simply need to remain in the solution at the proper concentration to maintain a protective film on the metal components. Their only loss is through blowdown and drift.

Most scaling inhibition is accomplished using polymer-based chemicals, organic phosphorous compounds (phosphonates), or by acid addition. The acid reacts with the alkalinity in the water to release CO_2 and is used up. The polymer and phosphonate scale inhibitors remain in the solution to delay scaling; their major loss is through blowdown and drift. Some polymers are designed to be biodegradable, easily broken down by bacteria in the environment, while others are not.

There are very wide assortments of biocides. A typical system will maintain an oxidizing biocide (e.g., bromine or chlorine) at a constant level and should be given a dose of a nonoxidizing biocide once a week. Chlorine and bromine have a high vapor pressure, and thus, easily volatilize from the water. Much of the chlorine and bromine added to the tower is stripped from the water into the air; a small quantity actually reacts with organics in the tower. Drift and blowdown will contain all of the nonoxidizing biocides, a small quantity of oxidizing biocides, and the reaction products of the biocides.

Drift

To promote efficient evaporation, cooling towers force intimate contact of outdoor air with warm water. Besides removing heat by evaporation, dust, pollen, and gas components of the air will become entrained in the water, while some high vapor pressure components of the water (e.g., bromine or chlorine) as well as entrained water drops, will migrate to the air. Airborne dust and pollen that are captured by the water can promote biological growth in the tower.

Small water droplets entrained and carried out with the air passing through the tower are called *drift*. Drift is always present when operating a cooling tower. Since drift is actually just small droplets of the cooling tower water, it contains all of the dissolved minerals, microbes, and water treatment chemicals in the tower water. Drift is a source of PM10 emission (particulate matter less than 10 μ in diameter) and is as well a suspected vector in *Legionella* transmission.

Drift is usually reported as a percentage of the recirculating water, though it may also be described in terms of parts per million (ppm) of the air passing through the tower. Tower designs use drift eliminators to capture some of this entrained water. A typical value for drift from cooling towers is 0.005% of the recirculating water; many tower designs have drift values as low as 0.001%.

To put these values in perspective, consider an example of a 400 ton cooling system with a particular treatment program. A 400 ton system would circulate approximately 1200 gpm (4543 L/min) through the tower and chiller. At nominal rates, 12 gpm (45.4 L/min) would be lost to evaporation, 0.06 gpm (0.23 L/min) would be lost to drift (0.005%), and, at four cycles of concentration, 4 gpm (15.1 L/min) would be intentionally bled from the system. Chlorine addition would be 0.09 gal/h (0.34 L/h) of 12.5% liquid bleach, equivalent to 0.06 lb/h or 0.027 kg/h of Cl_2. This chlorine addition should maintain about 0.2 ppm free chlorine and 0.2 ppm combined chlorine. The combined chlorine is from the reaction of chlorine with organic molecules and may include some hazardous by-products, such as chloroform. For corrosion and scale protection, the water would be maintained with 2 ppm zinc, 3 ppm triazole, and 20 ppm polyphosphate. Once a week, 4 lb of a 1.5% solution of isothiazolin, a nonoxidizing biocide, will be fed to the system for biofilm control. The drift and bleed will contain the same quantity of minerals and chemicals as are maintained in the recirculated water.

Table 11-1 shows monthly results for operating this tower at an assumed 75 h/week (300 h/month). From Table 11-1, it can be seen that most of the chlorine used in this tower is unaccounted for. Some of this loss is due to oxidation of organic and inorganic material in the cooling system, resulting in nonhazardous chloride ions; much of the chlorine is released into the atmosphere as chlorine gas. While it is hard to be quantitative, over the course of one year, this tower could release more than 100 lb (45.4 kg) of chlorine gas into the immediate building environment. If less-effective drift eliminators were used, 12,000 gal (45.4 kL) of contaminated tower water would also be released every year and 800,000 gal (3028 kL) of water containing heavy metals, phosphates,

Table 11-1: Material Release from Example Cooling Tower Operation

	Release Rate	Total per Month
Evaporation	12 gpm (45.4 L/min)	216,000 gal (817,560 L)
Bleed at four cycles	4 gpm (15.1 L/min)	72,000 gal (272,520 L)
Drift	0.060 gpm (0.23 L/min) (0.005%) 0.012 gpm (0.05 L/min) (0.001%)	1080 gal (4088 L) 216 gal (818 L)
Chlorine addition	0.06 lb/h (0.027 kg/h)	18.0 pounds Cl_2 (8.2 kg)
Chlorine in bleed	Free—about 0.2 ppm Combined—about 0.2 ppm	0.12 lb free (0.055 kg) 0.12 lb combined (0.055 kg)
Unaccounted chlorine		17.7 lb Cl_2 (8.0 kg)

and biocides would be sent to a POTW system. Most POTW systems are designed to handle only organic waste; much of these cooling tower chemicals would pass through the system untreated or would be released later as gaseous emissions at the POTW.

Over the lifetime of the building, these releases could be among the most significant impacts that the building would have on the local environment. This example highlights what could happen if the issue is not addressed.

Green Choices—Water Treatment

The water treatment plan illustrated above is not the only choice. There are many ways to treat the system that will have a less negative impact on the environment. Besides being rapidly stripped into the air, chlorine and bromine may react with organic molecules to produce very hazardous daughter products; however, other oxidizing biocides (e.g., hydrogen peroxide) do not have this issue. By continuous monitoring of the cooling system, chemical additions can be added only when needed. This technique can yield equivalent performance with less-added chemicals. Also, the U.S. Environmental Protection Agency (EPA) maintains a website on green chemistry (www.epa.gov//greenchemistry/), which contains criteria on how to evaluate the life-cycle environmental impact of a particular chemical.

Nonchemical water treatment has the potential to be a powerful method for water treatment; however, its success depends on the water chemistry, operating procedures, and degree of pollution of the specific system. There are several different nonchemical technologies available, including those based on pulsed electric fields, mechanical agitation, and ultrasound. Each of these technologies has developed a widespread following. ASHRAE has investigated the scale prevention effectiveness of some of these nonchemical technologies and has published the results from ASHRAE Research Project RP-1155 (Cho et al. 2003). These technologies offer an alternative to the storage and handling of toxic chemicals at the site, eliminating the risk caused by any spills or leaks of cooling tower water, the issues of the bleed water at the POTW, and much of the concern with drift. Bleed water, instead of being sent to the POTW, could be used on site for irrigation or other nonpotable needs. The pulsed-field technology is the subject of GreenTip #11-1.

Green Choices—Tower Selection

All cooling tower designs are efficient at removing heat from water through evaporation; however, not all designs perform as well environmentally. Some tower designs are more prone to splashout, spills, drift, and algae growth than others. Splashout involves tower water splashing from the tower. This happens most often in no-fan conditions (i.e., when tower water is circulated with the fans off) when there are strong winds. Cross-flow towers are more prone to this issue since, in a no-fan condition, some water will fall outside of the fill. On the other hand, cross-flow towers have a gravity feed system as opposed to pressurized nozzles for counterflow towers, making the pump head (and hence energy consumption) generally higher for counterflow towers. The designer is encouraged to analyze the requirements and constraints of each project in order to select the best tower configuration.

Spills can happen from the cold-water basin from overflowing at shutdown when all of the water in the piping drains into the basin. Proper water levels will prevent this. Some tower designs use hot-water basins to distribute water at the top of the tower, while other designs use a spray header pipe. The hot-water basin design can overflow if the nozzles clog; a spray header pipe never overflows.

Algae are a nuisance in basins and can contribute to microbial growth. Algae control requires harsher chemical treatment than typical biological control. Since algae are plants, they need sunlight to grow. Some tower designs are light-tight, virtually eliminating algae as an issue, while other designs are more open, and algae growth can be an issue.

The amount of drift varies extensively in tower design. Some tower designs have very little drift. The less the drift from a tower, the lower the amount of water containing minerals, water treatment chemicals, and microbes that will be released into the surrounding environment.

Fan-power draw also varies between designs. There are two general types of fans: axial and centrifugal. With axial fans, the fan blades are mounted perpendicular to the axis like propellers on a plane or boat. All induced-draft cooling towers and some forced-draft designs use axial fans. With centrifugal fans, the fan blades are mounted parallel to the axis, along the outside of the shaft hub. Centrifugal fans are only used in forced-draft towers and typically use twice the energy to achieve the same amount of airflow as an axial fan.

Maintenance

An often overlooked method for minimizing environmental impact is maintenance. Cooling towers operate outdoors under changing conditions. Wind damage to inlet air louvers, excessive airborne contamination, clogging of water distribution nozzles, and mechanical problems can best be prevented and quickly corrected with periodic inspections and maintenance.

Variable-Speed Fans

The designer is encouraged to include variable-frequency drives (VFDs) on the cooling-tower fans. With the decreasing cost of these drives, their life-cycle cost is likely to be favorable. VFDs on fans can save energy by reducing the fan speed to better match the load, reduce the fan on/off cycling (thus reducing inrush currents), and can accomplish other tasks such as fan rotation reversal (beneficial in cold climates).

DISTRICT ENERGY SYSTEMS

District energy (DE) systems involve the provision of thermal energy (heating and/or cooling) from one or more central energy plants to multiple buildings or facilities via a network of interconnecting thermal piping. Generally, district heating systems deliver heat as steam or hot water, while district cooling systems deliver cooling as chilled water or chilled secondary coolant (such as an aqueous glycol or an aqueous sodium nitrite solution) or even as a refrigerant.

DE systems often deliver multiple significant positive impacts to the local building environments that they serve. These typical impacts include the areas outlined in the following sections.

Energy Consumption

Heating and/or cooling buildings using DE systems can affect overall energy consumption in various ways, from modest increases or decreases to very dramatic decreases in fuel and energy consumption. The energy consumed within the boundaries of DE-served buildings will, of course, be dramatically reduced compared to a baseline building with its own dedicated boilers and chillers. This energy reduction within the buildings will be offset by the energy consumed in the central DE plants and in distributing the thermal energy from the central plants to the customer buildings. If the central DE plant uses similar technology (e.g., gas boilers and electric chillers) as otherwise used in the individual buildings, there may be little or no net reduction in energy use. However, the larger (and generally more efficient) DE plant equipment more than offsets extra energy consumed in the distribution of the thermal energy to the buildings for at least a slight net reduction in overall energy consumption. Reductions in overall fuel and energy consumption are achieved through the ability of DE plants to more readily and more economically use alternative technologies than is the case for individual buildings. These technologies include, but are not limited to, dual-fuel boilers; alternative fuel boilers (including renewable fuels such as low-Btu landfill gas, municipal solid waste, wood waste, etc.); high-efficiency boilers; high-efficiency chillers; alternative energy-efficient refrigerants (e.g., ammonia); nonelectric chiller plants (e.g., absorption chillers, engine-driven chillers, or turbine-driven chillers); hybrid chiller plants (with various combinations of electric and nonelectric chillers); energy-efficient series or series-parallel chiller configurations for high ΔT systems; thermal energy storage (TES); cogeneration of combined heating and power (CHP); trigeneration or combined cooling, heating, and power (CCHP); and the use of natural renewable thermal energy (such as geothermal heat for district heating systems and cold deep water sources [e.g., lakes or oceans] for district cooling systems). In addition, DE plants are better able to meet the changing loads of the system.

Central energy plants associated with DE systems (compared to the alternative of dispersed, multiple, smaller boilers and chiller plants within individual buildings), generally have higher levels of operational efficiency and reliability. This is because larger DE plants can more easily justify sophisticated design, automated optimized control systems, more attentive maintenance programs, and more highly trained and focused operations and maintenance personnel. Central energy plants are also better able to match the system load with central equipment versus part-load, especially at part-load throughout a campus.

Emissions

As is the case with energy consumption, DE serves to eliminate many emissions from the local building environment, such as boiler exhausts and chiller plant heat rejection. Some emissions are of course relocated to the site of the central DE plant. However, just as DE plants tend to have higher levels of energy efficiency, they tend to have lower levels of emissions vs. those associated with individually heated and cooled buildings. And for DE systems utilizing one or more of the alternative technologies (as cited above), the overall emissions can be significantly reduced in terms of air pollutants (i.e., SO_X, NO_X, and precipitates) and greenhouse gases (i.e., CO_2, NO_X, and some refrigerants).

Noise and Vibration

Through the avoidance of needing boilers and/or chillers in the buildings, the occupants of DE-served buildings experience a local building environment that is free from the potential noise and vibrations associated with such equipment.

Chemical Supplies and Blowdown

Due to the avoidance of needing boilers and/or chiller plants in each building, DE systems can eliminate or greatly mitigate the presence of potentially harmful fuel and chemicals to be handled within the occupied buildings. With the use of DE, the storage, handling, and disposal of fuel, boiler water treatment chemicals, refrigerants, condenser water treatment chemicals, and chilled-water treatment chemicals can all be removed to the location of the central DE plants. Thus, the potential for related chemical spills, disruptions, and associated hazards are avoided within the occupied building environments.

Space Utilization and Aesthetics

By not needing heating and/or chilling plants in the buildings, DE systems provide improved space utilization for the occupants of the individual buildings. Also, there is no longer a need for local boiler exhaust stacks and/or for local chiller plant heat rejection via, for example, roof-mounted cooling towers. In addition to the improved aesthetics of having buildings without such stacks and towers, multiple and sometimes unsightly exhaust plumes from stacks and towers are also removed from the local building environment.

Other Factors for Consideration

District energy systems do require additional infrastructure (e.g., a piping distribution network from the central plant to the various buildings within the network). A full-fledged, life-cycle assessment of the net benefits from a DE system should take items including construction of the additional piping network and additional space requirements of distributed equipment into consideration.

Where DE Systems are Used

DE systems are routinely used on university campuses; DE systems are also often used for other institutional applications (including schools, hospital and medical facilities, airports, military installations, and other federal, state, and local government facilities), for privately owned multibuilding commercial/industrial facilities, and for thermal utilities serving urban business districts. DE systems serve as few as two buildings or as many as many hundreds of buildings. The ideal times to consider utilizing DE for serving the heating and/or cooling needs of a building are either during master planning and new construction or during expansion or renovation of buildings or their HVAC systems.

WATER CONSUMPTION FOR COOLING SYSTEM OPERATION

Trade-offs between energy consumption and water consumption must be carefully considered. In many cases, site energy can be saved at the expense of using site water. Examples of this include using evaporative cooling as opposed to direct-expansion cooling. Likewise, chillers with water-based cooling towers are more efficient than air-based systems, but this energy saving comes with the necessity of using water. Further complicating the analysis, is the fact that water does not arrive at the site without treating and pumping. Much of the electrical power is produced with thermal electric plants that use evaporative cooling for the condenser part of the Rankine cycle. As additional environmental pressures are applied, more power plants are using water evaporation through towers, rather than discharging the heat to rivers, lakes, and oceans. Even hydroelectric power plants evaporate water because of the large lake surface areas compared to free-running rivers.

We have just begun to understand these interactions and know that the numbers vary considerably by climate and location. Many of the variations are because of variations in power plant designs. In the United States, a national average of 0.50 gal (1.9 L) of water is consumed for every kWh of electricity produced. An additional breakdown by region and for energy mixes is provided in Torcellini et al. (2004).

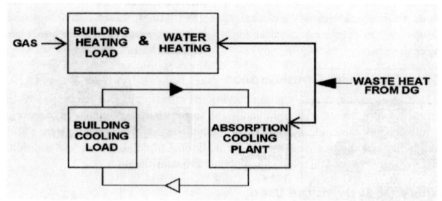

Image courtesy of Malcolm Lews, CTG Energetics, Inc.

Figure 11-1 Thermal uses of waste heat.

Because of the variations in climate and HVAC design, no rules of thumb exist to compare water and electrical consumption. Hourly computer simulations can be used to compare systems, with the water consumption calculated based on water and energy balances.

Other uses or sources of water should also be considered. For example, condensate from cooling coils can be collected and reused (see GreenTip #14-6 in Chapter 14). In some cases, blowdown from cooling towers can also be used, depending on the levels of dissolved solids and chemical treating.

DISTRIBUTED ELECTRICITY GENERATION

One opportunity for energy conservation in buildings is the use of on-site generation systems to provide both distributed electric power and thermal energy (otherwise wasted heat from the generation process), which can be used to meet the thermal loads of the building.

Distributed generation (DG) provides electricity directly to the building's electrical systems to offset loads that would otherwise have to be met by the utility grid (see Figure 11-1). The waste heat from that generation process goes through a heat recovery mechanism where it may be provided as heat to meet loads for conventional heating (such as space heating, reheat, and domestic water heating) or for specialized processes. Alternatively, that heat energy, if at a sufficiently high temperature, can be used to power an absorption chiller to produce chilled water to meet either space-cooling or process-cooling loads. This is shown in Figure 11-2.

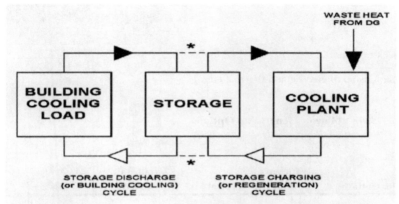

Image courtesy of Malcolm Lews, CTG Energetics, Inc.

Figure 11-2 Thermal energy storage and waste heat usage.

Any timing differences between the generation of the waste heat from the DG system and the thermal needs of the building can be handled utilizing a chilled-water TES system. This concept may also permit downsizing the absorption cooling system (so it does not need to be sized for the peak cooling load).

The overall usable energy value from the fuel input to the generation process is only about 30% (or less) if there is no waste heat recovery, but it can be more than 70% if most or all the waste heat is able to be used. This increased system efficiency can have a radical impact on the economics of the energy systems, because almost two-and-one-half times the useful value is obtained from the fossil fuel purchased. System sizing is generally done by evaluating the relative electrical and thermal loads over the course of the typical operating cycle and then selecting the system capacity to meet the lesser of the thermal or electrical loads. See Figure 11-3 showing comparative thermal and electrical energy for typical office buildings.

If the DG system is sized for the greater of the two, then there will be a net waste of energy produced, since it is seldom economical to sell electricity back to the grid. A key design issue that arises here is whether or not the system is being designed to improve efficiency, as discussed above, or as a baseload on-site generation system for purposes of improving the reliability of the electric and/or thermal energy supply. Either of these is a legitimate design criterion, although the goal, from a green design standpoint, almost always focuses on the energy-efficient strategy.

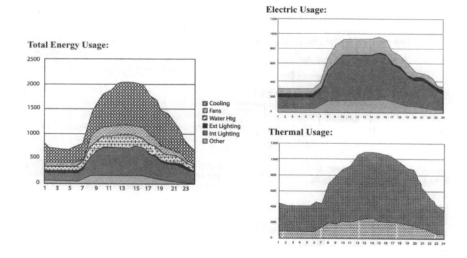

Image courtesy of Malcolm Lews, CTG Energetics, Inc.

Figure 11-3 Relationship between electric and thermal energy.

DG Technologies

Technologies that can be used for this type of generation include engine-driven generators, microturbine-driven generators, or fuel cells. Typically, each uses natural gas as the input fuel. There are advantages and disadvantages associated with these technologies, which are summarized briefly in Table 11-2.

Engine-Driven Generator (EDG). This technology has been around the longest of the three and is in many ways the least-expensive option. It produces a relatively high temperature of waste heat that can be more effectively used by the heat recovery systems. Disadvantages, however, include air pollution, acoustical impacts, and noise/vibration from the engines. EDG sets are available in sizes ranging from approximately 40 kW to several thousand kilowatts.

Microturbine Generator. At the moment, micro-turbine generators are somewhat more expensive in first cost than EDGs, but they have less air pollution and less severe acoustic and vibration impacts than EDGs. They also have longer operating lives and a projected lower cost. However, at the moment they are only available in sizes less than 100 kW per unit compared to EDGs, which can cost in the range of several hundred dollars per kilowatts.

Fuel Cells. Fuel cells are the most advanced form of power generation in terms of being a clean and green technology. They generate virtually no air pollution, have

Table 11-2: Comparison of Power Generation Options

Generation Option	Efficiency (%)	Typical Size	Installed Costs ($/kW)	Operations and Maintenance Costs ($/kWh)
Engine driven natural gas cogeneration with heat recovery	70%–80%	100kW–5MW	$3300–$1600	$0.007–$0.02
Turbine-driven natural gas cogeneration with heat recovery	70%–80%	1MW–10MW	$5000–$3000	$0.003–$0.01
Fuel cell natural gas cogeneration with heat recovery	56%–80%	200kW–1MW	$8000–$5000	$0.070–$0.090
Wind turbine	N/A	5kW–1MW	$8500–$2500	$0.001–$0.007
Photovoltaic	8%–20%	1kW–2MW	$4800–$3500	$0.003–$0.005

Note the following:
Fuel cells vary greatly in cost by type of technology (e.g. phosphoric acid fuel cell [PAFC] at low end, solid oxide fuel cell [SOFC] at high end), size of application, and gas cleanup requirements (e.g. biogas treatment).
Installed and O&M cost information for PV and wind turbine:
www.nrel.gov/analysis/tech_cost_dg.html.
Levelized cost of energy across multiple power options:
http://en.openei.org/apps/TCDB/.
Cogeneration information:
www.ashrae.org/File%20Library/docLib/Public/200412182611_326.pdf.

minimal acoustic and vibration impacts, and are considered the wave of the future. At this point, however, the cost of fuel cell equipment is high, so it's the least attractive of these options, from an economic standpoint. It is anticipated that this will change as continued development of the technology evolves over the next several years.

CCHP Systems

Combined heating and power is also referred to as *cogeneration*, and when combined with cooling they are called *trigeneration*. The larger, industrial-scale versions of such integrated energy systems have been in use through most of the twentieth century. More recently, with the focus shifting to distributed power generation for the reasons cited above (and with the California energy crisis of 2000), a new scale of CCHP has emerged, involving more integrated design of the components. These CCHP systems typically consist of one of the DG technologies described above (e.g., reciprocating engines or microturbines [possibly fuel cells, in the future] fired by natural gas, a heat exchanger to recover heat from the exhaust stream and/or jacket water, and, if there is a cooling demand, an absorption chiller). An intermediate medium such as hot water can be used to transfer heat between the exhaust stream of the prime-mover and the chiller (Figure 11-4), but lately a lot of development has gone into a dedicated heat recovery unit that also forms the generator of the absorption chiller (Rosfjord et al. 2004). Such integrated energy systems are admittedly in their infancy in the United States, but there is a significant drive by organizations including the U.S. Department of Energy (DOE) toward commercializing these systems as packaged systems.

The traditional application of CHP has been to directly use the waste heat for space heating and/or domestic hot-water heating. Larger central CCHP facilities use the waste heat to produce space or process cooling. The former is put together of off-the-shelf equipment, while the latter follows an integrated design approach with a better opportunity at optimization.

To illustrate the benefits of the DG-based CCHP systems, the following elementary view of the energy efficiencies, both at the component and system level, is offered. A typical reciprocating gas-fired engine has a thermal efficiency of about 35%; 65% of the energy in the fuel is ordinarily not used and wasted through the stack. The exhaust temperatures leaving the engine (>500°F or 260°C) are typically high enough to drive, at the very least, a single-effect absorption chiller with a coefficient of performance (COP) in the vicinity of 0.7. Factoring in the actual amount of available heat in the exhaust stream (temperature/enthalpy and flow rate), this translates to overall fuel utilization rates/efficiencies that are as high as 80%. The ratio of electrical to thermal load carrying capability shifts a bit when the prime mover is a microturbine or a fuel cell, with typical generation efficiencies being 25% (based on higher heating value) and 40%, respectively. However, the overall fuel utilization rate is relatively unchanged. The waste heat can also be directly used for desiccant dehumidification (regeneration).

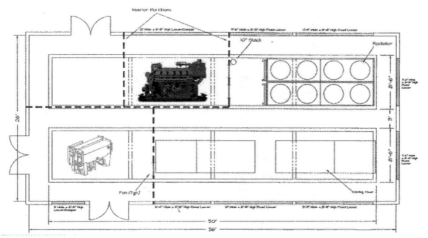

Image courtesy of Vikas Patnaik.

Figure 11-4 Schematic of actual CHP system consisting of a reciprocating engine and an indirect-fired absorption chiller, along with other ancillary equipment (Patnaik 2004).

Thus, the integrated energy systems have done the following:

- Brought the power generation closer to the point of application/load (through distributed generation), eliminating transmission and distribution losses, etc.
- Removed or reduced the normal electric and primary fuel consumption by independent pieces of equipment providing cooling, heating, and/or dehumidification (e.g., separate electric chiller and boiler), thereby substantially improving overall fuel utilization rates, inclusive of the power generation process
- Removed or reduced emissions of CO_2 and other combustion by-products associated with the operation of the cooling, heating, and/or dehumidification equipment

Challenges include the matching of electrical and thermal loads, given the diversity of energy usage patterns in buildings. The consultant/contractor can determine if the heating/cooling components of the CCHP system will play a primary or complementary role (the latter involving other, more conventional equipment to fill in the missing thermal load). Numerous studies were done in this regard, including one by Ryan (2003). Further resources are given in GreenTip #11-2.

REFERENCES AND RESOURCES

Published

ASHRAE. 1998. *Fundamentals of Water System Design*. Atlanta: ASHRAE.

ASHRAE. 2008. *ASHRAE Handbook—HVAC Systems and Equipment*, Chapter 12. Atlanta: ASHRAE.

Cho, Y.I., S. Lee, and W. Kim. 2003. Physical water treatment for the mitigation of mineral fouling in cooling-tower water applications. *ASHRAE Transactions* 109(1):346–57.

EDR. 2009. *Chilled Water Plant Design Guide*. Energy Design Resources. www.energydesignresources.com/Resources/Publications/DesignGuidelines/tabid/73/articleType/ArticleView/articleId/422/Default.aspx.

Patnaik, V. 2004. Experimental verification of an absorption chiller for BCHP applications. *ASHRAE Transactions* 110(1):503–507.

Rosfjord, T., T. Wagner, and B. Knight. 2004. UTC microturbine CHP product development and launch. Presented at the 2004 DOE/CETC Annual Workshop on Microturbine Applications, January 20–22, Marina del Rey, CA.

Ryan, W.A. 2003. CHP: The concept. Midwest CHP Application Center, Energy Resources Center, University of Illinois, Chicago. www.chpcentermw.org/presentations/WI-Focus-on-Energy-Presentation-05212003.pdf.

Torcellini, P.A., M. Longm, and R. Judkoff. 2004. Consumptive water use for U.S. power production. *ASHRAE Transactions* 110(1):96–100.

Trane. 2000. *Multiple-Chiller-System Design and Control Manual*. Lacrosse, WI: Trane Company.

Online

U.S. Environmental Protection Agency, Green Chemistry www.epa.gov/greenchemistry/.

ASHRAE GreenTip #11-1

Pulse-Powered, Chemical-Free Water Treatment

TECHNOLOGY DESCRIPTION

Pulse-powered physical water treatment uses pulsed, electric fields (a technology developed by the food industry for pasteurization) to control scaling, biological growth, and corrosion. This chemical-free approach to water treatment eliminates environmental and health and safety issues associated with water treatment chemicals. Pulse-powered systems do not require pumps or chemical tanks. Pulse-powered systems tend to be forgiving of operational upsets and promote cooling tower operation at higher cycles of concentration (so there's less blowdown and less water usage) than standard chemical treatment. Independent studies have shown that not only is the method effective for cooling towers, the performance of pulse-powered systems is superior to standard chemical treatment in biological control and water usage. The performance results of pulse-powered technology for chemical-free water treatment, as documented by various independent evaluations, support the objectives of green buildings and have earned LEED points for certification in a number of projects.

WHEN/WHERE IT'S APPLICABLE

Pulse-powered technology is applicable on the recirculating lines of cooling towers, chillers, heat exchangers, boilers, evaporative condensers, fluid coolers, and fountains.

The technology produces a pulsed, time-varying, induced electric field inside a PVC pipe that is fit into the recirculating water system. The electric signal changes the way minerals in the water precipitate, totally avoiding hard-lime scale by instead producing a nonadherent mineral powder in the water that does not adhere. The powder is readily filterable and easily removed. Bacteria are encapsulated into this mineral powder and cannot reproduce, thereby resulting in low bacteria populations. The water chemistry maintained by pulse-powered technology is noncorrosive, since the water is operating at the saturation point of calcium carbonate (a cathodic corrosion-inhibiting environment). The low bacteria count and reduction or elimination of biofilm reduces concern about microbial influenced corrosion. The absence of aggressive oxidizing biocides eliminates the risk of other forms of corrosion.

PRO

1. The potential for lower bacterial contamination while providing scale and corrosion control.
2. Lower energy and water use than in traditional chemical treatment.
3. Blowdown water is environmentally benign and recyclable.
4. Potentially lower life-cycle cost savings compared to chemical treatment.
5. Reduction or elimination of biofilm.
6. Removes health and safety concerns about handling chemicals.
7. Eliminates the environmental impact of blowdown, air emissions, and drift from toxic chemicals.

CON

1. It does not work effectively on very soft or distilled water, since the technology is based on changing the way minerals in the water precipitate.
2. Water with high chloride or silica content may limit the cycles of concentration obtainable to ensure optimum water savings, since the technology operates at the saturation point of calcium carbonate.
3. Energy usage is still required to operate.
4. It does not have a residual effect. Water in an isolated equipment (like an off chiller) will not have treatment and may corrode or become contaminated.

KEY ELEMENTS OF COST

The following economic factors list the various cost elements associated with traditional chemical treatment that are avoided with chemical-free water treatment. This is a general assessment of what might be likely, but it may not be accurate in all situations. There is no substitute for a detailed cost analysis as part of the design process.

- *Direct Cost of Chemicals.* This item is the easiest to see and is sometimes considered the only cost. For cooling towers in the United States, this direct cost usually runs between $8.00 and $20.00 per ton ($2.27 and $5.69 per kilowatt of refrigeration) of cooling per year.
- *Water Softener.* Water softeners have direct additional costs for salt, media, equipment depreciation, maintenance, and direct labor.
- *Occupational Safety and Health Administration (OSHA) and General Environmental Requirements.* Many chemicals used to treat water systems are OSHA-listed hazardous materials. Employees in this field are required to have documented, annual training on what to do in the event of a chemical release or otherwise exposed contamination.
- *General Handling Issues.* Chemical tanks, barrels, salt bags, etc., take space. A typical chemical station requires 100 ft (9.3 m) of space.
- *Equipment Maintenance.* Lower overall maintenance for the systems as a whole may be possible.
- *Water Savings.* Cooling towers are often a facility's largest consumer of water. Most chemically controlled cooling towers operate at two to four cycles of concentration. Cycles of concentration can often be changed to six to eight cycles with chemical-free technology, with an annual reduction in water usage costs and the associated environmental impacts.
- *Energy Savings.* Energy is required to operate the pulse-powered system, but overall energy usage can be lower. The reduction or elimination of biofilm (a slime layer in a cooling tower) results in energy savings versus chemical treatment due to improved heat transfer. Biofilm has a heat transfer resistance that is four times that of scale and is also the breeding ground for *Legionella* amplification. Thus, preventing this amplification saves money.

SOURCES OF FURTHER INFORMATION

Bisbee, D. 2003. Pulse-power water treatment systems for cooling towers. Energy Efficiency & Customer Research & Development, Sacramento Municipal Utility District.

PG&E. 2002. *Codes and Standards Enhancement Report, Code Change Proposal for Cooling Towers.* California: Pacific Gas & Electric.

HPAC. 2004. Innovative grocery store seeks LEED certification. *HPAC Engineering* 27:31.

Torcellini, P.A., N. Long, and R. Judkoff. 2004. Consumptive water use for U.S. power production. *ASHRAE Transactions* 110(1):96–100.

Trane. 2005. *Trane Installation, Operation, and Maintenance Manual: Series R® Air-Cooled Rotary Liquid Chillers*, RTAA-SVX01A-EN. Lacrosse, WI: Trane Company. www.trane.com/Commercial/Uploads/Pdf/1060/RTAA-SVX01A-EN_09012005.pdf.

ASHRAE GreenTip #11-2

CHP Systems

GENERAL DESCRIPTION

Other abbreviations that have been used to describe such integrated energy systems are CCHP (includes *cooling*) and BCHP (building cooling, heating, and power). The goal, regardless of the abbreviation, is to improve system efficiencies or source fuel utilization by availing of the low-grade heat that is a by-product of the power generation process for heating and/or cooling duty. Fuel utilization efficiencies as high as 80% were reported (Adamson 2002). The resulting savings in operating costs, relative to a conventional system, are then viewed against the first cost, and simple payback periods of less than four years have been anticipated (LeMar 2002). This is particularly important, from a marketing perspective, for both the distributed-generation and the thermal equipment provider. This is because, by themselves, a microturbine manufacturer and an absorption chiller manufacturer, for example, would find it difficult to compete with a utility and an electric chiller manufacturer, respectively, as the provider of low-cost power and cooling. Last, but by no means least, the higher fossil fuel utilization rates result in reduced emissions of CO_2, the greenhouse gas with a more than 55% contribution to global warming (Houghton et al. 1990).

Gas engines, microturbines, and fuel cells have been at the center of CHP activity as the need for reliable power and/or grid independence has recently become evident. These devices are also being promoted to reduce the need for additional central-station peaking power plants. As would be expected, however, they come at a first cost premium, which can range from $1000 to $4000/kW (Ellis and Gunes 2002). At the same time, operating (thermal) efficiencies have remained in the vicinity of those of the large, centralized power plants, even after the transmission and distribution losses are taken into account. This is particularly true of engines and microturbines (which have a 25% to 35% thermal efficiency), while fuel cells promise higher efficiencies (of ~50%), albeit at the higher cost premiums ($3000 to $4000/kW).

On the thermal side, standard gas-to-liquid or gas-to-gas heat exchanger equipment can be used for the heating component of the CHP system. This transfers the heat from the exhaust gases to the process/hydronic fluid or air, respectively. For the cooling component, the size ranges of distributed power generators offer a unique advantage, in terms of flexibility, in the selection of the chiller equipment. These can be smaller-end (relative to commercial), water-lithium bromide single- or double-effect absorption chillers or larger-end (relative to residential) single-effect or generator-absorber heat exchange ammonia-water absorption chillers (Erickson and Rane 1994). Such chillers have a typical coefficient of performance of 0.7, and, as a rule of thumb, for thermal-to-electrical load matching, for every 4 kW of power generated, 1 ton of cooling may be achieved (Patnaik 2004). Figure 11-5 illustrates typical operating conditions that an absorption chiller would see with a reciprocating engine.

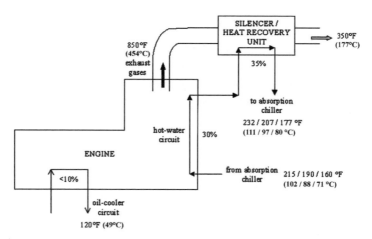

Image courtesy of Vikas Patnaik.

Figure 11-5 Schematic of CHP system consisting of a gas-fired reciprocating engine showing typical operating temperatures (Patnaik 2004).

WHEN/WHERE IT'S APPLICABLE

CHP is particularly suited for applications involving distributed power generation. Buildings requiring their own power generation, either due to a stringent power reliability and/or quality requirement or remoteness of location, must also satisfy various thermal loads (i.e., space heating or cooling, water heating, dehumidification). A conventional fossil-fuel-fired boiler and/or electric chiller can be displaced, to some extent if not entirely, by a heat recovery device (e.g., standard heat exchanger) and/or an absorption chiller driven by the waste heat from the power generator. Since the source of heating and/or cooling is waste heat that would ordinarily have been rejected to the surroundings, the operating cost of meeting the thermal demand of the building is significantly mitigated if not eliminated.

Economic analyses suggest that CHP systems are ideally suited for base-loaded distributed power generation and steady thermal (heating and/or cooling) loading. This is also the desirable mode of operation for the absorption chiller. Peak-loading is then met by utility power. Alternatively, if utility power is used for base-loading and the DG meets the peaking demand, the thermal availability may be intermittent and require frequent cycling of the primary thermal equipment (i.e., the boiler and/or chiller).

PRO

- One of the primary advantages of CHP systems is the reduction in centralized (utility) peak-load generating capacity. This is especially true, because one of the biggest contributors to summer peak loads is the demand for air conditioning. If some of this air-conditioning demand can be met by chillers fired by essentially free energy (waste heat), there is a double benefit.
- Additionally, DG-based CHP systems enable the following:
  - They bring the power generation closer to the point of application/load (i.e., distributed generation), eliminating transmission and distribution losses, etc.

- They remove or reduce the normal electric and primary fuel consumption by independent pieces of equipment providing cooling, heating, and/or dehumidification (e.g., a separate electric chiller and boiler), thereby substantially improving overall fuel utilization rates, inclusive of the power generation process.
- They remove or reduce emissions of CO_2 and other combustion by-products associated with the operation of the cooling, heating, and/or dehumidification equipment.

CON

- If the CHP system is to replace a conventional boiler/electric chiller system, the ratio of electrical to thermal load of the building must closely match the relative performance (i.e., efficiencies) of the respective equipment.
- Start-up times for absorption chillers are relatively longer than those for vapor-compression chillers, particularly when coupled to microturbines, which themselves have large time constants. Such systems would require robust and sophisticated controls that take these transients into account.

KEY ELEMENTS OF COST

The following provides a possible breakdown of the various cost elements that might differentiate a CHP system from a conventional one and an indication of whether the net cost for the alternative system is likely to be lower (L), higher (H), or the same (S). This assessment is only a perception of what might be likely, but it obviously may not be correct in all situations. There is no substitute for a detailed cost analysis as part of the design process. The listings below may also provide some assistance in identifying the cost elements involved.

First Cost

• Distributed power generator	H
• Heat recovery device/heat exchanger	L
• Absorption chiller	H
• Integrating control system	H

Recurring Costs

- Distributed power generator
  (engine/microturbine/fuel cell)   S/H/L
- Heat recovery device/heat exchanger   None
- Absorption chiller   None
- Integrating control system   H

SOURCES OF FURTHER INFORMATION

In keeping with the spirit of overlapping themes being promoted by ASHRAE, a number of technical sessions in recent meetings have been devoted to CHP, generally sponsored by technical committees on cogeneration systems (TC 1.10) and absorption/sorption heat pumps and refrigeration systems (TC 8.3). Presentations from these should be available on the ASHRAE website. The following is a link to recently sponsored programs by TC 8.3, including a couple of viewable presentations: http://tc83.ashraetcs.org/programs.html.

REFERENCES

Adamson, R. 2002. Mariah Heat Plus Power™ Packaged CHP, applications and economics. Second Annual DOE/CETC/CAN-DRA Workshop on Microturbine Applications, January 17–18, University of Maryland, College Park, MD.

Ellis, M.W., and M.B. Gunes. 2002. Status of fuel cell systems for combined heat and power applications in buildings. *ASHRAE Transactions* 108(1):1032–44.

Erickson, D.C., and M. Rane. 1994. Advanced absorption cycle: Vapor exchange GAX. ASME International Absorption Heat Pump Conference, January 19–21, New Orleans, LA.

Houghton, J.T., G.J. Jenkins, and J.J. Ephraums, eds. 1990. *Climate Change: The IPCC Scientific Assessment.* Intergovernmental Panel on Climate Change. New York: Cambridge UP.

LeMar, P. 2002. Integrated energy systems (IES) for buildings: A market assessment. Final report by Resource Dynamics Corporation for Oak Ridge National Laboratory, Contract No. DE-AC05-00OR22725.

Patnaik, V. 2004. Experimental verification of an absorption chiller for BCHP applications. *ASHRAE Transactions* 110(1):503–507.

ASHRAE GreenTip #11-3

Low-NO$_x$ Burners

GENERAL DESCRIPTION

Low-NO$_x$ burners are natural gas burners with improved energy efficiency and lower emissions of NO$_x$. When fossil fuels are burned, nitric oxide and nitrogen dioxide are produced. These pollutants initiate reactions that result in the production of ozone and acid rain. The NO$_x$ come from two sources: high-temperature combustion (thermal NO$_x$) and nitrogen bound to the fuel (fuel NO$_x$). For clean-burning fuels such as natural gas, fuel NO$_x$ generation is insignificant.

In most cases, NO$_x$ levels are reduced by lowering flame temperature. This can be accomplished by modifying the burner to create a larger (and therefore lower temperature) flame, injecting water or steam into the flame, recirculating flue gases, or limiting the excess air in the combustion process. In many cases a combination of these approaches is used. In general, reducing the flame temperature will reduce the overall efficiency of the boiler. However, recirculating flue gases and controlling the air-fuel mixture can improve boiler efficiency so that a combination of techniques may improve total boiler efficiency.

Natural-gas-fired burners with lowered NO$_x$ emissions are available for commercial and residential heating applications. One commercial/residential boiler has a burner with inserts above the individual burners; this design reduces NO$_x$ emissions by 30%. The boiler also has a wet base heat exchanger to capture more of the burner heat and reduce heat loss to flooring.

NO$_x$ production is of special concern in industrial high-temperature processes, because thermal NO$_x$ production increases with temperature. Processes include metal processing, glass manufacturing, pulp and paper mills, and cement kilns. Although natural gas is the cleanest-burning fossil fuel, it can produce emissions as high as 100 ppm or more.

A burner developed by the Massachusetts Institute of Technology and the Gas Research Institute combines staged-introduction combustion air, flue gas recirculation, and integral reburning to control NO$_x$ emissions. These improvements in burner design result in a low-temperature, fuel-rich primary zone, followed by a low-temperature, lean secondary zone; these low temperatures result in lower NO$_x$ formation.

In addition, any NO_x emission present in the recirculated flue gas is reburned, further reducing emissions. A jet pump recirculates a large volume of flue gas to the burner; this reduces NO_x emissions and improves heat transfer.

The low-NO_x burner used for commercial and residential space heating is larger in size than conventional burners, although it is designed for ease of installation.

WHEN/WHERE IT'S APPLICABLE

Low NO_x burners are best applied in regions where air quality is affected by high ground-level ozone and where required by law.

PRO

- Lowers NO_x and CO emissions, where that is an issue.
- Increases energy efficiency.

CON

- High cost.
- Higher maintenance.

KEY ELEMENTS OF COST

The following provides a possible breakdown of the various cost elements that might differentiate a low-NO_x system from a conventional one and an indication of whether the net cost for the alternative system is likely to be lower (L), higher (H), or the same (S). This is only a perception of what might be likely, but it may not be correct in all situations. There is no substitute for a detailed cost analysis as part of the design process. The listings below may provide some assistance in identifying the cost elements involved.

First Cost

- Conventional burner   L
- Low NO_x burner   H

Recurring Costs

- Maintenance   H
- Possible avoidance of pollution fines   L

SOURCE OF FURTHER INFORMATION

American Gas Association
 www.aga.org.

ASHRAE
GreenTip #11-4

Combustion Air Preheating

GENERAL DESCRIPTION

For fuel-fired heating equipment, one of the most potent ways to improve efficiency and productivity is to preheat the combustion air going to the burners. The source of this heat energy is the exhaust gas stream, which leaves the process at elevated temperatures. A heat exchanger, placed in the exhaust stack or ductwork, can extract a large portion of the thermal energy in the flue gases and transfer it to the incoming combustion air.

With natural gas, it is estimated that for each 50°F (10°C) the combustion air is preheated, overall boiler efficiency increases by approximately 1%. This provides a high leverage boiler plant efficiency measure, because increasing boiler efficiency also decreases boiler fuel usage. And, since combustion airflow decreases along with fuel flow, there is a reduction in fan-power usage as well.

There are two types of air preheaters: recuperators and regenerators. Recuperators are gas-to-gas heat exchangers placed on the furnace stack. Internal tubes or plates transfer heat from the outgoing exhaust gas to the incoming combustion air while keeping the two streams from mixing. Regenerators include two or more separate heat storage sections. Flue gases and combustion air take turns flowing through each regenerator, alternatively heating the storage medium and then withdrawing heat from it. For uninterrupted operation, at least two regenerators and their associated burners are required: one regenerator is needed to fire the furnace while the other is recharging.

WHEN/WHERE IT'S APPLICABLE

While theoretically any boiler can use combustion preheating, flue temperature is customarily used as a rough indication of when it will be cost-effective. However, boilers or processes with low flue temperatures but a high exhaust gas flow may still be good candidates and must be evaluated on a case-by-case basis. Financial justification is based on energy saved rather than on temperature differential. Some processes produce dirty or corrosive exhaust gases that can plug or attack a heat exchanger, so material selection is critical.

PRO

- Lowers energy costs.
- Increasing thermal efficiency lowers CO_2 emissions.

CON

- There are additional material and equipment costs.
- Corrosion and condensation can add to maintenance costs.
- Low specific heat of air results in relatively low U-factors and less economical heat exchangers.
- Increasing combustion temperature also increases NO_x emissions.

KEY ELEMENTS OF COST

The following provides a possible breakdown of the various cost elements that might differentiate a building with a combustion preheat system from one without and an indication of whether the net cost is likely to be lower (L), higher (H), or the same (S). This assessment is only a perception of what might be likely, but it obviously may not be correct in all situations. There is no substitute for a detailed cost analysis as part of the design process. The listings below may also provide some assistance in identifying the cost elements involved.

First Cost

- Equipment costs   H
- Controls   S
- Design fees   H

Recurring Costs

- Overall energy cost   L
- Maintenance of system   H
- Training of building operators   H

SOURCES OF FURTHER INFORMATION

DOE. 2002. *Energy Tip Sheet #1*, May. U.S. Department of Energy, Office of Industrial Technologies, Energy Efficiency, and Renewable Energy, Washington, D.C.

Fiorino, D.P. 2000. Six conservation and efficiency measures reducing steam costs. *ASHRAE Journal* 42(2):31–39.

ASHRAE GreenTip #11-5

Combination Space/Water Heaters

GENERAL DESCRIPTION

Combination space and water heating systems consist of a storage water heater, a heat delivery system (e.g., a fan coil or hydronic baseboards), and associated pumps and controls. Typically gas-fired, they provide both space and domestic water heating. The water heater is installed and operated as a conventional water heater. When there is a demand for domestic hot water, cold city water enters the bottom of the tank, and hot water from the top of the tank is delivered to the load. When there is a demand for space heating, a pump circulates water from the top of the tank through fan coils or hydronic baseboards.

The storage tank is maintained at the desired temperature for domestic hot water (e.g., 140°F [60°C]). Because this temperature is cooler than conventional hydronic systems, the space heating delivery system needs to be slightly larger than typical. Alternatively, the storage tank can be operated at a higher water temperature; this requires tempering valves to prevent scalding at the taps.

The water heater can be either a conventional storage-type water heater (either naturally venting or power vented) or a recuperative (condensing) gas boiler. Conventional water heaters have an efficiency of approximately 60%. By adding the space heating load, the energy factor increases because of longer runtimes and reduced standby losses on a percentage basis. Recuperative boilers can have efficiencies approaching 90%.

WHEN/WHERE IT'S APPLICABLE

These units are best suited to buildings that have similar space and water heating loads including dormitories, apartments, and condos. They are suited to all climate types.

PRO

- Reduces floor space requirements.
- Lowers capital cost.
- Improves energy efficiency.
- Increases tank life.

CON

- They are only available in small sizes.
- All space heating piping has to be designed for potable water.
- No ferrous metals or lead-based solder can be used.
- All components must be able to withstand prevailing city water pressures.
- Some jurisdictions require a double-wall heat exchanger for such a scheme to be acceptable.

KEY ELEMENTS OF COST

The following provides a possible breakdown of the various cost elements that might differentiate a combination space and water heating system from a conventional one and an indication of whether the net cost for the hybrid option is likely to be lower (L), higher (H), or the same (S). This assessment is only a perception of what might be likely, but it obviously may not be correct in all situations. There is no substitute for a detailed cost analysis as part of the design process. The listings below may also provide some assistance in identifying the cost elements involved.

First Cost

- Conventional heating equipment — L
- Combination space/domestic water heater — H
- Sanitizing/inspecting space heating system — H
- Piping and components able to withstand higher pressures — H
- Floor space used — L

Recurring Costs

- Heating energy — L
- Maintenance — L

SOURCES OF FURTHER INFORMATION

Sustainable Sources
    www.greenbuilder.com.

UG. 2010. Wise Energy Guide. Chatham, ON: Union Gas. www.uniongas.com/residential/energyconservation/education/wiseEnergyGuide/WEG_Booklet_Web_Final.pdf/.

U.S. Department of Energy, Energy Efficiency and Renewable Energy
www.eere.energy.gov/.

ASHRAE GreenTip #11-6

Ground-Source Heat Pumps (GSHPs)

GENERAL DESCRIPTION

A GSHP extracts solar heat stored in the upper layers of the earth; the heat is then delivered to a building. Conversely, in the summer season, the heat pump rejects heat removed from the building into the ground rather than into the atmosphere or a body of water.

GSHPs can reduce the energy required for space heating, cooling, and service water heating in commercial/institutional buildings by as much as 50%. GSHPs replace the need for a boiler in winter by utilizing heat stored in the ground; this heat is upgraded by a vapor-compressor refrigeration cycle. In summer, heat from a building is rejected to the ground. This eliminates the need for a cooling tower or heat rejection device and also lowers operating costs, because the ground is cooler than the outdoor air. (See Figure 11-6 for an example of a GSHP system.)

There are numerous types of GSHP loop systems. Each has its advantages and disadvantages. Visit the Geoexchange Geothermal Heat Pump Consortium website (www.geoexchange.org/about/how.htm) for a more detailed description of the loop options.

Water-to-air heat pumps are typically installed throughout a building with ductwork serving only the immediate zone; a two-pipe water distribution system conveys water to and from the ground-source heat exchanger. The heat exchanger field consists of a grid of vertical boreholes with plastic U-tube heat exchangers connected in parallel.

Simultaneous heating and cooling can occur throughout the building, as individual heat pumps, controlled by zone thermostats, can operate in heating or cooling mode as required.

Unlike conventional boiler/cooling, tower-type, water-loop heat pumps, the heat pumps used in GSHP applications are generally designed to operate at lower inlet water temperature. GSHPs are also more efficient than conventional heat pumps, with higher COPs and energy efficiency ratios. Because there are lower water temperatures in the two-pipe loop, piping needs to be insulated to prevent sweating. In addition, a larger circulation pump is needed because the units are slightly larger in the perimeter zones, requiring larger flows.

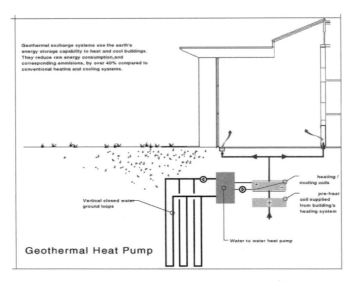

Image courtesy of Buro Happold.

Figure 11-6 Schematic example of GSHP closed-loop system.

GSHPs reduce energy use and, hence, atmospheric emissions. Conventional boilers and their associated emissions are eliminated, since no supplementary form of energy is usually required. Typically, single-packaged heat pump units have no field refrigerant connections and have significantly lower refrigerant leakage compared to central chiller systems.

GSHP units have life spans of 20 years or more. The two-pipe water-loop system typically used allows for unit placement changes to accommodate new tenants or changes in building use. The plastic piping used in the heat exchanger should last as long as the building itself.

When the system is disassembled, attention must be given to the removal and recycling of the hydrofluorocarbon (HFC) refrigerants used in the heat pumps and the antifreeze solution typically used in the ground heat exchanger.

WHEN/WHERE IT'S APPLICABLE

The most economical application of GSHPs is in buildings that require significant space/water heating and cooling over extended hours of operation. Examples are retirement communities, multi-family complexes, large office buildings, retail shopping malls, and schools. Building types not well-suited to the technology are buildings where space and water heating loads are relatively small or where hours of use are limited.

GSHPs are generally not suited for all climates, especially ones that are mostly very hot or very cold. In such climates where the total numbers of hours of cooling and heating per year are not close, the ground will not have a healthy charging/discharging cycle and its temperature will be affected over the years, resulting in system performance degradation.

PRO

- Requires less mechanical room space.
- Requires less outdoor equipment.
- Does not require roof penetrations, maintenance decks, or architectural blends.
- Relatively little operational noise.
- Reduces operation and maintenance costs.
- Requires simple controls only.
- Requires less space in ceilings.
- Loop piping, carrying low-temperature water, does not have to be insulated.
- Installation costs are lower than for many central HVAC systems.

CON

- Requires surface area for heat exchanger field.
- Higher initial cost.
- Requires additional site coordination and supervision.

KEY ELEMENTS OF COST

The following provides a possible breakdown of the various cost elements that might differentiate a GSHP system from a conventional one and an indication of whether the net cost for it is likely to be lower (L), higher (H), or the same (S). This assessment is only a perception of what might be likely, but it obviously may not be correct in all situations. There is no substitute for a detailed cost analysis as part of the design process. The listings may also provide some assistance in identifying the cost elements involved.

First Cost

- Conventional heating/cooling generators — L
- Heat pumps — H
- Outside piping system — H
- Heat exchanger field — H
- Operator training — H
- Design fees — H

Recurring Costs

- Energy cost (fossil fuel for conventional) — L
- Energy cost (electricity for heat pumps) — H
- Maintenance — L

SOURCES OF FURTHER INFORMATION

ASHRAE. 1995. *Commercial/Institutional Ground-Source Heat Pump Engineering Manual.* Atlanta: ASHRAE.

Earth Energy Society of Canada
www.earthenergy.ca.

Kavanaugh, S.P., and K. Rafferty. 1997. *Ground-Source Heat Pumps: Design of Geothermal Systems for Commercial and Institutional Buildings.* Atlanta: ASHRAE.

Natural Resources Canada, RETScreen
(software for renewable energy analysis)
www.retscreen.net.

ASHRAE GreenTip #11-7

Water-Loop Heat Pump Systems

GENERAL DESCRIPTION

A water-loop heat pump system consists of multiple water-source heat pumps serving local areas within a building that are tied into a neutral-temperature (usually 60°F to 90°F [15.5°C to 32°C]) water loop that serves as both heat source and heat sink. The loop is connected to a central heat source (e.g., a small boiler) and a central heat dissipation device (e.g., a closed-circuit evaporative condenser or open-circuit cooling tower isolated from the building loop via the heat exchanger). These operate to keep the temperature of the loop water within range.

The water-source heat pump itself is an electric-driven, self-contained, water-cooled heating and cooling unit with a reversible refrigerant cycle (i.e., a water-cooled air-conditioning unit that can run in reverse). Its components include a heat exchanger, heating/cooling coil, compressor, fan, and reversing controls, all in a common casing. The heat exchanger and coil are designed to accept hot and cold refrigerant liquid or gas. The units can be located either within the space (e.g., low, along the outside wall) or remotely (e.g., in a ceiling plenum or in a separate nearby mechanical room).

Piping all of the water-to-refrigerant heat exchangers together in a common loop yields what is essentially an internal source heat recovery system. In effect, the system is capable of recovering heat energy (through the cooling process) and redistributing it where it is needed.

During the cooling mode, heat energy is extracted from room air circulated across the coil (just like a room air conditioner) and rejected to the water loop. In this mode, the unit's heat exchanger acts as a condenser and the coil as an evaporator. In the heating mode, the process is reversed: specifically, a reversing valve allows the heat exchanger to function as the evaporator and the coil as the condenser so that heat extracted from the water loop is rejected to the air being delivered to the occupied space, thus heating the space.

In addition to the components mentioned above, the system includes equipment and specialties normally associated with a closed hydronic system (e.g., pumps, filters, air separator, expansion tank, makeup system, etc.).

WHEN/WHERE IT'S APPLICABLE

A water-source heat pump system is well qualified for applications where simultaneous heating and cooling needs/opportunities exist. (An example might be a building where, in certain seasons, south-side or interior rooms need cooling at the same time north-side rooms require heating.) Appropriate applications may include office buildings, hotels, schools, apartments, extended care facilities, and retail stores.

The system's characteristics may make it particularly suitable when a building is to be air conditioned in stages, perhaps due to cost constraints. Once the basic system is in, additional heat pumps can be added as needed and tied into the loop. Further, since it uses low-temperature water, this system is an ideal candidate for mating with a hydronic solar collection system (since solar hydronic systems are more efficient when generating lower water temperatures).

PRO

- It can make use of energy that would otherwise be rejected to the atmosphere.
- Loop piping, carrying low-temperature water, does not have to be insulated.
- When applied correctly, the system can save energy. (Note: that some factors tend to decrease energy cost, and some tend to increase it, which prevails will determine whether savings result.)
- It is quieter than a system utilizing air-cooled condensers (e.g., through-the-wall room air conditioners).
- Failure of one heat pump unit does not affect others.
- It can condition (heat or cool) local areas of a building without having to run the entire system.

CON

- Multiple compressors located throughout a building can be a maintenance concern because of their being noncentralized and sometimes difficult to access (e.g., above the ceiling).
- Effective water filtration is critical to proper operation of heat exchangers.

- There is an increased potential for noise within the conditioned space from heat pump units.
- Some of the energy used in the heating cycle is derived from electricity (used to drive the heat pump compressors), which may be more expensive than energy derived from fossil fuel.

KEY ELEMENTS OF COST

The following provides a possible breakdown of the various cost elements that might differentiate a water-loop heat pump system from a conventional one and an indication of whether the net incremental cost for the alternative option is likely to be lower (L), higher (H), or the same (S). This assessment is only a perception of what might be likely, but it obviously may not be correct in all situations. There is no substitute for a detailed cost analysis as part of the design process. The listings below may also provide some assistance in identifying the cost elements involved.

First Cost

- Equipment costs (will vary depending on what type of conventional system would otherwise be used)   S/L
- Controls   S
- Design fees   S

Recurring Costs

- Overall energy costs   L
- Maintenance of system   H
- Training of building operators   H

SOURCES OF FURTHER INFORMATION

Tri-State Generation and Transmission Association, Inc. www.tristategt.org/.

Trane. 1994. *Trane Water-Source Heat Pump System Design Application Engineering Manual*, SYS-AM-7. Lacrosse, WI: Trane Company.

ASHRAE GreenTip #11-8

TES for Cooling

GENERAL DESCRIPTION

There are several suitable media for storage of cooling energy, including the following:

- Chilled water
- Ice
- Calcium chloride solutions (brine)
- Glycol solutions
- Concentrated desiccant solutions

Active thermal storage systems use a building's cooling equipment to remove heat, usually at night, from an energy storage medium for later use as a source of cooling. The most common energy storage media are ice and chilled water. These systems decouple the production of cooling from the demand for cooling, (i.e., the plant output does not have to match the instantaneous building cooling load). This decoupling increases flexibility in design and operations, thereby providing an opportunity for a more efficient air-conditioning system than with an alternative that does not include storage. Before applying active thermal storage, however, the design cooling load should be minimized.

Although many operating strategies are possible, the basic principle of a TES system is to reduce peak building cooling loads by shifting a portion of peak cooling production to times when the building cooling load is lower. Energy is typically charged, stored, and discharged on a daily or weekly cycle. The net result is an opportunity to run a chiller plant at peak efficiency during the majority of its operating period. A system without storage, on the other hand, has to follow the building cooling load, and the majority of its operation is at part-load conditions. Part-load operation of chiller plants comes at the expense of efficiency.

Several buildings have demonstrated site energy reductions with the application of TES, as discussed in both the "Pro" and "Sources of Further Information" sections that follow.

In addition to the potential for site energy reduction, operation of TES systems can reduce energy resource consumption. This reduction is due to a shift toward using energy during periods of low aggregate electric utility demand. As a result, transmission and distribution losses are lower and power plant generating efficiencies can be higher because the load is served by base-load plants. Thermal storage can also have beneficial effects on CHP systems by flattening thermal and electric load profiles.

The ASHRAE *Design Guide for Cool Thermal Storage* (Dorgan and Elleson 1993) covers cool storage application issues and design parameters in some detail.

WHEN/WHERE IT'S APPLICABLE

TES systems tend to perform well in situations where there is variability in loads. Successful applications of TES systems have included commercial office buildings, schools, worship facilities, convention centers, hotels, health care facilities, industrial processes, and turbine inlet air cooling.

PRO

- Because TES allows downsizing the refrigeration system, the resulting cost savings (which may include avoiding having to add such equipment on an existing project) may substantially or entirely cover the added incremental cost of the storage system proper (also see the first con in the next section). However, if the first cost is more than another design option, there are still life-cycle cost benefits due to a significant reduction in utility costs.
- The addition of a TES system allows the size of refrigerating equipment to be reduced, since it will have to meet an average cooling load rather than the peak cooling load. Reduced refrigeration equipment size means less on-site refrigerant usage and a lower probability of environmental impacts due to direct effects.
- Because TES allows operation of the refrigeration system at or near peak efficiency during all operating hours, the annual energy usage may be lower than systems without storage that must operate at lower part-load ratios to meet instantaneous loads. In addition, since off-peak hours are usually at night when

lower ambient temperatures prevail, lower condensing temperatures required for heat rejection would tend to increase refrigeration efficiency. A number of carefully documented examples of energy savings can be found in the literature, including Bahnfleth and Joyce (1994), Fiorino (1994), and Goss et al. (1996).
- Because TES systems shift the consumption of site energy from on-peak to off-peak periods, the total energy resources required to deliver cooling to the facility will be lower (Reindl et al. 1995; Gansler et al. 2001). In addition, in some electric grids, the last generation plants to be used to meet peak loads may be the most polluting per kW of energy produced (Gupta 2002). In such cases, emissions would be further reduced by the use of TES.
- TES enables the practical incorporation of other high-efficiency technologies such as cold-air distribution systems and nighttime heat recovery.
- TES can be effective at preventing or delaying the need to construct additional power generation and transmission equipment.
- Liquid desiccant can be circulated in plastic pipes and does not need insulation.

CON

- Compared to a conventional system, the thermal storage element proper (i.e., the water tank or ice tank) and any associated pumping, piping accessories, and controls add to the incremental capital cost. If the system's refrigeration equipment can be reduced in size sufficiently (see the first pro listed in the previous section), this burden may be mitigated substantially or balanced out.
- The need to generate cooling at evaporator temperatures lower than conventional ones tends to decrease refrigeration efficiency. This reduction may be overcome, however, by factors that increase efficiency (see the third pro).
- Successful TES systems require additional efforts in the design phase of a project.
- TES systems will require increased site space usage. The impact of site space usage can be mitigated by considering ice storage technologies.

- Because a thermal storage system departs from the norm of system operation, continued training of facility operations staff is required, as are procedures for propagating system knowledge through a succession of facilities personnel.
- Ice requires special control of the melt rate to prevent uneven melting and to maximize performance.
- Calcium chloride brine needs management to prevent corrosion. Glycol needs management to prevent corrosion and toxicity.
- Liquid desiccant needs small resistant heating to be above 77°F (25°C) to prevent crystallization (similar to compressor sumps to prevent condensation).

KEY ELEMENTS OF COST

The following provides a possible breakdown of the various cost elements that might differentiate a TES system above from a conventional one and an indication of whether the net cost for the alternative option is likely to be lower (L), higher (H), or the same (S). This assessment is only a perception of what might be likely, but it obviously may not be correct in all situations. There is no substitute for a detailed cost analysis as part of the design process. The listings below may also provide some assistance in identifying the cost elements involved.

First Cost

- Storage element (e.g., chilled water, ice, glycol, and brine tanks) (Desiccant cost is higher than chilled water and brine but similar to glycol)   H
- Additional pumping/piping re-storage element   H
- Chiller/heat rejection system   L
- Controls   H
- Electrical (regarding chiller/heat rejection system)   S/L
- Design fees   H
- Operator training   H
- Commissioning   S/H
- Site space   H

Recurring Costs

- Electric energy — L
- Gas supply with low electrical demand — H
- Operator training (ongoing) — H
- Maintenance training — L

SOURCES OF FURTHER INFORMATION

Bahnfleth, W.P., and W.S. Joyce. 1994. Energy use in a district cooling system with stratified chilled-water storage. *ASHRAE Transactions* 100(1):1767–78.

CEC. 1996. Source energy and environmental impacts of thermal energy storage. Completed by Tabors, Caramanis & Assoc. for the California Energy Commission.

Dorgan, C., and J.S. Elleson. 1993. *Cool Storage Design Guide*. Atlanta: ASHRAE.

Duffy, G. 1992. Thermal storage shifts to saving energy. *Eng. Sys.* July/August: 32–38.

Elleson, J.S. 1996. *Successful Cool Storage Projects: From Planning to Operation*. Atlanta: ASHRAE.

Fiorino, D.P. 1994. Energy conservation with stratified chilled-water storage. *ASHRAE Transactions* 100(1):1754–66.

Galuska, E.J. 1994. Thermal storage system reduces costs of manufacturing facility (Technology Award case study). *ASHRAE Journal*, March.

Gansler, R.A., D.T. Reindl, T.B. Jekel. 2001. Simulation of source energy utilization and emissions for HVAC systems. *ASHRAE Transactions* 107(1):39–51.

Goss, J.O., L. Hyman, and J. Corbett. 1996. Integrated heating, cooling and thermal energy storage with heat pump provides economic and environmental solutions at California State University, Fullerton. *Proceedings of the EPRI International Conference on Sustainable Thermal Energy Storage, Minneapolis, MN*, pp. 163–67.

Lawson, S.H. 1988. Computer facility keeps cool with ice storage. *HPAC* 60(88):35–44.

Mathaudhu, S.S. 1999. Energy conservation showcase. *ASHRAE Journal* 41(4):44–46.

O'Neal, E.J. 1996. Thermal storage system achieves operating and first-cost savings (Technology Award case study). *ASHRAE Journal* 38(4).

Reindl, D.T., D.E. Knebel, and R.A. Gansler. 1995. Characterizing the marginal basis source energy and emissions associated with comfort cooling systems. *ASHRAE Transactions* 101(1):1353–63.

ASHRAE GreenTip #11-9

Double-Effect Absorption Chillers

GENERAL DESCRIPTION

Chilled-water systems that use fuel types other than electricity can help offset high electric prices, whether those high prices are caused by consumption or demand charges. Absorption chillers use thermal energy (rather than electricity) to produce chilled water. A double-effect absorption chiller using high-pressure steam (115 psig [793 kPa]) has a COP of approximately 1.20. Some double-effect absorption chillers use medium-pressure steam (60 psig [414 kPa]) or 350°F to 370°F (177°C to 188°C) hot water, but with lower efficiency or higher cost.

Double-effect absorption chillers are available from several manufacturers. Most are limited to chilled-water temperatures of 40°F (4.3°C) or above, since water is the refrigerant. The interior of the chiller experiences corrosive conditions; therefore, the manufacturer's material selection is directly related to the chiller life. The more robust the materials, the longer the life.

WHEN/WHERE IT'S APPLICABLE

Double-effect absorption chillers can be used in the following applications:

- When natural gas prices (used to produce steam) are significantly lower than electric prices.
- When the design team and building owner wish to have fuel flexibility to hedge against changes in future utility prices.
- When there is steam available from an on-site process; an example is steam from a turbine.
- When a steam plant is available but lightly loaded during the cooling season. Many hospitals have large steam plants that run at extremely low loads and low efficiency during the cooling season. By installing an absorption chiller, the steam plant efficiency can be increased significantly during the cooling season.
- At sites that have limited electric power available.
- In locations where district steam is available at a reasonable price (e.g., New York City).

PRO

- Reduces electric charges.
- Allows fuel flexibility, since natural gas, No. 2 fuel oil, propane, or waste steam may be used to supply thermal energy for the absorption chiller.
- Uses water as the refrigerant, making it environmentally friendly.
- Allows system expansion even at sites with limited electric power.
- When the system is designed and controlled properly, it allows versatile use of various power sources.

CON

- Cost of an absorption chiller will be roughly double that of an electric chiller of the same capacity, as opposed to 25% more for a single-effect absorption machine.
- Size of an absorption chiller is larger than an electric chiller of the same capacity.
- Although absorption chiller efficiency has increased in the past decade, the amount of heat rejected is significantly higher than that of an electric chiller of similar capacity. This requires larger cooling towers, condenser pipes, and cooling tower pumps.
- Few plant operators are familiar with absorption technology.

KEY ELEMENTS OF COST

The following provides a possible breakdown of the various cost elements that might differentiate an absorption chiller system from a conventional one and an indication of whether the net cost for the hybrid option is likely to be lower (L), higher (H), or the same (S). This assessment is only a perception of what might be likely, but it obviously may not be correct in all situations. There is no substitute for a detailed cost analysis as part of the design process. The listings below may also provide some assistance in identifying the cost elements involved.

First Cost

- Absorption chiller   H
- Cooling tower and associated equipment   H

- Electricity feed   L
- Design fees   H
- System controls   H

Recurring Costs

- Electric costs   L
- Chiller maintenance   S
- Training of building operators H

SOURCES OF FURTHER INFORMATION

ASHRAE. 2008. *ASHRAE Handbook—HVAC Systems and Equipment*, Chapter 50. Atlanta: ASHRAE.

ASHRAE. 2010. *ASHRAE Handbook—Refrigeration*, Chapter 43. Atlanta: ASHRAE.

Trane Co. 1999. *Trane Applications Engineering Manual, Absorption Chiller System Design*, SYS-AM-13. Lacrosse, WI: Trane Company.

ASHRAE GreenTip #11-10

Gas Engine-Driven Chillers

GENERAL DESCRIPTION

Chilled-water systems that use fuel types other than electricity can help offset high electric prices, regardless of whether those high prices are caused by consumption or demand charges. Gas engines can be used in conjunction with electric chillers to produce chilled water. Depending on chiller efficiency, a gas engine-driven chiller may have a cooling COP of 1.6 to 2.3.

Some gas engines are directly coupled to a chiller's shaft. Another option is to use a gas engine and switchgear. In such cases, the chiller may be operated either by using electricity from the engine or from the electric utility.

WHEN/WHERE IT'S APPLICABLE

A gas engine is applicable in the following circumstances:

- When natural gas prices are significantly lower than electric prices.
- When the design team and building owner wish to have fuel flexibility to hedge against changes in future utility prices.
- At sites that have limited electric power available.

PRO

- Reduces electric charges.
- Allows fuel flexibility if installed as a hybrid system (i.e., part gas engine and part electric chiller, so the plant may use either gas engine or electricity from utility).
- Allows system expansion even at sites with limited electric power.
- When the system is designed and controlled properly, these chillers allow for use of various fuel sources.
- May be used in conjunction with an emergency generator if switchgear is provided.

CON

- Added cost of gas engine.
- Additional space required for engine.

- Due to the amount of heat rejected being significantly higher than for similar capacity electric chiller, larger cooling towers, condenser pipes, and cooling tower pumps may be required.
- Site emissions are increased.
- Noise from engine may need to be attenuated, both inside and outside.
- Significant engine maintenance costs.

KEY ELEMENTS OF COST

The following provides a possible breakdown of the various cost elements that might differentiate a gas engine-driven chiller from a conventional one and an indication of whether the net cost is likely to be lower (L), higher (H), or the same (S). This assessment is only a perception of what might be likely, but it obviously may not be correct in all situations. There is no substitute for a detailed cost analysis as part of the design process. The listings below may also provide some assistance in identifying the cost elements involved.

First Cost

• Gas engine	H
• Cooling tower and associated equipment	H
• Electricity feed	L
• Site emissions	H
• Site acoustics	H
• Design fees	H
• System controls	H

Recurring Costs

• Electric costs	L
• Engine maintenance	H
• Training of building operators	H
• Emissions costs	H

SOURCE OF FURTHER INFORMATION

NBI. 1998. *Gas Engine Driven Chillers Guideline*. Fair Oaks, CA: New Buildings Institute and Southern California Gas Company. http://newbuildings.org/sites/default/files/AbsorptionChiller-Guideline.pdf.

ASHRAE GreenTip #11-11

Desiccant Cooling and Dehumidification

GENERAL DESCRIPTION

There are two basic types of open-cycle desiccant process: solid and liquid desiccant. Each of these processes has several forms, and these should be investigated to determine the most appropriate for the particular application. Both of these systems need to have the air being conditioned come in good contact with the desiccant, during which moisture is absorbed from the air and the temperatures of both the air and desiccant are coincidently raised.

The moisture absorption process is caused by desiccant having a lower surface vapor pressure than the air. As the temperature of the desiccant rises, its vapor pressure rises, and its useful absorption capability lessens. Some systems, particularly liquid types, have cooling of air and desiccant coincident with dehumidification. This can allow the need for less space and equipment.

The dehumidified air then has to be cooled by other means. Two supply air arrangements are available. One method uses a mixture of recycled return air and dehumidified outdoor ventilation air as supply air to the building. Moisture and contaminants such as volatile organic compounds can be absorbed by the desiccant and recycled; particles of solid or liquid desiccant may also be carried over into the ducts and to building occupants.

The other arrangement combines energy recovery from building exhaust air that is typically much cooler and less humid than outdoor ventilation air. By dehumidifying the exhaust air to a sufficiently low humidity ratio (i.e., moisture content), it can be used to indirectly cool outdoor air for supply to the building that has not contacted desiccant. Using the recovered energy, this arrangement can be used to process the total ventilation air requirement, even up to 100% from outside.

So that desiccant can be reused, it has to be re-dried by a heating process generally called *reactivation* or *regeneration* (for solid types) or *reconcentration* (for liquid types). The re-drying can be either direct (by contact with heated outdoor air with the desiccant) or indirect. Indirect may be preferable, particularly in high humid climates, because of the higher temperature needed to maximize the vapor difference and drying potential. The energy storage benefit for liquid desiccant was discussed in GreenTip #11-8.

Rotary desiccant dehumidifiers use solid desiccants such as silica gel to attract water vapor from the moist air. Humid air, generally referred to as *process air*, is dehumidified in one part of the desiccant bed while a different part of the bed is dried for reuse by a second airstream known as *reactivation air*. The desiccant rotates slowly between these two airstreams, so that dry, high-capacity desiccant leaving the reactivation air is available to remove moisture from the moist process air.

Process air that passes through the bed more slowly is dried more deeply, so for air requiring a lower dew point, a larger unit (and slower velocity) is required. The reactivation air inlet temperature changes the outlet moisture content of the process air. In turn, if the designer needs dry air, it is generally more economical to use high reactivation temperatures. On the other hand, if the leaving humidity need not be especially low, inexpensive, low-grade heat sources (e.g., waste heat or rejected cogeneration heat) can be used.

The process air outlet temperature is higher than the inlet temperature primarily because the heat of sorption of the moisture removed is converted to sensible heat. The outlet temperature rises roughly in proportion to the amount of moisture that is removed. In most comfort applications, provisions must be made to remove excess sensible heat from the process air following reactivation. Cooling is accomplished with cooling coils, and the source of this cooling affects the operating economics of the system.

WHEN/WHERE IT'S APPLICABLE

In general, applications that require a dew point at or below 40°F (4.3°C) may be candidates for active desiccant dehumidification. Examples of such candidates include facilities handling hygroscopic materials; film drying; the manufacturing of candy, chocolate, or chewing gum; the manufacturing of drugs and chemicals; the manufacturing of plastic materials; packaging of moisture-sensitive products; and the manufacturing of electronics. Supermarkets often use desiccant dehumidification to avoid condensation on refrigerated casework. And when there is a need for a lower dew point and a convenient source of low-grade heat for reactivation is available, rotary desiccant dehumidifiers can be especially economical.

PRO

- Desiccant equipment tends to be very durable.
- Often this is the most economical means to dehumidify below a 40°F (4.3°C) dew point.
- It eliminates condensate in the airstream, in turn, limiting the opportunity for mold growth.

CON

- Desiccant usually must be replaced, replenished, or reconditioned every five to ten years.
- In comfort applications, simultaneous heating and cooling may be required.
- The process is not especially intuitive, and the controls are relatively complicated.

KEY ELEMENTS OF COST

The following provides a possible breakdown of the various cost elements that might differentiate a building with a rotary desiccant dehumidification system from one without and an indication of whether the net cost is likely to be lower (L), higher (H), or the same (S).

First Cost

- Equipment costs   H
- Regeneration (heat source and supply)   H
- Ductwork   S
- Controls   H
- Design fees   S

Recurring Costs

- Overall energy cost   S/H
- Maintenance of system   H
- Training of building operators   H
- Filters   H

SOURCE OF FURTHER INFORMATION

ASHRAE. 2012. *ASHRAE Handbook—HVAC Systems and Equipment*. Atlanta: ASHRAE.

ASHRAE GreenTip #11-12

Indirect Evaporative Cooling

GENERAL DESCRIPTION

Evaporative cooling of supply air can be used to reduce the amount of energy consumed by mechanical cooling equipment. Two general types of evaporative cooling—direct and indirect—are available. The effectiveness of either of these methods is directly dependent on the extent that dry-bulb temperature exceeds wet-bulb temperature in the supply airstream.

Direct evaporative cooling introduces water directly into the supply airstream, usually with a spray or wetted media. As the water absorbs heat from the air, it evaporates. While this process lowers the dry-bulb temperature of the supply airstream, it also increases the air moisture content.

Two forms of indirect evaporative cooling (IEC) are described below.

1. **Coil/cooling Tower IEC.** This type uses an additional water-side coil to lower supply air temperature. The added coil is placed ahead of the conventional cooling coil in the supply airstream and is piped to a cooling tower where the evaporative process occurs. Because evaporation occurs elsewhere, this method of precooling does not add moisture to the supply air, but it is somewhat less effective than direct evaporative cooling. A conventional cooling coil provides any additional cooling required.

2. **Plate Heat Exchanger IEC.** This is composed of sets of parallel plates arranged into two sets of passages separated from each other. In a typical arrangement, exhaust air from a building is passed through one set of passages, during which it is wetted by water sprays. A stream of outdoor air is coincidently passed through the other set of passages and is cooled by heat transfer through the plates by the wetted exhaust air before being supplied to the building. Alternatively, the exhaust air may be replaced by a second stream of outdoor air. The wetted air is reduced in dry-bulb temperature to be close to its wet-bulb temperature. The stream of dry air is cooled to be close to the dry-bulb temperature

of the wetted exhaust air. In some applications, the cooled stream of outdoor air is passed through the coil of a direct expansion refrigeration unit, where it is further cooled before being supplied to the building. This system is an efficient way for an all outdoor air supply system. The plates in the heat exchanger can be formed from various metals and polymers. Consideration needs to be given to preventing the plate material from corroding.

WHEN/WHERE IT'S APPLICABLE

This may be used in climates with low wet-bulb temperatures where significant amounts of cooling are available. In such climates, the size of the conventional cooling system can also be reduced.

In more humid climates, indirect evaporative cooling can be applied during nonpeak seasons. It is especially applicable for loads that operate 24 h/day for many days of the year.

PRO

- Indirect evaporative cooling can reduce the size of the conventional cooling system.
- It reduces cooling costs during periods of low wet-bulb temperature.
- It does not add moisture to the supply airstream (in contrast, direct evaporative cooling does add moisture).
- It may be designed into equipment such as self-contained units.
- There is no cooling tower or condenser piping in the plate heat exchanger IEC that is described.

CON

- Air-side pressure drop (typically 0.2 to 0.4 in. of water [50 to 100 Pa]) increases due to an additional coil in the airstream.
- To make water cooler in the coil/cooling tower IEC, the cooling tower fans operate for longer periods of time and consume more energy.
- For the coil/cooling tower IEC, condenser piping and controls must be accounted for during the design process.

KEY ELEMENTS OF COST

The following provides a possible breakdown of the various cost elements that might differentiate an IEC system from a conventional one and an indication of whether the net cost for the hybrid option is likely to be lower (L), higher (H), or the same (S). This assessment is only a perception of what might be likely, but it obviously may not be correct in all situations. There is no substitute for a detailed cost analysis as part of the design process. The listings below may also provide some assistance in identifying the cost elements involved.

First Cost

- Indirect cooling coil   H
- Decreased conventional cooling system capacity   L
- Condenser piping, valves, and control   H

Recurring Costs

- Cooling system operating cost   L
- Supply fan operating cost   H
- Tower fan operating cost   H
- Maintenance of indirect coil   S

SOURCES OF FURTHER INFORMATION

ASHRAE. 2011. *ASHRAE Handbook—HVAC Applications*, pp. 52.2–5. Atlanta: ASHRAE.

ASHRAE. 2013. *ASHRAE Handbook—HVAC Systems and Equipment*, pp. 4.4; 41.3–5. Atlanta: ASHRAE.

ASHRAE GreenTip #11-13

Condensing Boilers

GENERAL DESCRIPTION

Condensing boilers are heat-producing equipment that recover heat from their wasted flue gases, cooling them to a point where water condensation occurs.

The flue of any boiler usually contains carbon dioxide, nitrous oxides, and water vapor with a dew-point temperature of about 138°F–140°F (58.8°C–60°C). Any return water temperature to the boiler less than this temperature will result in water vapor condensation (and hence corrosion). Conventional boilers avoid this problem by limiting return water temperatures and keeping the flue temperature hot enough. (Conventional boilers have flue temperatures in the range of 250°F–350°F [121°C–176°C]). This results in about 10%–15% of the heat produced by conventional boilers being lost through the flue, making their annual fuel utilization efficiency (AFUE) in the range of 70%–80%.

Condensing boilers, on the other hand, are designed to handle the corrosive nature of condensing water vapor. They usually include a second heat exchanger or an oversized single heat exchanger that will extract heat from the flue, thus cooling it below the dew-point temperature. Water vapor in the flue will condense and mix with the carbon dioxide and nitrous oxides to form carbonic and nitric acid, respectively. Flue temperatures of condensing boilers are generally in the 120°F–140°F (48.8°C–60°C) range. Return water temperatures to a condensing boiler are theoretically not bound by any minimum limit. However, typical values range from 130°F down to 80°F (54.4°C down to 26.6°C). The lower the return water temperature, the more condensation will occur and the more efficient the condensing boiler will be. Typical AFUEs of condensing boilers are in the 90%–98% range.

WHEN/WHERE IT'S APPLICABLE

Condensing boilers are widely used in Europe where certain countries, like the UK, have mandated efficiency levels that can only be met by them. The U.S. market, although late in adoption, is currently showing a tendency to outrun Europe in the efficiency race.

Although condensing boilers can theoretically be used on any new system or to retrofit any existing system, they are generally ideal and cost effective for low-temperature applications such as radiant floor heating, swimming pool heating, and water source heat pumping.

PROS AND CONS

Pro

- Very high efficiency
- Lower CO_2 and NO_x emissions due to lower combustion temperatures and the dilution of some of these gases with the condensing water
- Can accommodate a wider range of supply temperatures without the need for mixing valves
- No warm-up cycle required since the boiler can accommodate lower return water temperatures
- Can be coupled with other low-temperature equipment such as a heat recovery chiller
- Federal tax rebate (depending on year) for residential application. State and local incentives for residential and light commercial applications in many other jurisdictions

Con

- Higher capital cost
- A drain discharge is required which is generally corrosive
- Stack material must be able to withstand the corrosive nature of the flue
- In very cold climates, care should be exercised in running the flue stack outdoors to avoid freezing
- In retrofit applications, high-temperature coils may need to be upsized
- Lower equipment life

KEY ELEMENTS OF COST

The following provides a possible breakdown of the various cost elements that might differentiate a condensing boiler system from a conventional one and an indication of whether the net cost for an option is likely to be lower (L), higher (H), or the same (S). This assessment is only a perception of what might be likely, but it obviously may not be correct in all situations. There is no substitute for a detailed cost analysis as part of the design process. The listings below may also provide some assistance in identifying the cost elements involved.

First Cost

- Boiler — H
- Flue Stack — S/H
- Additional plumbing requirements — H

Recurring Cost

- Energy Consumption — L
- Testing and balancing (TAB) — S
- Maintenance — S/H
- Commissioning — S

SOURCES OF FURTHER INFORMATION

ASHRAE. 2012. *ASHRAE Handbook—HVAC Systems and Equipment*. Chapter 32, "Boilers." Atlanta: ASHRAE.

CHAPTER TWELVE

ENERGY/WATER SOURCES

RENEWABLE/NONRENEWABLE ENERGY SOURCES

There are often discussions about using renewable energy sources as a way to power the world, but little is done to actually implement it. This chapter focuses on ways to use renewable energy to offset nonrenewable energy sources. By definition, a renewable energy source (RES) is a fuel source that can be replenished in a short amount of time. The common renewable sources are solar, wind, hydro, and biomass. This is in contrast to the common nonrenewable energy sources such as coal, oil, natural gas, and nuclear.

The use of renewable energy is separated between using the energy source on site versus paying for the renewable resource. Today, many utility companies offer the ability to purchase renewable energy mixes from their generation portfolio. These utilities generate their own renewable energy with either large-scale wind or solar facilities. Note that quite often, large-scale hydroelectric plants are not considered a renewable resource because of the size of the environmental impact of such facilities. *Green energy*, as it is sometimes called, can also be purchased by third-party resellers of energy. The concept is that renewable energy is put into the utility grid and you, as the end user, can purchase that power somewhere else on the grid. The bottom line is that you can, in most areas, purchase green power, and you are not limited by the default utility offering. According to the U.S. Department of Energy's (DOE) National Renewable Energy Lab (NREL), "in 2009 total utility green power sales exceeded 6 billion kWh, a 60% increase since 2006. More than 850 utilities are participating in programs nationwide" (NREL 2010). The U.S. Environmental Protection Agency (EPA) publication, *Guide to Purchasing Green Power*, is a good overview of the green power market and steps to take for participating.

According to Europe's Energy Portal, in the European Union (EU), renewable energy contributed 10.3% of the total energy consumption in 2008 (www.energy.eu/#renewable). More than 4 million European consumers have already switched to green power.

To further strengthen the exploitation of RES, the EU officially adopted a 20-20-20 Renewable Energy Directive on December 17, 2008, setting the following ambitious targets for 2020: cutting greenhouse gas emissions by at least 20% of 1990 levels; cutting energy consumption by 20% through improved energy efficiency; and boosting the use of RES to 20% of total energy production (currently at about 8.5%). Under the terms of the directive, for the first time each member state has a legally binding renewable energy target for 2020 and by June 2010 each state was to have drawn up a National Action Plan detailing plans to meet their 2020 targets (EU 2008).

The designer has little say over how energy at the source will be provided. Electricity may be generated from coal, imported oil, natural gas, or uranium, and each source may have broad implications for national or industry interests, the environment, and/or economics. However, there is little a designer can do about it—at least in choosing between conventional nonrenewable energy sources. This subject is addressed in more detail in the Chapter 34 of the 2013 *ASHRAE Handbook—Fundamentals* (ASHRAE 2013), and the designer is referred to that source for more specific data.

What designers do have influence over is creating a building that is designed to minimize energy consumption, which is the focus of most of this book. Passive solar techniques—that is, those with minimal moving parts—can be incorporated into the design of a building. Daylighting, trombe walls, passive cooling, and natural ventilation are all methods for using the natural environment to help heat, light, and cool the building.

Other techniques can be used, such as solar water heating, solar ventilation preheat, photovoltaic (PV) systems, and wind systems, although these active systems tend to be a little more complicated to integrate into a building's operation.

In many cases, renewable energy can be considered free. The issue is that this free energy source usually needs some capital equipment to concentrate the diffuse nature of renewable energy into a useful form for the building. One way to illustrate this characteristic is to think of a gallon jug of, say, fuel oil compared to an array of hydronic solar collectors. The fuel oil could provide hot water, on demand, for hours in a simple water heater in a corner of a boiler room, but the fuel oil will consume fossil fuels and produce emissions. The equivalent job done by solar collectors would require an array of collectors on a roof, plus a tank, piping, and some controls (a simple thermosiphon-type solar collector can operate without controls) but would produce no emissions and would operate with free solar energy.

While consideration of renewables is a highly touted element of green design, the design team should be well aware of the key characteristics of whichever renewable is being considered and develop creative strategies accordingly. In the next sections, we focus on two main renewable energy sources (other than hydropower) in the world today.

SOLAR

Solar energy is the primary energy source that fuels the growth of the Earth's natural capital and drives wind and ocean currents that also can provide alternative energy sources. Since the beginning of time, solar energy has been successfully harnessed for human use. Early civilizations and some modern ones used solar energy for many purposes: food and clothes drying, heating water for baths, heating adobe and stone dwellings, etc. Solar energy is free and available to anyone who wishes to use it.

Solar thermal heating for domestic hot-water and space heating has grown considerably over the years and is well established in several countries. Global installed capacity is estimated at about 348×10^9 Btu/h (102 gigawatts of thermal energy [GW_{th}]) for glazed flat-plate (144×10^9 Btu/h [42.2 GW_{th}] or 649×10^6 ft² [60.3 $\times 10^6$ m²]) and evacuated-tube collectors (204×10^9 Btu/h [59.9 GW_{th}] or 921.4×10^6 ft² [85.6 $\times 10^6$ m²]), at 84×10^9 Btu/h (24.5 GW_{th}) for unglazed plastic collectors (376.7×10^6 ft² [35×10^6 m²]), and 4.1×10^9 Btu/h (1.2 GW_{th}) (17.2×10^6 ft² [1.6×10^6 m²]) for glazed and unglazed air collectors (Weiss et al. 2008). In North America (i.e., United States and Canada), swimming pool heating is dominant with an installed capacity of about 66.9×10^9 Btu/h (19.6 GW_{th}) of unglazed plastic collectors. In other countries, flat-plate and evacuated-tube collectors are used to generate hot water for sanitary use and space heating. These countries include China and Taiwan (225×10^9 Btu/h [66 GW_{th}]), Europe (51×10^9 [15 GW_{th}]), and Japan (17×10^9 Btu/h [5 GW_{th}]).

According to the European Solar Thermal Industry (ESTIF 2010), at the end of 2009, the total capacity of glazed collector area in operation in the European Union reached 73.7×10^9 Btu/h (21.6 GW_{th}) (about 334×10^6 ft² [31×10^6 m²]). Germany is the leader with 41% of the European market followed by Greece (13%), Austria (11.6%), and Spain (6.5%). In terms of capacity in operation per capita, Cyprus, where more than 90% of all buildings are equipped with solar collectors, leads Europe with 2.2×10^6 Btu/h (646 kW_{th}) per 1000 capita, followed by Austria at 1.0×10^6 Btu/h (301 kW_{th}) per 1000 capita, and Greece at 0.86×10^6 Btu/h (253 kW_{th}) per 1000 capita, with the European average at 0.15×10^6 Btu/h (44 kW_{th}) per 1000 capita.

According to the International Energy Agency (Weiss and Mauthner 2010), China dominates the world market (66%), with annual domestic sales of about 334×10^6 ft² (31×10^6 m²) in 2008 (74×10^9 Btu/h [21.7 GW_{th}]).

For many, a key impediment to increased solar use is economics. The cost of some solar technologies is perceived to be high compared to the fossil-based energy source it is offsetting. For example, while the simple payback of solar PV systems tends to be rather long (although much improvement has occurred over the past couple of decades), the recent increase in the cost of energy and advances in solar energy justify a fresh look at the applications and the engineering behind those applications. In addition, public policy in many areas encourages solar and other renewable energy applications through tax incentives and encouraging or requiring repurchase of excess electrical energy generated by the utility provider.

The applicability and, consequently, the economics and public policy incentives available with different solar energy system types and applications depend greatly on the location, which determines the technical factors (such as solar resources available), as well as the nontechnical (such as the country's political situation).

Solar Thermal Applications

Solar energy thermal applications range from low-temperature applications (e.g., swimming pool heating or domestic water) to medium- to high-temperature applications (e.g., space heating, absorption cooling or steam production for electrical generation). (See Figure 12-1 for examples of typical installations.)

The most common solar energy thermal application is for domestic hot water (DHW) production. However, the same solar collectors can be used to deliver thermal energy for space heating. A typical installation for the combined production of DHW and space heating (i.e., solar combi systems) includes the solar collectors, the heat storage tank, and a boiler used as an auxiliary heater. Combi systems require a larger collector area than a DHW system to meet the higher loads. It is possible to use a heat storage tank and a DHW storage vessel, but it may also be suitable to combine them in a single storage tank (with a high vertical stratification) to meet the different operating temperatures for space heating and DHW. To assess and compare performances of different designs for solar combi systems, the International Energy Agency (IEA) launched Task 26 to address issues in this area (http://task26.iea-shc.org/). Standardized classification and evaluation processes and design tools were developed for these systems, along with proposals for the international standardization of combi system test procedures.

The main drawback of solar combi systems has been the fact that during summer, the available high solar radiation and the heat produced from the solar collectors cannot be fully used, thus making the system financially less attractive and limiting its use to the low DHW summer demand. In addition, there are some technical problems related to stagnation (i.e. the condition when the medium in the solar collector loop vaporizes as a result of high solar radiation availability and low thermal demand). Since high building cooling loads generally coincide with high solar radiation, the readily available solar heat from the existing solar collectors can be exploited by a heat-driven cooling machine, thus extending the use of the solar field throughout the year (solar hot water [SHW] and space heating in winter and SHW and cooling in summer). Combining solar heating and cooling is usually referred to as a solar combi-plus system that can increase the total solar fraction.

Europe has the most sophisticated market for different solar thermal applications, including systems for hot-water production, plants for space heating of single- and multifamily houses and hotels, large-scale plants for district heating, and a growing number of systems for air-conditioning, cooling, and industrial applications.

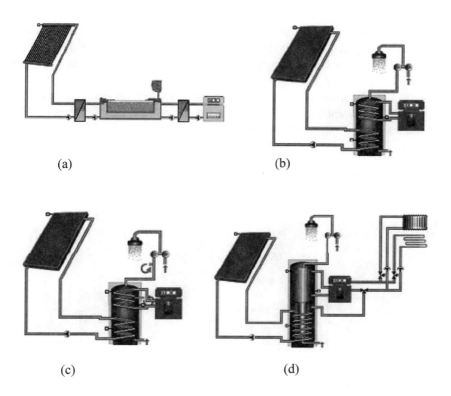

Figure 12-1 Examples of typical solar installations: (a) swimming pool, (b) domestic hot water, (c) domestic hot water for dwellings, and (d) solar combi systems for domestic hot water and space heating for houses. (Images generated using T-SOL 4.5 Expert; Valentin Energie software, www.valentin.de.)

The capital cost of solar thermal systems generally increases with higher working fluid temperatures. The higher the delivery temperature, the lower the efficiency, and the more solar collector area is generally required to deliver the same net energy. This is due to parasitic thermal losses that are inherent in solar collector design. Different solar collector types provide advantages and disadvantages, depending on the application, and there are significant cost differences among each solar collector type. Some common collector types are discussed here. (See Figure 12-2 for examples of hardware.)

Flat-Plate Solar Collectors. These are best suited for processes requiring low-temperature working fluids (80°F to 160°F or 27°C to 71°C) and can deliver 80°F (27°C) fluid temperatures, even during overcast conditions. The term *flat-plate collector* generally refers to a hydronic coil-covered absorber housed in an insulated box with a single- or double-glass cover that allows solar energy to heat the absorber. Heat is removed by a fluid running through the hydronic coils. Its design makes it more susceptible to parasitic losses than an evacuated tube collector, but more efficient in solar energy capture, because flat-plate collectors convert both direct and indirect solar radiation into thermal energy. This makes flat-plate collec-

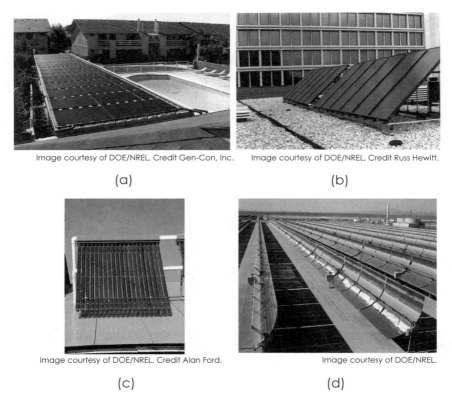

Figure 12-2 Examples of solar hardware: (a) unglazed plastic collector, (b) glazed flat-plate collector, (c) evacuated-tube collector, and (d) concentrating collectors.

tors the preferred choice for domestic hot-water and other low-temperature heating applications. Coupling a water-source heat pump with low fluid temperatures with solar collectors provides heating efficiencies that are higher than ground-source heat pump (GSHP) applications and standard natural gas furnaces.

Evacuated-Tube Collectors. These are a series of small absorbers consisting of small-diameter (approximately 3/8 in. or 10 mm) copper tubing encased in a clear, cylindrical, evacuated thermos bottle that minimizes parasitic losses even at elevated temperatures. Because of the relatively small absorber area, significantly more collector area is required than with the flat-plate or concentrating collector.

Typical flat-plate solar collector performance curves are illustrated in Figure 12-3. The collector's efficiency η is expressed as a function of the solar collector's working fluid inlet temperature T_{in}, the ambient temperature T_o, and the total solar radiation incident on the collector's surface G_T.

Concentrating Collectors. This refers to the use of a parabolic reflector that focuses the solar radiation falling within the reflector area onto a centrally located absorber. Concentrating collectors are best suited for processes requiring high-temperature working fluids (300°F to 750°F or 150°C to 400°C) and do not operate under overcast conditions. This type of collector converts only direct solar radiation, which varies dramatically with sky clearness and air quality (e.g., smog). Concentrating collectors rotate on one or two axes to track the sun and collect the available direct solar radiation.

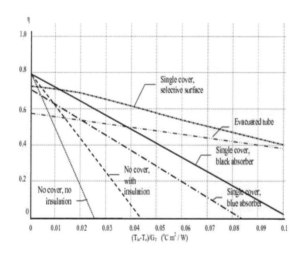

Image courtesy of C.A. Balaras.

Figure 12-3 Typical flat-plate solar collector performance curves.

Pros and cons of different solar collector types and other limitations of active solar systems can be found in a number of sources. Many of the key contributions came during the initial energy concerns of the 1970s, and a few are listed in the References and Resources section at the end of this chapter.

The percentage of energy a solar system can provide is known as the solar fraction. The F-Chart method (Klein and Beckman 2001) developed by Sanford Klein at the University of Wisconsin provides an accurate assessment of the amount of energy a solar thermal system will provide. This modeling provides the designer the ability to vary system parameters (e.g., collector area, storage volume, operating temperature, and load) to optimize system design. The method was originally developed to aid engineers who did not have computer resources available to do the complex analyses required for solar thermal calculations. This method has since been adapted for automated computer analysis using modern tools.

The cost-effectiveness of thermal solar systems is also dependent on having a constant load for the energy the solar system provides. Since space heating requirements are generally seasonal in most climates, it is advisable that energy from the solar system have more uses than space heating alone. Domestic water heating is usually a much steadier year-round load, though typically not very substantial. Solar cooling can extend utilization of solar collectors during the summer. The fact that peak cooling demand in summer is associated with high solar radiation availability offers an excellent opportunity to exploit solar energy with heat-driven cooling machines.

Solar hot water can be used as the sole source of domestic hot water or to preheat incoming supply water. This can be as simple as having an uninsulated tank in a hot attic in a southern climate to having a batch water heater on the roof of a building. Pool heating is the reason for the vast majority of current sales of hot-water-based systems in most markets in the United States. Water is taken from the pool and pumped through an unglazed collector and then returned back into the pool. In many climates, most of the energy for pool heating can be offset with this technique. Solar energy is a major source for domestic hot-water heating in several countries (i.e., China, Israel, Australia, and Greece) and is currently widely used in states such as Hawaii and Florida. Although it is an economically viable option for practically all areas of the United States, a barrier to increased market acceptance is the current lack of trained installation contractors and limited information available in most areas.

Besides preheating water, solar energy can be used to preheat incoming air—before it is introduced to a building—using systems integrated with the overall building design or collectors. This is a simple technological approach that can be economically viable and is well suited for northern climates or areas with high building heating load. The DOE has published a bulletin on these transpired solar collectors (DOE 2006).

The main heat-driven cooling technologies that can be used for solar cooling include:

- Closed-cycle systems (e.g., LiBr/H_2O and H_2O/NH_3 absorption systems [see Chapter 18 of 2010 *ASHRAE Handbook—Refrigeration* (ASHRAE 2010)] and adsorption [see Chapter 32 of 2013 *ASHRAE Handbook—Fundamentals* (ASHRAE 2013)] cycles). They produce chilled water that can be used in combination with any air-conditioning equipment, such as an air-handling unit, fan-coil systems, chilled ceilings, etc.
- Open cycles (e.g., desiccant systems [see Chapter 32 of 2013 *ASHRAE Handbook—Fundamentals* (ASHRAE 2013)]). The term *open cycle* is used to indicate that the refrigerant is discarded from the system after it provides the cooling effect. New refrigerant is supplied in its place in an open-ended loop.

Solar-assisted cooling systems employ solar thermal collectors connected to thermal-driven cooling devices. They consist of several main components (Figure 12-4), namely, the solar collectors, heat storage, heat distribution system, heat-driven cooling unit, optional cold storage, an air-conditioning system with appropriate cold distribution, and an auxiliary subsystem (which is integrated at different places in the overall system and is used as an auxiliary heater parallel

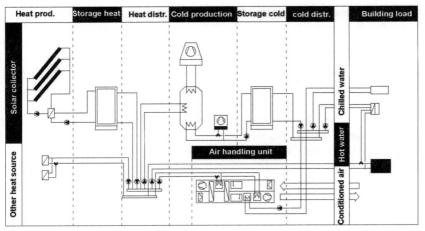

Image courtesy of Hans-Martin Henning.

Figure 12-4 Schematic description of a solar air-conditioning system describing the integration of different component options.

to the collector or the collector/storage, as an auxiliary cooling device, or both). International research and development efforts, along with monitoring data from several installations are available from Task 38 on solar air conditioning and refrigeration from the International Energy Agency (IEA) (http://task38.iea-shc.org/).

Due to their low first cost, common flat-plate solar collectors are used for reaching a driving temperature of 140°F to 194°F (60°C to 90°C) (Balaras et al. 2006). With selective-surface, flat-plate solar collectors, the driving temperature can be up to 248°F (120°C); however, the collector efficiency will be quite low at this temperature level. Stationary flat-plate, evacuated tube solar collectors are typically used for 175°F to 250°F (79°C to 121°C) applications, and can reach higher temperatures but at a lower collector efficiency. Compound parabolic concentrators can reach 207°F to 329°F (97°C to 165°C).

The average specific solar collector area averages 136 ft^2/ton (3.6 m^2/kW refrigeration), ranging from 19 to 208 ft^2/ton (0.5 to 5.5 m^2/kW refrigeration), depending on the employed technology. Adsorption and absorption systems typically use more than 76 ft^2/ton (2 m^2/kW refrigeration) and usually less than 189 ft^2/ton (5 m^2/kW refrigeration). Overall, H_2O/NH_3 systems require larger specific collector areas than $LiBr/H_2O$ systems, and, as a result, the installations are usually more expensive. The initial overall cost per installed cooling capacity in averages about 4000 Euro/kWR, excluding the cost for distribution networks among the system and the application and the delivery units.

Each technology has specific characteristics that match the building's HVAC design, loads, and local climatic conditions. A good design must first exploit all available solar radiation and then cover the remaining loads from conventional sources. Proper calculations for collector and storage size depend on the employed solar cooling technology. Hot-water storage may be integrated between the solar collectors and the heat-driven chiller to dampen the fluctuations in the return temperature of the hot water from the chiller. The storage size depends on the application: if cooling loads mainly occur during the day, then a smaller storage unit will be necessary than when the loads peak in the evening. Heating the hot-water storage by the backup heat source should be strictly avoided. The storage's only function is to store excess heat of the solar system and to make it available when sufficient solar heat is not available. An overview of European efforts on solar combi and combi-plus systems, summarizing their main design, and operational and performance characteristics, is available in Balaras et al. (2010).

Solar-Electric Systems

The direct conversion from sunlight to electricity is accomplished with PV systems. These systems continue to drop in price, and many states (as well as the federal government) offer tax incentives for the installation of these systems. The most common application is a grid-tied system where electricity is directly fed into the grid.

Other applications of PV may be attractive and provide additional value beyond the cost of offsetting utility power. Whenever power lines must be extended, PV should be investigated. It is cost-effective to use stand-alone PV applications for signs, remote lights, and blinking traffic lights. Many times, if an existing small grid line is available and additional power is needed, it is less expensive to add the PV system rather than increase the power line. This may be true of remote guard shakes, restroom facilities, or other outbuildings. PV should also be considered as part of the uninterruptible power system for a building. Every building has battery backup for certain equipment. Batteries can be centralized and the PV fed directly into this system. The cost can be less than other on-site generation and fuel storage.

Technical and legal factors have generally been worked out that have hindered the use of grid-tied solar PV systems in the past. In temperate climates, grid-tied PV systems provide a good method to reduce peak summer electrical demand, since peak solar gains generally correspond to peak air-conditioning demand. This was a major factor that led to the passing of the California Million Solar Roofs initiative in early 2006. The German 100,000 Roofs Program (HTDP), which began in early 1999, had a goal of 300 MW. It was successfully completed in 2003 with the parallel introduction of the German Renewable Energy Resources Act (EEG) that came into effect in 2000. The EEG and HTDP program secured commercially oriented PV investors (because they received a full payback of their investment).

According to the European Photovoltaic Industry Association (EPIA 2010), by the end of 2012 the cumulative installed capacity of PV systems around the world had reached about 31.1 GWp (peak gigawatts). Germany was the world's top PV market, with 7.6 GW of newly connected systems; China was second with an assumed 5 GW, followed closely by Italy (3.4 GW), the U.S. (3.3 GW), and Japan with an estimated 2 GW. Together, Germany, China, Italy, the U.S., and Japan accounted for nearly 21.3 GW, or two-thirds of the global market during 2012.

The economic viability of PV systems is dependent on many factors. Improvements in technology have lowered the cost per peak watt from $5.14 and $3.08 in 1989 (in nominal dollars) to $3.49 for modules and $1.94 for cells in 2008, according to the DOE (EIA [see the References and Resources section]).

The module cost represents about 50%–60% of the total installed cost of a solar energy system. Therefore, the solar module price is the key element in the total price of an installed solar system. All prices are exclusive of sales taxes, which, depending on the country or region, can add 8%–20% to the prices, with the highest sales tax rates being in Europe. The typical PV cell efficiencies (i.e., the ratio of electrical energy produced by a solar cell to the incident solar irradiance) currently range from less than 5% for the first generation of thin-film cells to more than 24% for the most advanced crystalline-silicon cells in laboratory conditions.

The integration of PV systems with building materials is a new development that may, in the long run, help make PV systems viable in most areas. One of the key ways to incorporate solar energy technologies is to incorporate them directly into the architecture of the building (see Figure 12-5). Trombe walls can replace spandrel panels and other glass facades; PV panels can become overhangs for the building or parking shading structures. Using these elements as part of the building doubles the value of the systems. An example of these technologies is the concept called *building-integrated PVs*. A picture of one concept is shown in Figures 12-5 and 12-6. These are being developed by the National Renewable Energy Laboratory (NREL), among others.

Finally, PV is very visible and can be used as a marketing tool for a building project. PV collectors are addressed in GreenTip #12-3.

Image courtesy of DOE/NREL, Credit Ben Kroposki.   Image courtesy of DOE/NREL, Credit Lawrence Berkeley Lab.

Figure 12-5 Building-integrated PVs.

Image courtesy of DOE/NREL, Credit United Solar Ovonic.

Figure 12-6 Building-integrated PV roofing.

WIND

Using prevailing breezes and wind energy is one of the most promising alternative technologies today. Wind turbine design and power generation have become more reliable over the last 30 years. Consistency and velocity of available breezes are essential to successful application of wind for electrical generation and natural ventilation in buildings. Information on velocities, durations, and direction of winds in a project area is generally available from the National Oceanic and Atmospheric Administration (NOAA).

Global wind energy installations reached about 282 GW at the end of 2012, according to the Global Wind Energy Council (GWEC 2010). By 2012 China was leading the world with about 75 GW of total installed grid connected capacity, followed by the United States which had 60 GW capacity. Germany and Spain were next and led the European countries

From a natural-ventilation perspective for passive cooling of buildings, other factors (e.g., temperature and relative humidity) must also be taken into consideration. Many areas can use outdoor air/natural breezes to provide passive cooling only during a limited period of the year, requiring designers wanting to use natural ventilation to carefully analyze both climatic conditions and cost to implement. The benefits of using natural ventilation must be weighed against the consistency of breezes, potential for higher relative humidity levels (which may affect indoor air quality), and the impact on occupant comfort and HVAC operation.

There are many excellent examples where natural ventilation provides the cooling needed throughout the year, but most are located in cool, arid climates. Even in moderate climates, such as that in Atlanta, Georgia, natural ventilation can provide passive cooling during some periods of the year, but in doing so, the designer must consider the impact on HVAC operation and building control. These elements may be very difficult to get to work effectively with operable windows. Additional programming and control points are required for successful operation. The potential of natural ventilation, its appropriate use, the design and dimensioning methodologies, the need for an integrated design approach, and how to overcome barriers are available in a handbook by Alvarez et al. (1998).

Incorporating one or more wind turbine generators at a building site is a more active approach. Initially, it should be recognized that there is a disadvantage of scale: one or two wind turbines alone are destined to be less cost-effective than a wind farm with hundreds. Wind turbines should first be evaluated by looking at the wind resource at the site. For most small-scale wind turbines, steady winds over 20 mph (9 m/s) are preferred to maximize the cost effectiveness. Other issues include noise, vibration, building geometry (if on the building), wind pattern interrupters, wildlife effect, periodic maintenance, safety, visual impact, and community acceptance; these must also be taken into account.

HYDRO

According to the International Hydropower Association (IHA) (http://hydropower.org/iha/sustainability/), hydropower is "the world's largest source of renewable energy." In all of its forms, from conventional power generation to emerging technologies like ocean and wave energy, hydrokinetic, and tidal power, there is a role of all types and scales of hydropower. According to the National Hydropower Association (NHA) (http://hydro.org), wave energy conversion devices capture mechanical power from the waves and use it to directly or indirectly power a turbine and a generator. The flow from the tides of an ocean or stream may also be harvested to produce power. Pumped storage may be combined with other RES systems to meet power demand variations. The water may be pumped from a lower reservoir to an upper reservoir when demand for electricity is low (the power for pumping may be supplied from RES systems or the grid). The water is then released from the upper reservoir back to the lower reservoir to generate power using the hydrokinetic energy of the water flow through a generator (this is done during peak power demand or low generating output from RES systems).

The ability to use hydro is limited on an individual building scale. On rare occasions, older building renovations can take advantage of existing river dams and generate hydroelectric on site. This is especially true in the northeastern United States, where the primary power supply in the 1800s was hydro. In some cases, small streams can be diverted and small turbines installed—this is called *low-impact hydro*, as a minimal dam is needed and water is returned to the same stream.

In some cases, pressure reduction values can be replaced with small turbines. Although this is not really free energy from the big picture, as the water was pressurized before being distributed to the site, it is better to recover some of the energy rather than essentially waste it during pressure reduction.

BIOMASS

Biomass is the conversion of plant and animal matter into useful energy. Several ideas should be investigated to be considered as an energy source in a building. According to the DOE Bioenergy Technologies Office web page (www1.eere.energy.gov/bioenergy/biobasics.html), biomass currently supplies about 3% of total U.S. energy consumption in the form of electricity, process heat, and transportation fuels. In the European Union, "bioenergy contributes only to 3.7% of the total primary energy supply; however, it plays a considerable role in several European countries, such as Finland and Sweden where its contribution amounts to respectively 20% and 16% of the gross inland consumption" according to the European Biomass Industry Association (www.eubia.org/index.php/about-biomass).

Wood chip conversion to heat (and possibly then electricity) can replace boiler systems in buildings. Automatic feed systems can now take wood chips and make hot water as conveniently as oil and natural gas systems can. In many locations, wood chips are free for the cost of the transportation, saving further cost. Pellet burners are a similar concept, except for the additional cost of the wood pellets. Burners are also available for corn cobs. In some cases, manufacturers will lease equipment for the energy savings, resulting in no additional cost to the end user.

The production and distribution of bio-oils and biodiesel fuels derived from bio-based materials is growing in momentum. Heating systems are available to burn used vegetable oils. These systems are useful near the point of production of used vegetable oil—usually the food service industry. Grocery stores and restaurants may benefit from such technologies. Most diesel-fueled machines can run on biodiesel without any (or minor) modifications. In addition, hauling and waste costs for the oil are eliminated, increasing the attractiveness of the technology.

REFERENCES AND RESOURCES

Published

Alvarez, S., E. Dascalaki, G. Guarracino, E. Maldonado, S. Sciuto, and L. Vandaele. 1998. *Natural Ventilation of Buildings—A Design Handbook*. F. Allard, ed. London: James & James Science Publishers Ltd.

ASHRAE. 2010. *ASHRAE Handbook—Refrigeration*. Atlanta: ASHRAE.

ASHRAE. 2013. *ASHRAE Handbook—Fundamentals*. Atlanta: ASHRAE.

Balaras, C.A., E. Dascalaki, P. Tsekouras, and A. Aidonis. 2010. High solar combi systems in Europe. *ASHRAE Transactions* 116(1).

Balaras, C.A., H.-M. Henning, E. Wiemken, G. Grossman, E. Podesser, and C.A. Infante Ferreira. 2006. Solar cooling—An overview of European applications and design guidelines. *ASHRAE Journal* 48(6):14–22.

Beckman, W., S.A. Klein, and J.A. Duffie. 1977. *Solar Heating Design*. New York: John Wiley & Sons, Inc.

DOE. 2006. Solar buildings. Transpired air collectors ventilation preheating. DOE/GO-102001-1288, U.S. Department of Energy, Washington, DC.

EU. 2008. Climate change: Commission welcomes final adoption of Europe's climate and energy package. Europa press release IP/08/1998, European Union, Brussels.

Henning. H-M., ed. 2004. *Solar Assisted Air-Conditioning in Buildings—A Handbook For Planners*. Wien: Springer-Verlag.

Kreith, F., and J. Kreider. 1989. *Principles of Solar Engineering*, 2d ed. Wasington, DC: Hemisphere Pub.

NREL. 2010. NREL highlights utility green power leaders. NREL press release NR-1910. Golden, CO: National Renewable Energy Laboratory.

Shiklomanov, I.A. 1993 World fresh water resources. In *Water in Crisis*. Gleick, P.H., ed. New York: Oxford University Press.

Online

American Solar Energy Society
www.ases.org.

American Wind Energy Association
www.awea.org.

Building Energy Efficiency Research Project at the Department of Architecture, The University of Hong Kong
www.arch.hku.hk/research/beer/.

Centre for Analysis and Dissemination of Demonstrated Energy Technologies
www.caddet-re.org.

EIA. Table 10.8, Photovoltaic cell and module shipments by type, trade, and prices, 1982–2008. U.S. Energy Information Administration. www.eia.doe.gov/aer/txt/stb1008.xls.

EPA. 2010. *Guide to Purchasing Green Power*. Washington, DC: U.S. Environmental Protection Agency. www.epa.gov/grnpower/buygreenpower/guide.htm.

EPIA. 2010. 2014 Global Market Outlook for Photovoltaics until 2014. European Photovoltaic Industry Association.
www.greenunivers.com/wp-content/uploads/2010/05/Global_Market_Outlook_for_Photovoltaics_until_20141.pdf.

ESTIF. 2010 Solar thermal markets in Europe—Trends and market statistics 2009.
www.estif.org/fileadmin/estif/content/market_data/downloads/2009%20solar_thermal_markets.pdf.

Europe's Energy Portal
www.energy.eu/#renewable.

European Biomass Association
www.aebiom.org.

European Biomass Industry Association
www.eubia.org/.

European Photovoltaic Industry Association
www.epia.org.

European Renewable Energy Centres Agency
www.eurec.be.

European Solar Thermal Industry
www.estif.org.

European Wind Energy Association
www.ewea.org.

EUROSOLAR—The European Association for Renewable Energies e.V.
www.eurosolar.org.

Global Wind Energy Council
www.gwec.net.

Green Energy In Europe
www.ecofys.com/.

Green Venture
  www.greenventure.ca.
GWEC. 2010. Global wind 2009 report. Global Wind Energy Council, Brussels, Belgium. www.gwec.net/index.php?id=167.
International Energy Agency, Bioenergy
  www.ieabioenergy.com.
International Energy Agency, Photovoltaic Power Systems Programme
  www.iea-pvps.org.
International Energy Agency, Solar Heating And Cooling Programme
  www.iea-shc.org.
International Energy Agency, Task 38
  www.ieabioenergy-task38.org/.
International Hydropower Association
  www.hydropower.org.
International Hydropower Association Fact Sheet on the Hydropower Sustainability Assessment Protocol.
  www.internationalrivers.org/campaigns/the-hydropower-sustainability-assessment-protocol.
International Solar Energy Society
  www.ises.org.
Klein, S., and W. Beckman. 2001. *F-Chart Software.* The F-Chart Software Company., Madison, WI.
  www.fchart.com.
National Hydropower Association
  www.hydro.org.
National Oceanic and Atmospheric Administration
  www.noaa.gov.
National Renewable Energy Laboratory, Solar Energy Research Facility www.nrel.gov/pv/facilities_serf.html.
North Carolina Solar Center, DSIRE
  www.dsireusa.org/.
North Carolina Cooperative Extension Service, Water Quality and Waste Management
  www.bae.ncsu.edu/programs/extension/publicat/wqwm.
Oak Ridge National Laboratory, Bioenergy Feedstock Information Network
  http://bioenergy.ornl.gov/.
Solar Energy Industries Association
  www.seia.org.
U.S. Department of Energy, Energy Efficiency and Renewable Energy, Bioenergy Technologies Office web page
  www1.eere.energy.gov/biomass/for_consumers.html.
Weiss, W., I. Bergmann, and G. Faninger. 2008. *Solar Heat Worldwide.* Solar Heat & Cooling Programme, International Energy Agency, Paris, France.

https://www.iea-shc.org/data/sites/1/publications/Solar_Heat_Worldwide-2008.pdf.

Weiss, W., and F. Mauthner. 2010. Solar heat worldwide—Markets and contribution to the energy supply 2008.
https://www.iea-shc.org/data/sites/1/publications/Solar_Heat_Worldwide-2010.pdf.

World Meteorological Organization, World Radiation Data Centre www1.eere.energy.gov/bioenergy/.

The World's Water
www.worldwater.org.

ASHRAE GreenTip #12-1

Passive Solar Thermal Energy Systems

GENERAL DESCRIPTION

Passive solar thermal energy systems use solar energy (mainly for space heating) via little or no use of conventional energy or mechanisms other than the building design and orientation. All above-grade buildings are passive solar. Making buildings collect, store, and use solar energy wisely then becomes the challenge for building designers. A building that intentionally optimizes passive solar heating can visually be a solar building, but many reasonably sized features that enhance energy collection and storage can be integrated into the design without dominating the overall architecture.

To be successful, a well-designed passive solar building needs (1) an appropriate thermal load (e.g., space heating); (2) aperture (e.g., clear, glazed windows); (3) thermal storage to minimize overheating and to use heat at night; (4) control, either manual or automatic, to address overheating; and (5) night insulation of the aperture so that there is not a net heat loss.

HIGH-PERFORMANCE STRATEGIES

The following strategies are general in nature and are presented as guidelines to help maximize the performance of a passive thermal solar design:

- Think about conservation first. Minimizing the heating load will reduce conventional and renewable heating systems' sizes and yields the best economics. Insulate, including the foundation, and seal the building well. Use quality exterior windows and doors.
- In the northern hemisphere, the aperture must face due south for optimal performance. If this is not possible, make it within ±10 degrees of due south. In the southern hemisphere, this solar aperture looks north.
- Minimize use of east- and west-facing glazing. They admit solar energy at nonoptimal times, at low angles that minimize storage, and are difficult to control by external shading. Also, reduce north glazing in the colder regions of the northern hemisphere, because this can lead to high heat loss rates.

- Use optimized and/or moveable external shading devices, such as overhangs, awnings, and side fins. Internal shading devices should not be relied upon for passive solar thermal control—they tend to cause overheating.
- Use high-mass direct gain designs. The solar collector (windows) and storage (floors and potentially walls) are part of the occupied space and typically have the highest solar savings fraction, which is the percent of heating load met by solar. Directly irradiated thermal masses are much more effective than indirect, thus floors or trombe walls are often best.
- Use vertical glazing. Horizontal or sloped windows and skylights are hard to control and insulate.
- Calculate the optimal thermal mass—it is often around 8 in. thick (about 200 mm)—high-density concrete for direct-gain floors over conditioned basements. The optimization should direct the designer to a concept that will capture and store the highest amount of solar energy without unnecessarily increasing cost or complexity. For all direct-gain surfaces, make sure they are of high-absorptivity (dark color) and are not covered by carpet, tile, much furniture, or other items that prevent or slow solar energy absorption. Be sure the thermal mass is highly insulated from the outdoor air or ground.
- Seeking a very high annual solar savings fraction f_s often leads to disappointment and poor economics, so keep expectations reasonable. Even a 15% annual fraction represents a substantial reduction in conventional heating energy use. A highly optimized passive solar thermal single-family house, in an appropriate climate, often only has about a 40% solar fraction. Combining passive with active solar, PVs, wind, and other renewable energy sources is often the most satisfactory way to achieve a very high annual solar savings fraction.
- An old solar saying, of unknown origin, is "the more passive a building, the more active the owner." Operating a passive solar building to optimize collection has thus been called *solar sailing* and requires time and experience. Be sure that passive solar and the building operator are good matches for each other.

KEY ELEMENTS OF COST

Passive solar energy systems must be engineered; otherwise, poor performance is likely.

Window sizes are typically larger than for conventional design. Some operable windows and/or vents, placed high and low, are needed for overheat periods. Proper solar control must also be foreseen.

Concrete floors and walls are commonly thicker in the storage portion of a passive solar building. A structural engineer's services are likely required.

Conventional backup systems are still needed. They will be used during cloudy and/or cold weather, so select high-efficiency equipment with low-cost fuel sources.

Night insulation must be used consistently or else there will be a net heat loss. Making the nightly installation and removal automatic is recommended, but costly.

SOURCES OF FURTHER INFORMATION

ASHRAE. 2011. *ASHRAE Handbook—HVAC Applications*, Chapter 35. Atlanta: ASHRAE.

Balcomb, J.D. 1984. *Passive Solar Heating Analysis: A Design Manual*. Atlanta: ASHRAE.

Crosbie, M.J., ed. 1998. *The Passive Solar Design and Construction Handbook*, Steven Winter Associates. New York: John Wiley & Sons, Inc.

ASHRAE GreenTip #12-2

Active Solar Thermal Energy Systems

GENERAL DESCRIPTION

Both passive and active solar thermal energy systems rely on capture and use of solar heat. Active solar thermal energy systems differ from passive systems in the way that they use solar energy, as they use it primarily for space and water heating. This can allow for greatly enhanced collection, storage, and use of solar energy.

To be successful, a well-designed active solar system needs (1) an appropriate thermal load (e.g., potable water, space air, or pool heating); (2) collectors such as flat-plate solar panels; (3) thermal storage to use the heat at a later time; and (4) control, typically automatic, to optimize energy collection and storage and for freeze and overheat protection.

The working fluid or coolant that moves heat from the collectors to the storage device is typically water, a water/glycol solution, or air. The heat storage medium is often water but could be a rock-bed or a high-mass building for air-coolant systems.

The solar energy collectors are most often of fixed orientation and nonconcentrating, but they could be tracking and/or concentrating. Flat-plate collectors are most common and are typically installed as fixed and nonconcentrating. Large surface areas of collectors, or mirrors and/or lens for concentration, are needed to gather heat and to achieve higher temperatures.

There are many different types of active solar thermal energy systems. For example, one type often used for potable water heating is flat-plate, pressurized water/glycol coolant, and two-tank storage. An internal double-wall heat exchanger is typically employed in one tank, known as the preheat tank, and the other tank, plumbed in series, is a conventional water heater. These preheat tanks are now widely available due to nonsolar use as indirect water heaters. One-tank systems typically have an electric-resistance heating element installed in the top of the special tank.

HIGH-PERFORMANCE STRATEGIES

The following strategies are general in nature and are presented as a guideline to helping maximize the performance of an active thermal solar design:

- Think about conservation first. Minimizing the heating load will reduce conventional and renewable heating systems' sizes, and yield the best economics.
- In the northern hemisphere, the solar collectors must face due south for optimal performance. If not possible, make them within ±10 degrees of due south. In the southern hemisphere, the solar collectors look north.
- When using flat-plate collectors for space heating, mount them at an angle equal to the local latitude plus 15 to 20 degrees. For water heating, use the local latitude plus 10 degrees.
- Calculate the optimal thermal storage—about one day's heat storage (or less) often yields the best economics. Place the thermal storage device within the heated space, and be sure it is highly insulated, including under its base.
- Seeking a very high annual solar savings fraction f_s often leads to disappointment and poor economics, so keep expectations reasonable. Even a 25% annual fraction represents a substantial reduction in conventional energy use. A highly optimized, active solar thermal domestic water-heating system, in an appropriate climate, often has about a 60% solar fraction. Combining active solar with passive, PVs, wind, and other renewable energy sources is often the most satisfactory way to achieve a very high annual solar savings fraction.

KEY ELEMENTS OF COST

Active solar energy systems must be engineered; otherwise, poor performance is likely. Fortunately, good design tools, such as F-Chart software (Klein and Beckman 2001), are readily available.

Well-designed, factory-assembled collectors are recommended, but are fairly expensive. Site-built collectors tend to have lower thermal performance and reliability.

Storage tanks must be of high quality and be durable. Water will eventually leak, so proper tank placement and floor drains are important. For rock storage, moisture control and air-entrance filtration are important for mold growth prevention.

Quality and appropriate pumps, fans, and controls can be somewhat expensive. Using surge protection for all the electrical components is recommended. Effective grounding, for lightning and shockmitigation, is normally required by building code. For liquid coolants in sealed loops, expansion tanks and pressure relief valves are needed. For domestic water heating, a temperature-limiting mixing valve is required for the final potable water to prevent scalding; even nonconcentrating systems can produce 180°F (82°C) or so water at times. All thermal components require insulation for safety and reduced heat loss.

Installation requires many trades: a contractor to build or install the major components, a plumber to do the piping and pumps (i.e., water systems), an HVAC contractor to install ducts (i.e., air systems) and/or space-heating heat exchangers, an electrician to provide power, and a controls specialist.

Conventional backup systems are still needed. They will be used during cloudy and/or cold weather, so select high-efficiency equipment with low-cost fuel sources.

SOURCES OF FURTHER INFORMATION

ASHRAE. 1988. *Active Solar Heating Systems Design Manual.* Atlanta: ASHRAE.

ASHRAE. 2011. *ASHRAE Handbook—HVAC Applications*, Chapter 35. Atlanta: ASHRAE.

Howell, J., R. Bannerot, and G. Vliet. 1982. *Solar-Thermal Energy Systems: Analysis and Design.* New York: McGraw-Hill.

Klein, S., and W. Beckman. 2001. *F-Chart Software.* The F-Chart Software Company., Madison, WI. www.fchart.com.

ASHRAE GreenTip #12-3

Solar Energy System—PV

GENERAL DESCRIPTION

Light shining on a PV cell, which is a solid-state semiconductor device, liberates electrons, which are collected by a wire grid, to produce direct-current electricity.

The use of solar energy to produce electricity means that PV systems reduce greenhouse gas emissions, electricity costs, and resource consumption. Electrical consumption can be reduced. Because the peak generation of PV electricity coincides with peak air-conditioning loads (if the sun shines), peak electricity demands (from the grid) may be reduced, though it is unlikely without substantial storage capacity.

PV can also reduce electrical power installation costs where the need for trenching and independent metering can be avoided. The public appeal of using solar energy to produce electricity results in a positive marketing image for PV-powered buildings and, thus, can enhance occupancy rates in commercial buildings.

While conventional PV design has focused on the use of independent applications in which excess electricity is stored in batteries, grid-connected systems are becoming more common. In these cases, electricity generated in excess of immediate demand is sent to the electrical grid, and the PV-powered building receives a utility credit. Grid-connected systems are often integrated into building elements. Increasingly, PV cells are being incorporated into sunshades on buildings for a doubly effective reduction in cooling and electricity loads.

PV power is being applied in innovative ways. Typical economically viable commercial installations include the lighting of parking lots, pathways, signs, emergency telephones, and small outbuildings.

A typical PV module consists of 33 to 40 cells, which is the basic block used in commercial applications. Typical components of a module are aluminum, glass, tedlar, and rubber. The cell is usually silicon, with trace amounts of boron and phosphorus.

Because PV systems are made from a few relatively simple components and materials, the maintenance costs of PV systems are low. Manufacturers now provide 20-year warranties for PV cells.

PV systems are adaptable and can easily be removed and reinstalled in other applications. Systems can also be enlarged for greater capacity through the addition of more PV modules.

WHEN/WHERE IT'S APPLICABLE

PV is well suited for rural and urban off-grid applications and for grid-connected buildings with air-conditioning loads. The economic viability of PV depends on the distance from the grid, electrical load sizes, power line extension costs, and incentive programs offered by governmental entities or utilities.

PV applications include prime buildings, outbuildings, emergency telephones, irrigation pumps, fountains, lighting for parking lots, pathways, security, clearance, billboards, bus shelters or signs, and remote operation of gates, irrigation valves, traffic signals, radios, telemetry, or instrumentation.

Grid-connected PV systems are better suited for buildings with peak loads during summer cooling operation but are not as well suited for grid-connected buildings with peak wintertime loads.

Note that a portion of a PV electrical system is direct current, so appropriate fusing and breakers may not be readily available. A PV system is also not solely an electrical installation; other trades, such as roofing and light steel erectors, may be involved with a PV installation. When a PV system is installed on a roof or wall, it will likely result in envelope penetrations that will need to be sealed.

PRO

- Reduces greenhouse gas emissions.
- Reduces nonrenewable energy demand, with the ability to help offset demand on the electrical grid during critical peak cooling hours.
- Enhances green-image marketing.
- Lowers electricity consumption costs and may reduce peak electrical demand charges.

- Reduces utility infrastructure costs.
- Increases electrical reliability for the building owner. May be used as part of an emergency power backup system.

CON

- Relatively high initial capital costs.
- Requires energy storage in batteries or a connection to electrical utility grid.
- May encounter regulatory barriers.
- High-capacity systems require large-building envelope areas that are clear of protuberances and have uninterrupted access to sunshine.
- Capacity to supply peak electrical demand can be limited, depending on sunshine during peak hours.

KEY ELEMENTS OF COST

The following provides a possible breakdown of the various cost elements that might differentiate a PV system from a conventional one and an indication of whether the net cost for this system is likely to be lower (L), higher (H), or the same (S). This assessment is only a perception of what might be likely, but it obviously may not be correct in all situations. There is no substitute for a detailed cost analysis as part of the design process. The listings below may also provide some assistance in identifying the cost elements involved.

First Cost

- PV modules   H
- Wiring and various electrical devices   H
- Battery bank   H
- Instrumentation   H
- Connection cost (if grid-connected)   H

Recurring Costs

- Electricity   L

SOURCES OF FURTHER INFORMATION

California Energy Commission, Renewable Energy Program www.energy.ca.gov/renewables/.

Canadian Renewable Energy Network (CanREN) http://users.encs.concordia.ca/~raojw/crd/essay/essay002164.html.

ecoENERGY for Renewable Power www.ecoaction.gc.ca/ecoenergy-ecoenergie/power-electricite/index-eng.cfm.

National Renewable Energy Laboratory, Building-Integrated PV www.nrel.gov/pv/building_integrated_pv.html.

NRC. *Photovoltaic Systems Design Manual*. Ottawa, Ontario, Canada: Natural Resources Canada, Office of Coordination and Technical Information.

NRC. RETScreen (Renewable Energy Analysis Software). Natural Resources Canada, Energy Diversification Research Laboratory, Ottawa, Ontario, Canada. www.retscreen.net.

Photovoltaic Resource Site www.pvpower.com.

School of Photovoltaic and Renewable Energy Engineering, University of New South Wales www.pv.unsw.edu.au.

Solar Energy Industries Association www.seia.org.

Sustainable Sources www.greenbuilder.com.

Watsun. 1999. *WATSUN-PV—A computer program for simulation of solar photovoltaic systems, User's Manual and Program Documentation*, Version 6.1. Watsun Simulation Laboratory, University of Waterloo, Ontario, Canada.

ASHRAE GreenTip #12-4

Solar Protection

GENERAL DESCRIPTION

Shading the building's transparent surfaces from solar radiation is mandatory during summer and sometimes even necessary during winter. This way, it is possible to prevent solar heat gains when they are not needed and to control daylighting to minimize glare problems. Depending on the origin of solar radiation (i.e., direct, diffuse, reflected), it may be possible to select different shading elements that provide more effective solar control.

Depending on the specific application and type of problem, there may be different options for selecting the optimum shading device. The decision can be based on several criteria, from aesthetics to performance and effectiveness or cost. Different types of shading elements are suitable for a given application, result in varying levels of solar control effectiveness, and have a different impact on indoor daylight levels, natural ventilation, and overall indoor visual and thermal comfort conditions.

There are basically three main groups of solar control devices. The first group is external shading devices, which can be fixed and/or movable elements. They have the most apparent impact on the aesthetics of the building. If properly designed and accounted for, they can become an integral part of the building's architecture, as they are integrated into the building envelope. Fixed types are typically variations of a horizontal overhang and a vertical side fin, with different relative dimensions and geometry. When properly designed and sized, fixed external shading devices can be effective during summer, while during winter they allow the desirable direct solar gains through the openings. This is a direct positive outcome given the relative position of the sun and its daily movement in winter (when there is low solar elevation) and summer (when there is high solar elevation). Movable types are more flexible, since they can be adjusted and operated either manually or automatically for optimum results and typically include various types and shapes of awnings and louvers.

The second group is interpane shading devices, which are usually adjustable and retractable louvers, roller blinds, screens, or films that are placed within the glazing. This type of a shading device is more suitable for solar control of scattered radiation or sky diffuse radiation. Given that the incident solar radiation is already absorbed by the glazing, thus increasing its temperature, one needs to take into account the heat transfer component to the indoor spaces.

The third group, which is internal shading devices, is very common because of indoor aesthetics. They offer privacy control and easy installation, accessibility, and maintenance. Although on the interior they are very practical and, most of the time, necessary, their overall thermal behavior needs to be carefully evaluated, since the incident solar radiation is trapped inside the space and will be absorbed and turn into heat if not properly controlled (i.e., reflecting solar radiation outward through the opening). Numerous types or combinations of the various shading devices are also possible, depending on the application.

HIGH-PERFORMANCE STRATEGIES

- Use natural shading. Deciduous plants, trees, and vines offer effective natural shading. It is critical for their year-round effectiveness not to obstruct solar radiation during winter (in order to increase passive solar gains). Plants also have a positive impact on the immediate environment surrounding the building (i.e., the microclimate) because of their evaporative cooling potential. However, the plants need some time to grow, may cause moisture problems if they are too close to opaque elements, and can suffer from various diseases. The view can be restricted and some plants, especially large leafless trees, can still obstruct solar radiation during winter and may reduce natural ventilation. In general, for deciduous plants, the shading effect is best for east and west orientations, along with southeast and southwest.
- Incorporate louvers into the design. These are also referred to as *venetian blinds* and can be placed externally (which is preferable) or internally (which allows for easier maintenance and installation in existing buildings). The external louvers can be

fixed in place with rotating or fixed tilt of the slats. The louvers can also be retractable. The slats can be flat or curved. Slats from semitransparent material allow for outdoor visibility. The louvers can be operated manually (i.e., slat tilt angle, up or down movement) or they can be electrically motor driven. Adjusting the tilt angle of the slats or raising/lowering the panel can change the conditions from maximum light and solar gains to complete shading. Louvers can also be used to properly control air movement during natural ventilation. Slat curvature can be used to redirect incident solar radiation before entering into the space. Slat material can have different reflective properties and can also be insulated. During winter, fully closed louvers with insulated slats can be used at night for providing additional thermal insulation at the openings.

- Incorporate awnings into the design. External or internal awnings can be fixed in place, operated manually, or driven electrically by a motor that can also be automated. Light-colored materials are preferable, as they allow for high solar surface reflectivity. Awnings are easily installed on any type and size of opening and may also be used for wind protection during winter to reduce infiltration and heat losses.

KEY ELEMENTS OF COST

Natural shading usually is a cost-effective strategy that reduces glare, and, depending on the external building façade, can improve aesthetics. Plants should be carefully selected to match local climatic conditions (in order to optimize watering needs).

External electrically driven and automated units have a higher cost and it takes money to maintain the motors, but they are more flexible and effective. Louvers are difficult to clean on a regular basis. Nonretractable louvers somewhat obstruct outward vision.

Awning fabric needs periodic replacement, depending on local wind conditions. Electrically driven and automated units have a higher cost and it takes money to maintain the motors.

SOURCES OF FURTHER INFORMATION

Argiriou, A., A. Dimoudi, C.A. Balaras, D. Mantas, E. Dascalaki, and I. Tselepidaki. 1996. *Passive Cooling of Buildings.* M. Santamouris and A. Asimakopoulos, eds. London: James & James Science Publishers Ltd.

Baker, N. 1995. *Light and Shade: Optimising Daylight Design.* European Directory of sustainable and energy efficient building. London: James & James Science Publishers Ltd.

Lawrence Berkeley National Laboratory. 2000. *Daylight in Buildings: A Source Book on Daylighting Systems and Components.* Berkeley, California: Lawrence Berkeley National Laboratory. http://gaia.lbl.gov/iea21/ieapubc.htm.

Givoni, B. 1994. *Passive and Low Energy Cooling of Buildings.* New York: Van Nostrand Reinhold.

Stack, A., J. Goulding, and J.O. Lewis. Shading systems: Solar shading for the European climates. DG TREN, Brussels, Belgium. http://erg.ucd.ie/down_thermie.html.

CHAPTER THIRTEEN

LIGHTING SYSTEMS

OVERVIEW

This chapter is intended to familiarize the reader with the process of sustainable lighting design. Lighting has a significant impact on building loads and energy usage. In the United States, energy powering lighting accounts for roughly 20% of the total energy used in buildings. Percentages differ in other countries; for example, in Canada (where heating takes up a greater percentage of total energy use), only 4% of residential energy and about 12% of commercial energy is used for lighting. Lighting also adds heat to the conditioned space, which increases cooling loads and decreases heating loads. Therefore, it is important that the HVAC designer understand the chosen lighting systems and their effects on HVAC loads. Architectural configurations and interior design decisions dictate the potential to offset electrical lighting needs with daylighting. Consequently, the productive collaboration of all design professionals including the lighting designer; architect; interior designer; as well as the HVAC, energy, and electrical engineers can enhance the overall energy and the environmental performance of buildings.

ELECTRIC LIGHTING

Energy-Effective Lighting Design

The lighting design profession is increasingly focused on providing high-quality visual environments, using energy efficient strategies. Lighting technologies have become more efficient, and expectations for interior illuminance (measured in foot-candles [I-P] or lux [SI] at the task level) have moderated. It is now possible to design high-quality lighting at connected power levels that are much lower than 30 years ago. Energy efficient lighting is now regulated in many parts of the world. A number of governmental entities have incorporated regulatory restrictions limiting the lighting power density (LPD), which is the watts of power allocated for lighting per square foot (or square meter) area in a building, and some have limited or banned outright the manufacture of lamps with low efficacies (lumen/watt ratio) (e.g., incandescent lamps that convert only one-tenth of their consumed energy to supply light have been banned in Europe and a phaseout has been planned in the United States).

Most energy codes have building LPD limits between 0.8 and 1.4 W/ft^2 (about 8.5 and 15 W/m^2). All buildings can be efficiently and comfortably illuminated using carefully selected, standard lighting equipment. Successful lighting systems with LPDs of 0.7 to 1.0 W/ft^2 (8 to 11 W/m^2) can be applied to most building types by following basic application design criteria set forth by the Illuminating Engineering Society (IES) and lighting equipment manufacturers. For offices and schools, lighting quality is generally defined as being relatively uniform, visually comfortable, with low glare, and designed to appropriate illuminance levels for the tasks performed. This includes balancing both the horizontal and vertical brightness.

ANSI/ASHRAE/IES Standard 90.1-2013 lists LPDs for various types of buildings, but excludes some types of specialty lighting from a required maximum LPD (e.g., those used for museum displays, medical procedures, retail display, exit signs, performance theater). Users may use one of two methods to comply: (1) the space-by-space method, which assigns LPDs to individual space types based on the specific tasks which will occur within, or (2) the building area method, which averages all of a building's space types to a single building LPD. When using the space-by-space method, some spaces within a building are allowed to have much higher or much lower LPD levels than specified for the building through the building area method. The way the lighting systems are controlled also may be a factor in determining the final required LPD. ANSI/ASHRAE/USGBC/IES Standard 189.1-2011, *Standard for the Design of High-Performance Green Buildings*, specifies reductions in LPDs from those specified in ANSI/ASHRAE/IES Standard 90.1-2013 for more aggressive energy savings targets and will be discussed later in this chapter.

The lighting designer will select the appropriate luminaires, sources, and control configurations to create appropriate illuminance levels (measured in footcandles or lux). These selections will be based primarily on the room's intended activity while conforming to the maximum LPD required by code. The fundamental guide used by lighting professionals is the 10th edition of the *IES Lighting Handbook* (IES 2011). It is a highly recommended resource for designers and engineers involved in the design and specification of lighting systems. This handbook comprehensively describes recommended design approaches, illuminance and uniformity levels, power density targets, daylighting integration strategies, luminaire efficiency, and related information necessary for appropriate lighting design.

Efficient Lighting Systems

The lighting industry has made great strides in energy efficiency in the last three decades and continues to undergo a major shift with the advent of high-efficiency lamp/ballast technologies and solid-state light-emitting diode (LED) technology. While LED lighting already offers several energy efficient alternates to traditional solutions and promises to continue to progress in terms of energy efficiency, the higher first cost often makes implementation prohibitive. In general, a lighting designer should perform a detailed life-cycle cost analysis to ensure the investment makes sense. Selected applications for LED lighting that have been proven to be cost-effective solutions over more established technologies are included in this chapter, but generally the approaches discussed herein focus on economical solutions that require reduced or minimized up-front cost.

Tables 13-1, 13-2, and 13-3 contain a listing of one state's standard fluorescent lighting systems and related criteria. These selections will generally satisfy current IES light level recommendations and comply with energy codes while also meeting a modest project budget. Table 13-1 lists two common lighting systems that can be used for a wide variety of project types at very low costs. Table 13-2 lists lighting systems suitable for private and open office areas and similar spaces (e.g., exam rooms in clinics and hospitals). Table 13-3 lists a variety of common lighting systems that can be used in industrial and commercial applications.

Table 13-1: General Use Systems

Primary Application	Luminaire Type	Lamps or Total Lamp Watts	Spacing between Luminaries (in Plan View)	Lamp Ballast System
General use spaces of all types	Nominal 4 ft (1.2 m) recessed or surface-mounted fluorescent troffer (e.g., basket, lensed, etc.) with high-efficiency electronic ballast	Two F32-T8 lamps	No less than 8 ft (2.4 m) OC	Maximum 56 W, 45–48 W, low-ballast-factor ballast preferred
General use spaces of all types	Nominal 2 ft (0.6 m) recessed or surface-mounted fluorescent troffer (e.g., basket, lensed, etc.) with electronic ballast	Three F17-T8 lamps	No less than 8 ft (2.4 m) OC	Maximum 52 W

OC = on center

Table 13-2: Lighting for Offices (including Commercial, Academic, and Institutional)

Primary Application	Luminaire Type	Lamps or Total Lamp Watts	Spacing between Luminaires (in Plan View)	Lamp Ballast System
Open offices	Suspended linear fluorescent fixtures consisting of nominal 4 ft (1.2 m) sections in continuous rows with electronic ballast(s) (ceiling heights 9 ft [2.7 m] or above)	One F54T-5HO, two F32-T8, or two F28T-5	Continuous rows no closer than 15 ft (4.6 m) apart	Maximum 60 input W/4 ft (1.2 m) unit
	Nominal 4 ft (1.2 m) recessed or surface-mounted fluorescent troffers (ceiling heights below 9 ft [2.7 m])	Two F32-T8 lamps	Regular grid 8 ft (2.4 m) OC	Maximum 48 input W/ luminaire
Very small private offices < 105 ft² (10 m²)	One recessed or suspended 4 ft (1.2 m) linear fluorescent fixture (minimum ceiling height 9 ft [2.7 m] for suspended fixtures)	Three F32-T8 lamps	One luminaire per office	Maximum 90 input W
Small private offices 105–125 ft² (10–12 m²)	Two recessed or suspended 4 ft (1.2 m) linear fluorescent fixtures (minimum ceiling height 9 ft [2.7 m] for suspended fixtures)	Two F32-T8 lamps per fixture	No less than 6 ft (1.8 m) OC	Maximum 48 input W to each luminaire

OC = on center

Table 13-2: Lighting for Offices (including Commercial, Academic, and Institutional) *(Continued)*

Primary Application	Luminaire Type	Lamps or Total Lamp Watts	Spacing between Luminaires (in Plan View)	Lamp Ballast System
Small private offices 125–160 ft^2 (10–15 m^2)	Two recessed or suspended 4 ft (1.2 m) linear fluorescent fixtures (minimum ceiling height 9 ft [2.7 m] for suspended fixtures)	Two F32-T8 lamps per fixture	No less than 6 ft (1.8 m) OC	Maximum 56 input W to each luminaire
Medium private offices 160–200 ft^2 (15–19 m^2)	Two recessed or suspended 4 ft (1.2 m) linear fluorescent fixtures Three recessed or suspended 4 ft (1.2 m) linear fluorescent fixtures Four recessed 2 ft (0.6 m) linear fluorescent fixtures (Minimum ceiling height 9 ft [2.7 m] for suspended fixtures)	Three F32-T8 lamps/fixture Two F32-T8 lamps/fixture Two F17-T8 lamps/fixture	No less than 6 ft (1.8 m) OC	Maximum 72 input W to each luminaire Maximum 48 input W to each luminaire Maximum 36 input W per fixture
Executive offices and conference rooms 200-250 ft^2 (19–23 m^2)	Four recessed or suspended 4 ft (1.2 m) linear fluorescent fixtures Four recessed 2 ft (0.6 m) linear fluorescent fixtures (Minimum ceiling height 9 ft [2.7m] for suspended fixtures)	Two F32-T8 lamps/fixture Two F32-T8U lamps/fixture	No less than 8 ft (2.4 m) OC	Maximum 48 input W to each luminaire

OC = on center

Table 13-3: Other Common Lighting Systems

Primary Application	Luminaire Type	Lamps or Total Lamp Watts	Spacing between Luminaires (in Plan View)	Lamp Ballast System
Lobbies, atriums, etc. Industrial spaces	Metal halide, induction, or multiple compact fluorescent lamps (of equivalent lamp watts with electronic ballasts), downlights, pendants, etc.	100 W or less 150 W or less 250 W or less 400 W or less	No less than 12 ft (3.7 m) OC No less than 15 ft (4.6 m) OC No less than 18 ft (5.5 m) OC No less than 22 ft (6.7 m) OC	Mounting height at least 12 ft (3.7 m) above-finished door; only recommended for high-bay spaces
Corridors, lobbies, meeting rooms, etc.	Compact fluorescent (including twin tube, quad tube, or triple tube) or metal halide downlights, wallwashers, monopoints, and similar directional luminaires Wall sconces using any of the above light sources	40 W or less 60 W or less 80 W or less 100 W or less	No less than 6 ft (1.8 m) OC No less than 8 ft (2.4 m) OC No less than 10 ft (3.0 m) OC No less than 12 ft (3.7 m) OC	Any space height

OC = on center

Table 13-3: Other Common Lighting Systems *(Continued)*

Primary Application	Luminaire Type	Lamps or Total Lamp Watts	Spacing between Luminaires (in Plan View)	Lamp Ballast System
Undercabinet and under-shelf task lighting	Hardwired undercabinet or undershelf fluorescent or LED luminaires, nominal 2, 3, or 4 ft (0.6, 0.9, or 1.2 m) in length, with electronic ballast (fluorescent sources)	No greater than 8.5 W/ft (27.09 W/m) of luminaire	No typical spacing requirements. Must be addressed specifically for each application.	Luminaires may be mounted end-to-end if needed to accommodate cabinet length
Lobby, executive office, and conference room accent lighting	Low-voltage downlights, accent lights, or monopoint lights with an integral transformer	Rated at 50 W or less	No less than 8 ft (2.4 m) OC	For accent lighting only; should not be used for general lighting
Copyroom, storeroom, etc.	Nominal 4 ft (1.2 m) recessed or surface-mounted fluorescent troffer, wraparound, strip lights, etc., with electronic ballast	One or two lamps totaling 64 W or less	No less than 8 ft (2.4 m) OC	Maximum 60 input W to each luminaire
Small utility, storage, and closet spaces	Single-lamp fluorescent with electronic ballast (strip, wrap, industrial, or other fixture)	32 W	One luminaire in a closet, electric room, or other small space	Maximum 35 input W to each luminaire

OC = on center

Table 13-3: Other Common Lighting Systems *(Continued)*

Primary Application	Luminaire Type	Lamps or Total Lamp Watts	Spacing between Luminaires (in Plan View)	Lamp Ballast System
Storage and utility spaces	Industrials, wraparounds, strip lights, etc., consisting of nominal 4 ft (1.2 m) sections	Two F32-T8	Individual 12-lamp luminaires 8 ft (2.4 m) OC Continuous single-lamp rows no closer than 8 ft (2.4 m) apart	48 input W per two lamps
Bathroom vanities and stairwells	Two-lamp fluorescent with electronic ballast (wrap, cove, troffer, corridor, vanity, valence, or other fixture)	Two F32-T8	One luminaire per vanity in a toilet or locker room	Maximum 48 input W to each luminaire
Exit signs	LED	No greater than 3.5 W per sign	One luminaire per landing in a stairwell	Maximum 48 input W to each luminaire

OC = on center

For most spaces, designers should use lighting layouts that conform to these criteria. Spacing measurements are taken from the plan-view center of the luminaire. Luminaires should be mounted at least one-third of the indicated mounting distance away from any ceiling-high partition. To allow integration with daylighting, luminaires in rows should ideally be parallel to windowed walls so that controls can be used to step lamps on or off based on the amount of daylight penetration into the room.

The lamp/ballast combination and luminaire type need to be carefully considered to provide visual acuity for the intended task, visual comfort, glare control, efficacy (lumen/watt output), and total system efficiency for any particular space type. The proper specification of lighting equipment is necessary to optimize lighting quality and minimize energy use.

The following luminaire types have cost effective energy alternatives and, therefore, should be avoided:

- Luminaires using Edison base (standard screw-in) sockets or halogen lamps using any sockets rated over 150 W
- Luminaires designed for incandescent or halogen low-voltage lamps exceeding 75 W
- Track lighting systems using any kind of incandescent or halogen
- Line-voltage monopoints allowing the later installation of incandescent or halogen track luminaires.

However, an exception is allowed: for every 20 luminaires meeting these requirements, a single hardwired luminaire of any type (except track), rated not more than 150 lamp W may be placed as desired. This allows architects, interior designers, and lighting designers the ability to add lighting for aesthetic effects or décor without an unreasonable energy burden. Note that if more than one such luminaire is permitted, any number of them may be located in any of the project's spaces. In other words, in a project with 100 types of luminaires from Tables 13-1, 13-2, or 13-3, five decorative luminaires would be permitted, and they could all be installed over a receptionist's desk.

These lists are not intended to be comprehensive but rather straightforward and instructive approaches to basic sustainable lighting design. There are certainly other satisfactory and energy-efficient designs that are not listed. Professional lighting design assistance may be needed to reach optimum performance in complex spaces and for special conditions.

In all cases, the reader should be certain to review the subsequent sections on other aspects of lighting for detailed information on product specifications and additional energy-saving alternatives.

When/Where Applicable

The systems described in the tables above are generally good for most conventional spaces with ordinary ceiling systems. These include the following:

- Typical private offices
- Typical open office areas

- Office area corridors
- Conference rooms and classrooms
- Meeting and seminar rooms
- Most laboratories
- Equipment, server, and cable rooms
- Building lobbies
- Elevator lobbies
- Building core and circulation areas
- Industrial areas, shops, and docks
- Big-box and grocery stores

For commercial buildings, these recommendations assume standard acoustical tile or drywall ceilings (with a reflectance of 80%). For industrial buildings, open bar joist construction is assumed. Ceiling heights and wall and floor finishes are assumed have standard reflectances as well (50% and 20%, respectively). Note that in many newer designs, commercial structures may also have ceilings exposing duct work and piping, which do not meet the reflectance of 80%. Adjustments must be made to account for these conditions. Deep, dark ceilings have lower reflectance and require higher-powered or additional fixtures to meet designed illumination levels.

Efficient Lamps and Ballasts

The lighting industry has made significant improvements in lamp and ballast technology over the last 30 years. Recent improvements continue to permit quality lighting even at greatly reduced power levels. Proper specification of lighting equipment is an important part of actualizing these results.

Specifications. The following specifications are recommended to ensure the latest technology is being used:

- Ballasts for all fluorescent lamps and for high-intensity discharge (HID) lamps rated 150 W and less should be electronic. Harmonic distortion should be less than 20%.
- Four-foot (1.2 m) T8 fluorescent lighting systems should use high-efficiency electronic ballasts. Because instant-start ballasts are the most efficient and least costly, they should be used in all longer duty cycle applications where the lights are turned on and off infrequently. Fluorescent systems controlled by motion sensors in spaces where the lights will be turned on and off frequently should have programmed-start ballasts.
- Designers are strongly encouraged to use ballasts with low-ballast-factors (BF < 0.80) wherever possible. Low-BF ballasts affect light output, so it is important to take this into consideration to reach target illuminance. T8 low-BF and normal-BF ballasts should be high-efficiency electronic, not exceeding 28 input W/4 ft

(1.2 m) lamp at BF > 0.85 and not exceeding 24 input W for BF > 0.70 (American National Standards Institute free air rating). In lieu of the above, electronic dimming ballasts may be used as required by the function of the space.
- Metal halide ballasts 150 W and less should be electronic. Metal halide systems greater than 150 W should use linear-reactor, pulse-start ballasts wherever 277 volt power is available. For other voltages, pulse-start lamps and ballasts should be used.

The lamps listed in Table 13-4 represent the best common lamp types to use. Note that this list is not comprehensive; a better choice for a particular project may be other than those listed here. However, for the majority of applications, this list is a good guide.

The lamps recommended for primary lighting systems are both energy efficient and have long lamp life, representing excellent cost benefits. Lamps for other applications can be less efficient, have shorter lives, or both. Also, designers should strive to minimize the number of different types of lamps on a project to reduce maintenance costs and improve the efficiency of inventory management.

As previously mentioned, LED lamps have entered the market to replace some of the lamps cited in Table 13-4. The reader is encouraged to seek out LED alternatives that produce high-quality light with both high efficacy and long life.

Lighting Power Density Determination Using ASHRAE Standards

ANSI/ASHRAE/USGBC/IES Standard 189.1-2011 specifies reductions in the LPD values given in ANSI/ASHRAE/IES Standard 90.1-2013 and these lower LPD values are encouraged to become the basis of design for high-performance buildings. As discussed, either of two methods are acceptable to use: building area method and the space-by-space method can be used to comply with this standard. Examples are given below for both methods for an office building.

Under the building area method, the ANSI/ASHRAE/IES Standard 90.1 LPD assigned to an office building is listed as 0.90 W/ft^2 (9.65 W/m^2). This LPD is then multiplied by the gross (interior) area of the building to arrive at the maximum allowable installed lighting wattage for the building. This total must be further multiplied by a 95% office LPD factor to conform to the more aggressive Standard 189.1-2011.

An example of determining LPD values by the space-by-space method for an office building follows. Typical spaces found in office buildings were assumed for this example. A hardwired lighting system would be installed with the lighting power density (LPD) requirements shown in the last column of Table 13-5. The area of each space is multiplied by the LPD for that space to determine the maximum connected (hardwired) lighting wattage. For example, for an open office area of 400 ft^2 (37.2 m^2), the maximum wattage for lighting would be (400 ft^2) × (0.83 W/ft^2) = 332 W or (37.2 m^2) × (8.93 W/m^2) = 332 W.

Table 13-4: Common Lamp Types

Generic Lamp Types	Applications	Requirements
4 ft (1.2 m) T8 lamps F28-T8	Primary lighting systems in commercial, institutional, and low-bay industrial spaces	TCLP-compliant (low-mercury) lamps with barrier coat and high-lumen phosphor (minimum 3100 initial lumens) Premium long-life-rated lamp
Fluorescent T5 and T5HO lamps F14-T5, F21-T5, and F28-T5; F24-T5HO, F39-T5HO, and F54-T5HO	Primary lighting systems in commercial, institutional, and low and high bay industrial spaces	Standard T5 and T5HO lamps
Metal halide pulse-start lamps over 250 W	Primary lighting systems in large spaces with very high ceilings and/or special lighting requirements	Pulse-start lamps only—be certain to specify pulse-start ballasts Use linear reactor ballasts on 277 volt systems
Fluorescent T8 lamps F17-T8, F25-T8	Secondary and specialized applications in commercial, institutional, and low bay industrial spaces	TCLP-compliant (low-mercury) lamps with barrier coat and 800 series phosphor Premium long-life-rated lamps
Compact fluorescent long lamps F40T-T5, F50T-T5, and F55T-T5	Specialized applications in commercial, institutional, and low-bay industrial spaces	Standard long twin-tube lamps

Table 13-4: Common Lamp Types *(Continued)*

Generic Lamp Types	Applications	Requirements
Compact fluorescent 4-pin lamps CF13, CF18, CF26, CF32, CF42, CF57, and CF70	Downlighting, wallwashing, sconces, and other common space and secondary lighting systems in commercial, institutional, and low-bay industrial spaces	Standard twin-, quad-, triple-tube, and four-tube lamps.
Halogen MR16 lamps	Accent lighting for art and displays only; do not use for general lighting	Halogen IR 12 volt compact reflector lamps
Ceramic PAR and T HID Lamps PAR20, PAR30, PAR38, ED17, and T-6	Downlighting, accent lighting, and other special, limited applications in commercial, institutional, and low-bay industrial spaces, retail display lighting	39, 70, 100, and 150 W ceramic lamps
Halogen infrared reflecting PAR30 and PAR38 lamps	Applications requiring full-range dimming, retail display lighting	Halogen IR 50, 60, 80, and 100 W reflector lamps

Additional lighting, such as that hardwired to a furniture system, may be acceptable, but should not be specified in a space unless a more complete analysis and design are undertaken. These systems need to be carefully coordinated with the permanent lighting systems of the building. An exception is portable plug-in lamps and manufacturer-installed luminaires on modular furniture, overhead cabinets, bins, or shelves.

Review appropriate LPD requirements for all types of buildings using both ANSI/ASHRAE/USGBC/IES Standard 189.1-2011 and ANSI/ASHRAE/IES Standard 90.1-2013. It is important to note that there are additional requirements regarding lighting control that are not discussed in this chapter.

Table 13-5: LPD Values for Office Building Example (Space-by-Space Method)

Type of Area	Connected LPD limit per Std. 90.1-2013		Percent Reduction of LPD per Std 189.1-2011	Connected LPD limit	
	(W/ft^2)	(W/m^2)		(W/ft^2)	(W/m^2)
Private Office Areas	1.11	11.90	95%	1.05	11.30
Open Office Areas	0.98	10.51	85%	0.83	8.93
Restrooms	0.98	10.51	100%	0.98	10.51
Conference/ Meeting/ Multipurpose Areas	1.23	13.19	90%	1.11	11.87
Lobby	0.9	9.65	95%	0.86	9.17
Elevator Lobby	0.64	6.86	85%	0.54	5.83
Hallways and Corridors	0.66	7.08	85%	0.56	6.01
Lounge/ Recreation Area	0.73	7.83	85%	0.62	6.65
Mechanical and Electrical Rooms	0.95	10.18	100%	0.95	10.18
Stairway	0.69	7.40	100%	0.69	7.40

Application Notes

Always check required illuminance levels using IES guidelines and LPD restrictions using ANSI/ASHRAE/IES Standard 90.1 or another applicable standard or code to ensure compliance.

Open Offices. For general illumination in spaces with ceilings that are 9 ft (2.7 m) or higher, consider suspended linear fluorescent indirect, direct/indirect, or semi-indirect lighting systems, supplemented by task lights. Use high-performance luminaires and T8, T5, T5HO (high-output) lamps. If troffers are preferred, consider using T8 lamps as specified with high-performance ballasts. Task lighting should be added, as required.

Private Offices. Where ceiling heights allow, suspended linear fluorescent lighting should be a first consideration for private offices, although recessed troffers can also be used. Luminaires should use T8 or T5 lamps. Task lights can also be integrated where appropriate.

Executive Offices, Board, and Conference Rooms. Executive offices can be designed similarly to private offices. If desired, a premium approach using compact fluorescent or LED downlights, wallwashers, and/or accent lights can be used, but care should be taken not to exceed the required LPD. Look for new LED lamps coming on the market for downlights and accent lighting to further reduce the wattage; products are improving at a rapid rate. If the number of executive offices is high, lighting power levels should be reduced to match the recommendations for private offices.

Classrooms. Ideally, classroom lighting should be designed using direct/indirect lighting systems using high-efficiency T8 lamps and ballast combinations. Care needs to be taken to insure that all desks are evenly illuminated.

Corridors. Typically, corridors strategies include compact fluorescent or LED sconces, downlights, ceiling-mounted or close-to-ceiling decorative diffuse fixtures, or similar equipment. LED lamps will provide opportunities for changing lighting color as well as having the highest efficacy. Note that these luminaires may be connected to a generator and/or equipped with emergency battery backup when needed as an alternative to less attractive "bug eye" type emergency lighting.

High-Bay Spaces. Industrial, grocery, and retail space without ceilings or with very high ceilings (usually 15 ft [4.6 m] and above) need special lighting fixtures. For mounting heights up to 20 to 25 ft (6.1 to 7.6 m), first consider LED or fluorescent industrial luminaires with T8 lamps, keeping in mind that two super 4 ft (1.2 m) T8 lamps and a high-light-output overdrive ballast at 77 W produce as much light (mean lumens) as a 100 W metal halide lamp that, with ballast, consumes 120 W. Four super T8 lamps with overdrive ballasts at 154 W produce as much light as a 175 W pulse-start metal halide (195 to 205 W) or a standard 250 W metal halide lamp (286 to 295 W with ballast). A study by the Lighting Research Center (LRC) of the Rensselaer Polytechnic Institute indicated that T8 fluorescent lamps provide energy savings over high-bay metal halide lamps in typical warehouse applications.

For mounting heights above 25 ft (7.6 m), consider T-5HO high-bay luminaires. Similar savings relative to metal halide are possible. High-wattage metal halide should be reserved for very high mounting (above 50 ft [15.2 m]) and for special applications (e.g., sports lighting).

Other Applications. The following luminaire types are generally recommended for these areas:

- *Artwork, bulletin/display surfaces, etc.:* use compact fluorescent wall washers or low-voltage monopoint lights. Make sure to pay attention to the color rendition index (CRI) of the specific lamps when illuminating art work. A higher CRI will render colors better.
- *Utility spaces (e.g., cable and equipment rooms):* use two-lamp strip lights, industrials, or surface luminaires equipped with high-efficiency lamp/ballast combinations.
- *Lobby spaces, cafeterias, and other public spaces:* use appropriate selections from among these luminaires as much as possible.

DAYLIGHT HARVESTING

General Description

Most buildings are designed to have some type of natural light that is transmitted through windows and/or skylights. The majority of commercial, industrial, and institutional buildings have windows and, in some cases, skylights, tubular daylighting devices, clerestories, or more extensive fenestration systems.

From an energy perspective, the optimal use of daylight is to reduce the load of the electric lighting system by dimming or switching off luminaires when natural light provides ample illuminance for the tasks performed in the space. It is important to note that the incoming light is only usable if it is controlled to reduce glare and illuminates a space in a way that is comfortable to the occupant (including the consideration of thermal comfort). This process of reducing electric lighting in the presence of daylight, which is known as *daylight harvesting* (see Figure 13-1), is discussed in this section because of its significant energy saving potential. The prediction of daylight harvesting savings is a complicated process that involves a comprehensive understanding of the site, building orientation, weather conditions, materials, and system integration. There are added capital costs for daylight harvesting elements such as dimming ballasts and photoelectric controls. It is important to justify these costs by accurately predicting the potential energy savings of daylight harvesting techniques. However, when the challenges of daylighting are appropriately addressed, significant energy savings are possible.

Image courtesy of Cannon Design.

Figure 13-1 Example of daylight harvesting rendering.

From a lighting design perspective, daylight can be treated similarly to any other light source, so it can be used to compose lighting design solutions that take illuminance, luminance, contrast, color rendition, and other lighting design elements into consideration. However, the lighting designer is challenged to deal with the fact that the light source varies daily and throughout the year. The designer can use blinds, shades, curtains, moveable shutters, light shelves, light conveyors, and other mechanical forms of attenuation and shielding to control daylight.

It is possible to simulate the performance of daylight to determine the amount, and to a certain extent, the quality of available daylight under varying conditions of season, time of day, and weather. However, this is exhaustive analytical work of a highly specialized nature, and it is recommended that appropriate experts perform such studies. In the meantime, some buildings can benefit tremendously from some simple daylight-harvesting considerations.

Basic Toplighting

Basic toplighting typically involves using simple skylights or tubular daylighting devices in the roof. (This is not to say that other toplighting configurations [e.g., the clerestory, roof monitor, or sawtooth roof] are not workable; indeed, they often have advantages over horizontal skylights.) There are many architectural elements, including structure, waterproofing, and other details, that should be considered when exploring the appropriateness of skylights. When used in a manner similar to light fixtures and laid out to provide uniform illumination, toplights are an acceptable way to illuminate single-story, large spaces or the top floor of a multi-story building. Toplighting is best when a number of smaller skylights are used, in much the same way effective lighting systems have many light fixtures rather than one larger light source in the middle of the room. Skylights should be diffuse or prismatic rather than clear to avoid harsh glare from direct sunlight. When layouts adhere to these guidelines, skylights generally do not have to incorporate mechanical or electronic light control louvers, because the optimum size of the skylight is chosen for passive daylighting (i.e., typically, there are no active or moving elements needed to regulate the amount of interior light). If installed in an area requiring the ability to "black out" or darken the space (e.g., for showing movies or videoconferencing), some type of shading should be provided along with controls.

ANSI/ASHRAE/USGBC/IES Standard 189.1-2011 requires toplighting for buildings of three stories or less having a large enclosed space of 20,000 ft^2 (2000 m^2) located "directly under a roof with finished heights greater than 15 ft. (4 m) and that have a lighting power allowance for general lighting equal to or greater than 0.5 W/ft^2 (5.5 W/m^2)." Section 8.3.3.3 of the standard details the minimum amount of toplighting required.

To determine the optimum size of skylights, one can download a program called SkyCalc available at www.energydesignresources.com/resource/129/. This program, which is optimized for California and the northwestern region of the United States, can be applied (with appropriate care) anywhere in North America. It takes into account location, utility rates, and other basic data, and yields recommended skylight area.

Note that the ideal amount of fenestrated roof is generally about 5%, but designers should always review governing standards and codes. All too often, architects, although well-intentioned, design skylights that are too big. An HVAC&R designer's input here, especially when backed up by calculations, can help ensure that energy is not wasted.

When/Where It's Applicable

Daylight harvesting is best suited for skin-load-dominated (SLD) structures, especially single-story or narrow, multistory buildings. It is also an effective strategy for the top story of an internal-load-dominated (ILD) building. In structures where merchandise is sold, it has been found that sales increase for products displayed under skylights. Therefore, large-volume, single-story structures often used for big-box retail stores can still benefit from daylighting, because vertical glazing would be ineffective in bringing light to the deep interior.

Buildings serving several functions can benefit from daylighting, including the following types of buildings:

- Gyms
- Industrial work spaces
- Big box retail stores
- Grocery stores

- Exhibition halls
- Storage
- Warehouses
- Office buildings (if skin load dominated)
- Classrooms

It is essential to use automatic lighting controls programmed to dim or extinguish electric luminaires to efficiently integrate daylighting Without them, the energy savings will not be realized. (See the later section on Lighting Controls.)

Pros of Daylighting

- Daylight harvesting offers significantly reduced energy consumption (exceeding 60%) and reduced HVAC load (as long as solar gains do not outweigh electric lighting reductions).
- Daylight extends the electric lighting maintenance cycle (lamps can last two to three times as long in calendar years with proper selection of fluorescent lamp/ballast combinations).
- Daylight has been shown to lead to improved human factors and increased enjoyment of space.
- Merchandise daylit under skylights sells at a faster rate.
- Daylight assists in maintaining proper circadian rhythm (which is especially important to sufferers of seasonal affective disorder [SAD])

Cons of Daylighting

- Sunlight is intermittent and variable.
- Daylight harvesting requires intensive architectural, structural, and lighting design coordination.
- There is a potential for glare and, therefore, glare control devices are often required.
- The user must be educated on the proper use and maintenance of the components.
- There is no assurance that the design will meet exact project lighting requirements.
- There can be higher construction costs and higher maintenance.
- There is a risk of poor design or installation workmanship, which can result in roof leaks.
- Daylighting may not be suitable for uncommon room shapes, sizes, and/or finishes.
- There is net decreased roof insulation when skylights are installed.
- Even when using high-performance glazing, there will be significantly larger heat loss than that lost through a solid insulated wall in winter. Additionally, occupants in close proximity to glazing during colder periods may experience discomfort.
- Heat gain may occur at undesirable times.

LIGHT CONVEYORS (TUBULAR DAYLIGHTING DEVICES)

Light conveyors use a specialized technique whereby light from a source is transmitted some distance from the source to illuminate spaces, either along its length or some distance away. Some devices gather light and transmit it down a fiber-optic tube bundle to light an area at a distance; with sensors and lamps in the room it serves to seamlessly ensure constant illuminance. The source can be either natural light or an artificial source. Light conveyors are further described in Green-Tip #13-1.

LIGHTING CONTROLS

General

While all modern energy codes require automatic shutoff controls for commercial buildings, implementing automatic controls in all building projects is a sound money- and energy-saving idea. There are two ways to reduce lighting energy use through controls:

- Turn lights off when not needed (which reduces hours).
- Reduce lighting power to minimum need (which reduces kilowatts).

With very few exceptions, each interior space enclosed by ceiling-high partitions must have a separate local switching or dimming device and some form of automated "off" control (e.g., occupancy-sensing, time-based scheduling, or other) to meet code. In addition, wherever possible, providing separate switching/dimming for lights in daylit zones can further reduce energy usage or provide occupant control in situations where overrides to provide more light are required. To comply with code requirements and ensure maximum energy savings, specify the most appropriate lighting control option(s) as outlined in Table 13-6 and described in the following paragraphs.

Control Options

Below are some considerations for useful lighting control system components that correlate with recommendation in Table 13-6:

1. *Ceiling-mounted motion sensor with transformer/relay, auxiliary relay, and series switch.* The sensor should be located to look down on the work area, so as to detect anything from a small hand motion to major movements. The sensor may be mounted to the upper wall if a ceiling location is not workable. More than one sensor can be used for a large room or a room with obstructions (e.g., a library or server room). In such situations, sensor coverage zones should be slightly overlapped to ensure comprehensive coverage. The main transformer relay should control the overhead lighting system

(usually 277 volt) and the auxiliary relay should control at least one-half of a receptacle to switch task lights and other applicably controlled plug loads. Note that the light switch is in series so that it can only turn lights off in an occupied room; it cannot override the motion sensor's "off" control.

2. *Ceiling-mounted motion sensors connected to programmable time controller.* During programmed "on" times, the lights remain on. During programmed "off" times, motion within the space initiates lights on for a time out period. The controller should be programmable according to the day of the week and should have an electronic calendar to permit programming holidays.
3. *Workstation motion sensor connected to a plug strip or task light with auxiliary receptacle.*
4. *One or more ceiling-mounted motion sensors with transformer/relay (minimum of two luminaires controlled.)*
5. *Switchbox motion sensor (one or more luminaires controlled).*
6. *Programmable time controller with manual override switch(es) located in a protected or concealed location.* There are separate zones for retail and similar applications where displays can be controlled separately from general lighting. This may also control dimmers.
7. *Automatic daylighting sensor connected to dimming ballast(s) in each luminaire in the daylit zone (in addition to any of the above).*
8. Motion sensor connected to a high-low lighting system.

When using controls such as motion sensors or daylight sensors, be very thorough and carefully read the manufacturer's literature. Different sensors work for different applications, and their sensing systems are optimized. For instance, avoid wallbox motion sensors except in spaces where their sensing field is appropriate (e.g., small private offices, individual toilets, etc). For spaces with small-motion work, a lookdown sensor (situated on the ceiling) generally works much better than a lookout sensor (situated on a wallbox). Consider the sensing technology when specifying sensors. Infrared, ultrasonic, and microphonic sensor technologies present opportunities to fine-tune lighting control system design and can enhance the overall effectiveness of the lighting system.

Applicability

The controls mentioned in the previous section are applicable to most commercial, institutional, and industrial buildings. Use common sense in special spaces, keeping in mind safety and security. Never switch path-of-egress lighting systems except with properly designed emergency transfer controls.

Table 13-6: Recommended and Optional Lighting Controls

Type of Space	Minimum Recommended Control	Optional Control(s)
Private Office, Exam Room	1	2 2 + 4 1 + 8 2 + 4 + 8 2 + 8
Open Office	3	3 + 4 3 + 8 3 + 4 + 8
Conference Rooms, Teleconference Rooms, Boardrooms, Classrooms	2	1 2 + 8 1 + 8
Server Rooms, Computer Rooms, Other Clean Work Areas	5	
Toilet Rooms, Copy Rooms, Mail Rooms, Coffee Rooms	5 or 6	
Individual Toilets, Janitor Closets, Electrical Rooms, and Other Small Spaces	6	
Public Corridors	3	
Corridors, Hallways, Lobbies (Private Spaces Only)	3	3 + 8
Public Lobbies	7	7 + 8
Industrial Work Areas	7	7 + 8

Table 13-6: Recommended and Optional Lighting Controls

Type of Space	Minimum Recommended Control	Optional Control(s)
Warehousing and Storage	9 (high-intensity discharge [HID] systems) 3 or 5 (fluorescent systems)	3 or 5 + 8 (fluorescent)
Stores, Newsstands, Food Service	7	
Mechanical Rooms	Manual switching only	
Stairs	None	Motion sensors can be used to reduce light levels to minimum egress lighting levels only
Hotels		Standard 189.1-2011 requires automatic controls to turn power to lights off within 30 minutes of guests leaving their guest rooms.

Pro

- There is low to moderate cost for most space types.
- There is virtually no maintenance.
- These controls will generally lower energy use.

Con

- If controls are not properly commissioned, unacceptable results may occur until they are fixed.
- Poorly installed and/or commissioned controls negatively affect system performance.
- Substitutions and value engineering can cause bad results.

COST CONSIDERATIONS

Lighting Systems

The systems described above are generally low-to-moderate cost lighting systems. On average, they also use low-maintenance lamps and ballasts. The combination of low first cost, low maintenance, and low energy use leads to lighting choices that are among the most economical available.

Users should consider life-cycle costs when evaluating the use of LEDs. Although high in first cost, LED lamps have significantly higher efficacies and longer life, which can offset first cost in the course of the life cycle. Note, however, that LEDs installed in areas where they will overheat will experience shorter life.

Lamps and Ballasts

Premium lamps and ballasts can be used instead of conventional lamps and ballasts for as much as 100% of the bill of materials. Depending on the selections, this can increase the cost of a lighting system by 20% to 30%, although many utilities, at least in the United States, offer rebates for using these higher-efficiency products. These premium lamps, especially the T8 4 ft (1.2 m) lamps shown, offer the following specific benefits:

- *Increased lamp life by as much as 50%.* In a T8 application, this can be 5000 to 10,000 h. The cost of replacing a lamp is about 75% labor and 25% material. Relamping cost savings alone pay the difference.
- *Reduced lamp energy use when used with the correct ballast.* In a typical T8 application, this means achieving energy savings of around 6 W/lamp. At 3000 annual hours and $0.085/kWh, the combination saves more than $1.50/yr in energy costs per lamp. The premium for a two-lamp ballast and lamps is about $12.00. The energy savings pay for the added costs in about four years. (Costs are as of year 2003.)

Daylight Harvesting

Daylight harvesting is a potentially complex undertaking in which the first cost of lighting remains the same, the cost of lighting controls increases, and the added cost of skylights and/or structural changes/complications are incurred as well. To be cost-effective, this needs to be offset by a combination of HVAC energy savings, lighting energy savings, HVAC system first-cost reduction, and perhaps savings from utility incentives or tax credits. Expect daylight-harvesting systems to yield a 4 to 5 yr simple payback with a utility incentive, 6 to 8 yr or more without.

Controls

The lighting control systems described previously are generally of low to moderate cost. However, using better quality sensors and separate transformer/relay packs with remote sensors costs much more than wallbox devices. Savings can range from modest to considerable, depending on the building and occupants.

REFERENCES AND RESOURCES

Published

ASHRAE. 2011. ANSI/ASHRAE/USGBC/IES Standard 189.1-2011, *Standard for the Design of High Performance Green Buildings*. Atlanta: ASHRAE.

ASHRAE. 2013. ANSI/ASHRAE/IES Standard 90.1-2013, *Energy Standard for Buildings Except Low-Rise Residential Buildings*. Atlanta: ASHRAE.

Benya, J. and Peter Schwartz. 2010. *Advanced Lighting Guidelines*. Fair Oaks, CA: New Buildings Institute. http://newbuildings.org/advanced-lighting-guidelines-0.

IEA. 2000. *Daylight in Buildings: A Source Book on Daylighting Systems and Components*. Berkeley, CA: International Energy Agency (IEA) Solar Heating and Cooling Programme, Energy Conservation in Buildings & Community Systems http://gaia.lbl.gov/iea21/ (Free download).

IES. 2011. *Lighting Handbook*. New York: Illuminating Engineering Society of North America.

Online

Advanced Lighting Guidelines Online
www.advancedbuildings.net.

Architectural Energy Corporation
www.archenergy.com.

BetterBricks
www.betterbricks.com.

California Energy Commission
www.energy.ca.gov.

Collaborative for High Performance Schools
www.chps.net.

Energy Design Resources
www.energydesignresources.com/.

Energy Trust of Oregon
www.energytrust.org.

EIA, U.S. Energy Information Administration
www.eia.gov.

Lighting Research Center at Rensselaer Polytechnic Institute
   www.lrc.rpi.edu/programs/NLPIP/publications.asp.
Natural Resources Canada
   http://oee.nrcan.gc.ca/node/17725#bb.
New Buildings Institute
   www.newbuildings.org.
RealWinWin, Inc.
   www.realwinwin.com.
Rising Sun Enterprises
   www.rselight.com.
Savings by Design
   www.savingsbydesign.com.
SkyCalc
   www.energydesignresources.com/resource/129/.
U.S. Department of Energy
   www.energy.gov.

ASHRAE GreenTip #13-1

Light Conveyors (Tubular Daylighting Devices)

GENERAL DESCRIPTION

A light conveyor is pipe or duct with reflective sides that transmits artificial or natural light along its length. There are three types of such light-directing devices: plastic square or round duct with curves or bends that transmits light along its length; a relatively straight tube having an opaque, highly reflective interior coating which transmits daylight or high-intensity electric light to an interior space; and a bundle of multiple fiber-optic tubes which transmit light from the outside (horizontally or vertically) to the interior.

The first type is a square duct or round pipe made of plastic. Based on how the inside of the duct or pipe is cut and treated, light entering one end of the pipe is reflected off these configurations (similar to the way light is refracted through a prism) and transmitted through. The reflected light continues to travel down the pipe and out the other end. Because some light is absorbed and escapes along the length of the pipe (i.e., it is lost), the maximum distance that light can be piped into a building is generally about 90 ft (27.4 m). There are a few installations where sun-tracking mirrors concentrate and direct natural light into a light pipe. In most applications, however, a high-intensity electric light is used as the light source. Having the electric light separate from the space where the light is delivered isolates the heat, noise, and electromagnetic field of the light source from building occupants. In addition, the placement of the light source in a maintenance room separate from building occupants simplifies replacement of the light source.

The second type of light-directing device is a straight tube with a highly reflective interior coating. The device is mounted on a building roof and has a clear plastic dome at the top end of the tube and a translucent plastic diffusing dome at the bottom end. The tube is typically 12 to 16 in. (300 to 400 mm) in diameter. Natural light enters the top dome, is reflected down the tube, and is then diffused throughout the building interior. The light output is limited by the amount of daylight falling on the exterior dome.

A second type of tubular daylighting device uses the technology of fiber optics to transmit light into interior zones of a building. The most complex designs incorporate sensors, controls, and LED lamps to maintain relatively constant light output.

WHEN/WHERE IT'S APPLICABLE

Light-conveyor systems are best suited to building applications where there is a need to isolate electric lights from the interior space (e.g., operating rooms or theaters) or where electric light replacement is difficult (for example, above swimming pools or in roadway tunnels). For the latter, reflective-tube system, each device can light only a small area (10 ft^2 [1 m^2]) and is best suited to small interior spaces with access to the roof, such as interior bathrooms and hallways.

PRO

- A light conveyor transports natural light into building interiors.
- The first type of light conveyor isolates the electric light source from the lighted space.
- The first type of light conveyor reduces lighting glare.
- It lowers lighting maintenance costs.

CON

- A light conveyor may have greater capital (first) cost than traditional electric lighting.
- The tube type may increase roof heat loss.
- The tube type runs the risk of poor installation, resulting in leaks.
- The effectiveness may not be worth the additional cost.

KEY ELEMENTS OF COST

Because of the specialized nature of these techniques, it is difficult to address specific cost elements. As an alternative to conventional electric lighting techniques, it could add to or reduce the overall cost of a lighting system—and the energy costs required—depending on specific project conditions. A designer should not incorporate any such system without thoroughly investigating its benefits and applicability, and should preferably observe such a system in actual use.

SOURCES OF FURTHER INFORMATION
McKurdy, G., S.J. Harrison, and R. Cooke. Preliminary evaluation of cylindrical skylights. 23rd Annual Conference of the Solar Energy Society of Canada Inc., Vancouver, British Columbia.

CHAPTER FOURTEEN

WATER EFFICIENCY

Water efficiency and conservation continues to become a critical factor in green-building design. Buildings consume 20% of the world's available water, a resource that becomes scarcer each year, according to the United Nations Environmental Program (Epoch Times 2009). Efficient practices and products provide opportunities to save significant amounts of water. The reduction of energy use and operating costs, and the expectation of increased government regulation will continue to drive faster adoption of water-efficient products and methods.

In a typical commercial building when the HVAC system employs a water-cooled chiller, the HVAC system can account for approximately one third of water consumption. Therefore, minimizing the water needed to operate HVAC systems while not significantly increasing energy usage should be a major consideration in green-building design.

Plumbing and fire protection systems are normally not considered within the purview of the HVAC&R designer's expertise. Nevertheless, both subsets of designers, in practice, must work closely in putting together a functional building mechanical system. Indeed, frequently the designing firm for the HVAC systems and plumbing systems are one and the same. For detailed design guidelines and information on these systems, please refer to the National Fire Protection Association (NFPA) and the American Society of Plumbing Engineers (ASPE).

Recently, ASHRAE partnered with the American Water Works Association (AWWA), U.S. Green Building Council (USGBC), and ASPE to develop a new proposed standard titled Standard 191, *Standard for the Efficient Use of Water in Building, Site and Mechanical Systems.* The target date for the final standard to be published is late 2013. You can follow the progress of this proposed standard at http://spc191.ashraepcs.org/.

In green-building design, it is important for the practitioners of each design discipline to be familiar with what the other disciplines may bring to an effective green design. This holds true as well with plumbing design. The editors of this guide have chosen to include a discussion of some key aspects of plumbing design that impact green design, including several ASHRAE GreenTips. Several of these GreenTips may have an impact in other areas as well. For instance, point-of-use hot-water heaters (GreenTip #14-3) not only save heating energy and distribution energy, but they also often result in the use of less water.

THE ENERGY-WATER BALANCE

The continued security and economic health of the population depends on a sustainable supply of both energy and water. These two critical resources are inextricably and reciprocally linked. A nation's ability to continue providing both clean and affordable energy and water is being seriously challenged by a number of emerging issues.

Energy production requires a reliable, abundant, and predictable source of water, a resource that is already in short supply throughout much of the United States and the world. Electricity generation is second only to agriculture as the largest user of water in the United States, although some water listed as consumption is returned to waterways, albeit at a higher temperature, which does lead to a greater evaporation rate. Electricity production from fossil fuels and nuclear energy in 2005 required 201,000,000 million gallons (761,000 m^3) of water per day, accounting for 41% of all freshwater withdrawals in the nation (U.S. Geological Survey).

According to the World Health Organization, approximately 1.1 billion people do not have access to improved water supply sources. Two primary solutions exist: shipping water over long distances or cleaning nearby, but dirty, water supplies. Both of these options require large amounts of energy. Even if you have ready access to a good, clean water supply nearby, there is a significant amount of embodied energy in the water we use to drink, cook, flush toilets, and bathe.

In areas where water supply is not plentiful, an engineer focusing on green design should take into account the total energy as well as water consumption required to operate a cooling plant, including the embodied energy of the water. For example, if a life-cycle analysis is performed to compare an air-cooled, chilled-water plant to a water-cooled, chilled-water plant, the total energy of the air-cooled plant may end up being approximately the same as the total energy that would be used by a water-cooled plant if, for example, the water-cooled plant was consuming desalinated water that was delivered to the site.

Therefore, when evaluating predicted energy (and carbon) loads for alternative building system design options, it is important to look at the embodied energy of the water in areas where water is not a plentiful resource.

WATER SUPPLY

This basic resource is obviously essential at every building site, but what has changed over the last several decades is the realization that it is fast becoming a precious resource. While the total amount of water in its various forms on the planet is finite, the amount of fresh water, of a quality suitable for the purposes for which it may be used, is not uniformly distributed (e.g., 20% of the world's fresh

ASHRAE GreenGuide | 401

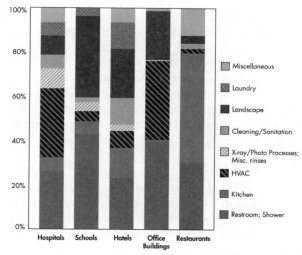

Image courtesy of Environmental Building News 17:2.

Figure 14-1 Water use in buildings.

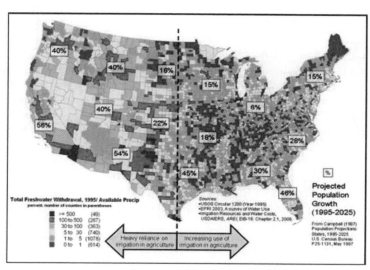

Image courtesy of Universities Council on Water Resources
Journal of Contemporary Water Research and Ed/Sandia National Laboratories.

Figure 14-2 United States water availability map.

water is in the U.S. Great Lakes, while elsewhere it is often nonexistent or in very meager supply). Nevertheless, water must be allocated somehow to the world's populated lands, many of which are undergoing rapid development. In short, it is becoming more and more difficult to provide for the adequate and equitable distribution of the world's water supply to those who need it.

This trend has implications for not only how prudently we use the water we have, but what we do to avoid contaminating water supplies. While many of the measures to protect and preserve the world's freshwater supplies are beyond the design engineer's purview, there are a number of simple things related to building sites that can be done as part of a green design effort.

COOLING TOWER SYSTEMS

Since cooling towers are commonly used in larger-scale building systems and they consume water through evaporation, drift, and blowdown losses, the following section provides details on cooling tower operations, particularly as they affect green-building design and water efficiency concern.

Cooling towers are efficient devices for removing heat from water through evaporation. Cooling towers remove heat by evaporation and can cool close to the ambient wet-bulb temperature. The wet-bulb temperature is always lower than the dry-bulb; thus, water cooling allows more efficient condenser operation (as much as a 50% energy savings) than air cooling. However, cooling towers use a significant amount of water. A typical tower operation will evaporate 3 gpm of water per 100 tons of cooling capacity. In addition to evaporation, some recirculated water must be bled from the system to prevent soluble and semisoluble minerals from reaching too high a concentration. This bleed or blowdown is usually drained to the sanitary plumbing system and ends up being treated at a sewage waste treatment plant. By treating the water with chemical-free systems, by appropriately addressing splash-out and spill issues, and by using filtration systems with low water usage for the blowdown requirement, it is possible to maximize both the energy and the water efficiency of the cooling tower system.

It is important, from a sustainability perspective, that a proper tower design is selected. The tower that is selected must be energy efficient as defined below and designed to minimize splash losses, spills, drift, and algae growth. For more information, consult Chapter 40 in the 2012 *ASHRAE Handbook—HVAC Systems and Equipment* (ASHRAE 2012).

Drift

To promote efficient heat transfer through evaporation, cooling towers force intimate contact between outdoor air and warm water. During this heat exchange process dust, pollen, and gas, components of the air become entrained in the water, while some lower-vapor pressure components in the water (e.g., bromine or chlorine), as well as entrained water drops, migrate to the air. Airborne dust and pollen that are captured by the water can promote biological growth in the tower.

Small water droplets entrained and carried out with the air passing through the tower are called *drift*. Drift is always present when operating a cooling tower. Because drift is generated by small droplets of the cooling tower water, it contains all of the dissolved minerals, microbes, and water treatment chemicals in the tower water. Drift is a source of PM10 emissions (i.e., particulate matter that is less than 10 μm in diameter) and, since it can contain biological growth suspended in the water, it is a suspected vector in *Legionella* transmission.

Drift is usually reported as a percentage of the recirculating water flow rate, though it is more accurately described in terms of the parts per million (ppm) of the air passing through the tower. Tower designs use drift eliminators to capture virtually all of this entrained water. A typical value for drift high-efficiency eliminator performance in modern cooling towers is 0.001 to 0.005% of the recirculating water rate.

It should further be noted that since drift is liquid water being aspirated from the cooling tower, is it simply one component of tower blowdown. Because of this, the drift rate has no impact on total water consumption since the amount of drift will result in a commensurate reduction in blowdown.

Water Consumption of Typical Cooling Tower

In times of increasing pressures on water supplies, consider as an example a 400 ton (1407 kW of refrigeration) rated cooling system. This system would circulate approximately 1200 gpm (4542 L/min) through the tower and chiller at full cooling load. At nominal rates, 12 gpm (45.4 L/min), or about 1% of the total system flow, would be lost to evaporation, while 0.06 gpm (0.23 L/min) would be lost to drift (0.005%), and, at four cycles of concentration, 4 gpm (15.1 L/min) would be intentionally bled from the system.

Table 14-1 shows monthly results for operating this 400 ton rated tower at an assumed 75 hr/week or 300 hr/month. In this table, it is assumed that the tower operates at an average of 60% of its rated capacity, or 240 tons cooling average for 300 hours per month. These values also assumes that the water flow rate is reduced proportionally using a variable-speed pumping system.

Depending on the water treatment system used, this cooling tower would also be sending a blowdown waste stream potentially containing heavy metals, phosphates, and biocides to a publicly owned treatment works (POTW) system. Most POTW systems are designed to handle only organic waste; much of these cooling tower chemicals will pass through the system untreated or will be released later as gaseous emissions at the POTW. In addition, harmful air emissions may result from the water treatment chemicals used, again depending on the type of water treatment.

Table 14-1: Water Consumption from Example Cooling Tower Operation

	Flow at Rated Capacity	Total/Month (Assuming 60% Average Capacity)
Evaporation	12 gpm (45.4 L/min)	129,600 gal (490,536 L)
Bleed at Four Cycles	4 gpm (15.1 L/min)	43,200 gal (163,512 L)
Drift	0.060 gpm (0.23 L/min) (0.005%) 0.012 gpm (0.05 L/min) (0.001%)	648 gal (2453 L) 130 gal (491 L)

Over the lifetime of the building, these releases could be among the more significant impacts on the local environment that the building will cause. This example highlights the magnitude of what can happen if this issue is not addressed.

Green Choices—Water Treatment

The water in the evaporative cooling loop must be treated to minimize biological growth, scaling, and corrosion. Typically, a combination of biocides, corrosion inhibitors, and scale inhibitors are added to the system.

Corrosion inhibitors are usually phosphate- or nitrogen-based (e.g., fertilizers) or molybdenum- or zinc-based (e.g., heavy metals). These inhibitors are more effective when added in combinations. These materials have low vapor pressure and are not used by the system. The inhibitors simply need to remain in the solution at the proper concentration to maintain a protective film on the metal components. Their only loss is through bleed and drift.

Most scaling inhibition is done by polymer-based chemicals, organic phosphorous compounds (e.g., phosphonates), or by acid addition. The acid reacts with the alkalinity in the water to release CO_2 and is used up. The polymer and phosphonate scale inhibitors remain in the solution to delay scaling; their major loss is through bleed and drift. Some polymers are designed to be biodegradable (i.e., easily broken down by bacteria in the environment) while others are not.

There are very wide assortments of biocides. A typical system will maintain an oxidizing biocide (e.g., bromine or chlorine) at a constant level and slug feed a nonoxidizing biocide once a week. Chlorine and bromine have a high vapor pressure in water. Much of the chlorine and bromine added to the tower is stripped from the water into the air; a small quantity actually reacts with organics in the tower. Drift and bleed will contain all of the nonoxidizing biocides and a small quantity of oxidizing biocides and the reaction products of the biocides.

There are, however, many ways to treat the system that will have a less negative impact on the environment. Besides being rapidly stripped into the air, chlorine and bromine may react with organic molecules to produce very hazardous daughter products. However, other oxidizing biocides (e.g., hydrogen peroxide) do not have this issue. With continuous monitoring of the cooling system, chemical additions can be added only when needed. This technique can yield equivalent performance with less-added chemicals. Also, the U.S. Environmental Protection Agency (EPA) maintains a website on green chemistry (www.epa.gov//greenchemistry/) which contains criteria on how to evaluate the life-cycle environmental impact of a particular chemical.

Nonchemical water treatment has the potential to be a powerful method for water treatment. However, its successful use depends on the water chemistry, operating procedures, and degree of pollution of the specific system. There are several different nonchemical technologies available including those based on pulsed electric fields, mechanical agitation, and ultrasound. Each of these technologies has developed a widespread following. These technologies offer the potential to substantially reduce or eliminate the storage and handling of toxic chemicals at the site. By not adding chemicals to the cooling tower water system, there is a reduced risk from spills or leaks of cooling tower water, the bleed water will be better to send to the local POTW, and any concern associated with drift will be reduced. Bleed water, instead of being sent to the POTW, could possibly be used on site for irrigation or other needs where potable water is not required.

ASHRAE has conducted several research projects that investigated the water treatment options and their effectiveness. For example, the effectiveness of scale prevention of some of these nonchemical water treatment technologies was the focus of ASHRAE Research Project RP-1155 (Cho et al., 2003). Others, such as ASHRAE Research Project RP-1361 raise question regarding the overall efficacy of alternative methods. Technologies for water treatment continue to be developed and the reader is encouraged to investigate thoroughly any option chosen and evaluate all the trade-offs.

Green Choices—Tower Design and Selection

All cooling tower designs are efficient at removing heat from water through evaporation. However, not all designs perform as well environmentally. Some tower designs are more prone to splashout, spills, drift, and algae growth than others. Splashout involves tower water splashing from the tower. This happens most often in no-fan conditions (i.e., when circulating tower water with the fans off) when there are strong winds. Cross-flow towers are more prone to this issue since, in no-fan condition, some water will fall outside of the fill.

Spills can happen from the cold-water basin from overflowing at shutdown when all of the water in the piping drains into the basin. Proper water levels will prevent this. Some tower designs use hot-water basins to distribute water at the top of the tower, while other designs use a spray header pipe. The hot-water basin design can overflow if the nozzles clog; a spray header pipe never overflows.

Algae are a nuisance in basins and can contribute to microbial growth. Algae control requires harsher chemical treatment than typical biological control. Since algae are plants, they need sunlight to grow. Some tower designs are light-tight, which completely eliminates algae as an issue in cooling towers, while other designs are more open and algae growth can be an issue.

The amount of drift varies extensively in tower design. Some tower designs have very little drift, and the less the drift from a tower, the lower the amount of water containing minerals, water treatment chemicals, and microbes that will be released into the surrounding environment.

Fan-power draw also varies between designs. All induced-draft cooling towers and some forced-draft designs use axial fans. With centrifugal fans, the fan blades are mounted parallel to the axis. Centrifugal fans are only used in forced-draft towers due to the higher overall power consumption for the same airflow.

Green Operation and Maintenance

An often overlooked method to minimize environmental impact is maintenance. Cooling towers operate outdoors under changing conditions. Wind damage to inlet air louvers, excessive airborne contamination, clogging of water distribution nozzles, and mechanical problems can best be prevented and quickly corrected with periodic inspections and maintenance. Continuous filtration of the circulating water and basin sweeping systems should be considered to keep the basins free from all visible dirt.

Green Choices—Filtration for Water Cooled Systems

A critical consideration with respect to green operation of water-cooled systems is to choose to design, install, and maintain a filtration system capable of continually removing visible solids (40 µm and larger) from the water-cooled system. Filtration of water-cooled systems is essential for sustainable operation. Keeping the system free from dirt, corrosion by-products, and scale maintains optimal heat transfer efficiency and reduces under-deposit corrosion. Filtration systems that use centrifugal action and require no media or moving parts have the advantage of reduced maintenance, no disposal of used media, better water efficiency, and offer minimal or zero liquid loss that reduces chemical usage and disposal.

The filtration system and sweeper piping should be able to provide adequate pressure and flow (constant 20 psi at the nozzles, 1 gpm/ft² [138 kPa at the nozzles, 40.8 L/min per m²]) to remove visible solids from the water and provide for efficient handling of the solids. It is not necessary to remove particles that are less than 40 microns (drinking water is filtered to 5 microns), and the extra energy and water loss to achieve this level of filtration for a water-cooled system is wasteful.

While it is obvious that water purity must meet the health and safety standards prescribed by the authorities, there are other techniques that can contribute not only to using lesser quantities of the highest-quality water needed at a site, but to requiring less energy to process that water (i.e., distribute it, heat it, and dispose of it once used).

Regarding the quality of the water provided for a given building/building site, the design team should do the following:

- Supply local or municipal water that exceeds the EPA's requirements or more stringent local requirements for potable and heated water.
- Meet the EPA's national primary drinking water regulations, including maximum contaminant level, by testing or by installing appropriate treatment systems. The quality of the municipal water supply must be evaluated at applicable points (e.g., restrooms/showers, kitchen/pantry areas, drinking fountains, architectural fountains, and/or indoor water features).
- Exceed national primary drinking water regulation maximum contaminant level goals and secondary standards by testing or by installing appropriate treatment systems. The quality of the municipal water supply must be evaluated at applicable points (e.g., restrooms/showers, kitchen/pantry areas, drinking fountains, architectural fountains, and/or indoor water features).

Some of the techniques for reducing the demand for domestic water in a building are well known and in fairly widespread use. These include water restrictors (e.g., flow-control shower heads) and spring-closing or timed lavatory faucets, especially in public or semipublic washrooms. GreenTip #14-1 addresses other water-conserving fixtures that can be used.

GreenTip #14-2 would also impact the amount of pure fresh water used at a site by making use of used water (known as *graywater*) for purposes where the use of potable water is not a requirement.

DOMESTIC WATER HEATING

One of the earliest techniques used for lowering the energy required for domestic water heating, going back to the mid-1970s, was reducing the temperature of the water supplied. The previous norm before these energy crises for domestic hot water was 140°F (60°C), and the energy-saving recommendation thereafter was 105°F to 110°F (40°C to 43°C). While this can save heating energy, where hot/cold mixing valves are used (as with some shower controls), it may also cause more hot water to be used. In addition, because of the dangers of *Legionella*, which thrives at the lower-heated water ranges, the prudent recommendation for the hot-water supply temperature has been revised to be 105°F to 120°F (40°C to 49°C). Protection against possible *Legionella* can be achieved by generating 140°F (60°C) centrally (in a well-insulated heater/storage tank) and then mixing to the lower temperature through a mixing valve.

One other energy-reducing technique is point-of-use water heating, which can also reduce the water quantity used. (This is covered in GreenTip #14-3.) Another technique to reduce water-heating costs is to combine the function of domestic and space water heating, where allowed by codes. (See Chapter 11 for GreenTip #11-5 on combination space and water-heating systems). Yet another water-heating technique, with more limited and specialized application, is covered in GreenTip #14-4.

Solar thermal water heating is often used for preheating or full heating of domestic hot-water systems.

Strategies should be considered to preheat and/or fully heat domestic hot water using waste heat (e.g., flash steam, etc.) from the main building heating system.

SANITARY WASTE

See GreenTip #14-2, which deals with sanitary wastewater and a strategy to conserve potable water.

RAINWATER HARVESTING

GreenTip #14-5 also deals with a strategy for conserving potable water, though the water source differs.

FIRE SUPPRESSION SYSTEMS

These systems are designed for life safety of the occupants. However, there may be times when there is an opportunity to use the water source, or the water in the sprinkler and standpipe piping, as a heat sink for heat pump systems.

WATER RECOVERY AND REUSE

For HVAC&R engineers looking for water recovery and reuse opportunities, condensate collection should be considered on most projects, generally in more humid climates, although other factors to be considered would include the amount of outdoor ventilation air used and the ease for using this water. Additional factors that determine whether condensate collection should be considered include location (i.e., climate), building type (particularly relating to amount of outdoor air required), the size, number, and accessibility of air handlers that condition outdoor air, location of potential uses for the condensate, etc. Location determines both the potential to collect a significant amount of condensate as well as the value of the water to the local community. As water demand continues to grow, even regions that might have been considered water-rich will experience tight supplies of fresh, clean water. ANSI/ASHRAE/USGBC/IES Standard 189.1-2011 requires that condensate collection be done on all air-conditioning units with a cooling capacity greater than 65,000 Btu/h (19 kW) and in regions where the ambient mean coincident wet-bulb temperature at 1% design cooling conditions is 72°F (22°C) or greater.

Building or space occupancy type determines the amount of outdoor air required, and thus, the amount of moisture in the incoming air. As the air passes across cold cooling coils, if the coil surface temperature is less than the dew point of the airstream passing by, then the potential exists for water condensation on the coils. (See Digging Deeper sidebar titled "How Much Water Will Collect at Design Conditions?") Water that condenses on the coils will collect and drop to the drain pan below. The actual amount of water collected depends on parameters such as the absolute humidity level, total airflow, and coil bypass ratio. A major source of the moisture being condensed is from outdoor air brought in through outdoor air intakes or through infiltration exchange with the outdoors. A building or space that requires a lot of outdoor air on an ongoing basis (e.g., a laboratory) is an ideal candidate for condensate collection. Other obvious candidates include spaces with indoor water features, natatoriums, gymnasiums, and shower rooms, although these may face special challenges. Dedicated outdoor air systems, or DOAS units, are also prime candidates for consideration. More information on the subject of condensate collection can be found in GreenTip #14-6.

REFERENCES AND RESOURCES

Published

ASHRAE. 2012. *ASHRAE Handbook—HVAC Systems and Equipment*, Ch. 6. Atlanta: ASHRAE.

ASHRAE. 2011. ANSI/ASHRAE/USGBC/IES Standard 189.1-2011, *Standard for the Design of High-Performance Green Buildings*. Atlanta: ASHRAE.

ASHRAE. 2013. ANSI/ASHRAE/IES Standard 90.1-2013, *Energy Standard for Buildings Except Low-Rise Residential Buildings*. Atlanta: ASHRAE.

Cho, Y.I., S. Lee, and W. Kim. 2003. Physical water treatment for the mitigation of mineral fouling in cooling-tower water applications. *ASHRAE Transactions* 109(1):346–57.

Epoch Times. 2009. Water efficiency a priority for green buildings. www.theepochtimes.com/n2/content/view/17751.

Kenny, J.F., Barber, N.L., Hutson, S.S., Linsey, K.S., Lovelace, J.K., and Maupin, M.A. 2009. Estimated use of water in the United States in 2005: U.S. Geological Survey Circular 1344, 52. http://pubs.usgs.gov/circ/1344/pdf/c1344.pdf.

Online

ASHRAE Standard Project Committee 191
   http://spc191.ashraepcs.org/.
U.S. Environmental Protection Agency, Green Chemistry
   www.epa.gov/greenchemistry/.
U.S. Geological Survey
   www.usgs.gov.
World Health Organization.
   www.who.int/water_sanitation_health/hygiene/en/.

ASHRAE GreenTip #14-1

Water-Conserving Plumbing Fixtures

GENERAL DESCRIPTION

Water conservation strategies save building owners money when it comes to both consumption and demand charges. Further, municipal water and wastewater treatment plants save on operating and capital costs for new facilities. As a general rule, water conservation strategies are very cost-effective when properly applied.

The Energy Policy Act of 1992 set reasonable standards for the technologies then available. Now, there are plumbing fixtures and equipment capable of significant reduction in water usage. For example, a rest stop in Minnesota that was equipped with ultralow-flow toilets and waterless urinals has recorded a 62% reduction in water usage.

Tables 14-2, 14-3, and 14-4 list the maximum water usage standards established by the Energy Policy Act of 1992 for typical fixture types. Also listed is water usage for flush-type and flow-type fixtures. Listing of conventional fixture usage allows comparison to the low-flow and ultralow-flow fixture usage. These values would be used to calculate the baseline water consumption if, for example, you are doing a LEED project. ANSI/ASHRAE/USGBC/IES Standard 189.1 requires more efficient fixtures in the design, equivalent to the levels set under the U.S. Environmental Protection Agency's WaterSense program. The corresponding WaterSense levels are shown in the right-hand side column of Tables 14-2.

WHEN/WHERE IT'S APPLICABLE

Applicable state and local codes should be checked prior to design as some of them have approved fixture lists; some code officials have not approved the waterless urinal and low-flush toilet technologies. Waterless urinals and low-flow lavatory fixtures can have a rapid payback period. Toilet technology continues to evolve rapidly, so be sure to obtain test data and references before specifying. Some units work very well, while others perform marginally.

Options that should be considered in the design of water-conserving systems include the following:

- Infrared faucet sensors
- Delayed-action shutoff or automatic mechanical shutoff valves (metering faucets at 0.25 gal/cycle [0.95 L/cycle])

Table 14-2: Energy Policy and Conservation Act (EPACT) Maximum Flows

Fixture Type	Energy Policy Act of 1992 Maximum Water Usage or Use Rates	Corresponding U.S. EPA WaterSense or Standard 189.1 Values
Water Closets, gpf (L/f)	1.6 (6.1)	1.28 (4.8)
Urinals, gpf (L/f)	1.0 (3.8)	0.5 (1.9)
Shower Heads, gpm (L/s)	2.5 (0.16)	2.0 (0.13)
Faucets, gpm (L/s) Residential Public Lavatory	2.5 (0.16)	2.2 (0.14) 0.5 (0.03)
Replacement aerators, gpm (L/s)	2.5 (0.16)	
Metering facets, gal/cycle (L/cycle)	0.25 (0.95)	0.25 (1.0)

Note: gpf = gallons per fixture (L/f = liters per fixture); gpm = gallons per minute (L/s = liters per second). The gpm (L/s) value is at flowing water pressure of 80 psi (552 kPa).

- Low-flow or ultralow-flow toilets
- Lavatory faucets with flow restrictors
- Low-flow kitchen faucets
- Domestic dishwashers that use 10 gal (38 L)/cycle or less
- Commercial dishwashers (conveyor type) that use 120 gal (455 L)/h
- Waterless urinals
- Closed cooling towers (to eliminate drift) and filters for cleaning the water

Table 14-3: Flush-Fixture Flows

Flush-Fixture Type	Water Use, gpf (L/f)
Conventional water closet	1.6 (6.1)
Low-flow water closet	1.1 (4.2)
Ultralow-flow water closet	0.8 (3.0)
Composting toilet	0.0
Conventional urinal	1.0 (3.8)
Waterless urinal	0.0

Note: gpf = gallons per fixture (L/f = liters per fixture)

Table 14-4: Flow-Fixture Flows

Flow-Fixture Type	Water Use, gpm (L/s)
Conventional lavatory	2.5 (0.16)
Low-flow lavatory	1.8 (0.11)
Kitchen sink	2.5 (0.16)
Low-flow kitchen sink	1.8 (0.11)
Shower	2.5 (0.16)
Low-flow shower	1.8 (0.11)
Janitor sink	2.5 (0.16)

Note: gpm = gallons per minute (L/s = liters per second)

A comparison of water consumption for a typical office building using the baseline Energy Policy Act specifications compared to the more efficient WaterSense program values is given in the Digging Deeper sidebar after this GreenTip.

PRO

- Water conservation reduces a building's potable water use, which reduces demand on the municipal water supply and lowers costs and energy use associated with water.
- It reduces a building's overall waste generation, thus putting fewer burdens on the existing sewage system.
- It may save capital costs since some fixtures (e.g., waterless urinals and low-flow lavatories) may be less expensive to install initially.

CON

- Some states and municipalities have approved fixture lists that may not include certain newer and more efficient fixtures. However, the design engineer would likely have the option to go to a review process in order to get new fixture technologies put on the approved fixture list.
- Maintenance of these fixtures is different and requires special training of staff.

KEY ELEMENTS OF COST

The following provides a possible breakdown of the various cost elements that might differentiate a building using water-conserving plumbing fixtures from one that does not and an indication of whether the net cost is likely to be lower (L), higher (H), or the same (S). This assessment is only a perception of what might be likely, but it obviously may not be correct in all situations. There is no substitute for a detailed cost analysis as part of the design process. The listing below may also provide some assistance in identifying the cost elements involved.

First Cost

- Low-flow and ultralow-flow flush water closets   S/H
- Waterless urinals   S/L

- Low-flow shower heads   S
- Metering faucets   S
- Electronic faucets   S/H
- Dual-flush water closets   S/H
- Water-conserving dishwashers   S/H

Recurring Costs

- Potable water   L
- Sewer discharge   L
- Maintenance   L/S
- Training of building operators   S/H
- Orientation of building occupants   S
- Commissioning   S

SOURCES OF FURTHER INFORMATION

American Society of Plumbing Engineers
 www.aspe.org.

Del Porto, D., and C. Steinfeld. 1999. *The Composting Toilet System Book*. New Bedford, MA: The Center for Ecological Pollution Prevention.

EPA. 2010. *How to Conserve Water and Use It Effectively*. U.S. Environmental Protection Agency, Washington, DC. www.epa.gov/owow/nps/chap3.html.

U.S. EPA. WaterSense Program.
 www.epa.gov/watersense/.

Green Building Council
 www.usgbc.org.

EXAMPLE CALCULATION TO COMPUTE A BASELINE PREDICTED WATER CONSUMPTION FOR A BUILDING

Often the design project team may want to, or need to, compute the estimated water consumption of a building and compare that to some baseline level of performance. For example, this type of comparison would be necessary for a building project that was working to demonstrate the minimum prerequisite water efficiency for a LEED project and to determine water savings credit points. This procedure involves estimating the water consumption using flow rate or usage rate values specified for the baseline performance in the LEED program description. These baseline levels correspond generally to the 1992 EPACT values listed in Table 14-2, with one exception for residential faucets.

It is fairly easy to determine the values to use for both the baseline level and the proposed building fixture flow or usage rates; these are taken from the table listed in the corresponding LEED program description and the specifications for the building project. Total water consumption is estimated based on occupancy expectations and thus requires some estimates or assumptions on occupancy levels and patterns. This is because it is the number of people in a building, not necessarily the number of fixtures in the building that determines how much water is consumed.

One common method for determining water consumption is to base it on the number of full-time occupants in the building on daily basis. For this example, let us assume an office building that contains space for 100 employees, with a 50/50 breakdown of males and females. We assume that 10% of the employees are not at this office on any given normal workday, that this facility has 14 outside visitors a day (7 male, 7 female), and that each visitor stays an average of two hours per visit. On weekends and holidays, let us also assume on average 5% of the employees are in the office for a total of four hours for each person each day, with no visitors. Basing the full-time occupancy estimate on an eight-hour workday, the full-time equivalent occupancy of this building therefore is estimated as follows.

Weekdays:

$$FTE_{male} = \frac{100-10}{2} \times \frac{8}{8} + \frac{14}{2} \times \frac{2}{8} = 46.75 \approx 47 \text{ people}$$

Because we have assumed an equal distribution male and female, this is also the female FTE for weekdays. In this example, we are also rounding up the population to a whole number for each case.

Weekends, holidays:

$$FTE_{male} = FTE_{female} = \frac{5}{2} \times \frac{4}{8} = 2.5 \approx 3 \text{ people}$$

Next and assumption is needed for how many times per day each occupant uses the facilities. We will assume that each person visits them three times during each full, eight-hour day. During each visit we assume that the males use the water closet once and a urinal twice. For each visit, regardless of gender, we assume that the lavatory sink is used a total of 15 seconds. We will also assume that this building contains a break room with a sink where the water is run a total of 30 minutes per day on weekdays and 5 minutes per day on weekends and holidays.

We now have all the information needed to compute an estimate of the water consumption, which is illustrated in the following tables.

Weekdays:

Fixture Type	Number of People	Rate	Units	Daily Use per Person	Length (min)	Water Used (gal)
W.C. Female	47	1.6	gpf	3		225.6
W.C. Male	47	1.6	gpf	1		75.2
Urinal	47	1	gpf	2		94
Sink		2.2	gpm		30	66

Weekends, holidays:

Fixture Type	Number of People	Rate	Units	Daily Use per Person	Length (min)	Water Used (gal)
W.C. Female	3	1.6	gpf	3		14.4
W.C. Male	3	1.6	gpf	1		4.8
Urinal	3	1	gpf	2		6
Sink		2.2	gpm		30	11

These give an estimated total water consumption of 460.8 gal (1744 L) per day during each weekday and 36.2 gallons (137 L) on weekends and holidays.

The final step in the calculation is to apply this to the number of days per year for each day type, which gives us an estimated total annual water consumption of 123,609 gal (467,911 L) per year.

Day Type	Number per Year	Daily Use (gal)	Total
Weekday	260	460.8	119,808
Weekend	165	36.2	3,801
Total			123,609

A similar procedure would be conducted for estimating the water consumption in the proposed building design that includes water conservation features. Note that these numbers provide an estimate for the amount of potable water used. If alternative sources of water were used to supplement, for example, the toilet flushing, then this would have to be factored into the water use per fixture, based on specific information for that project.

ASHRAE GreenTip #14-2

Graywater Systems

GENERAL DESCRIPTION

Graywater is generally wastewater from lavatories, showers, bathtubs, and sinks that is not used for food preparation. Graywater is further distinguished from blackwater, which is wastewater from toilets and sinks that contains organic or toxic matter. Local health code departments have regulations that specifically define the two kinds of waste streams in their respective jurisdictions.

Where allowed by local code, separate blackwater and graywater waste collection systems can be installed. The blackwater system would be treated as a typical waste stream and piped to the water treatment system or local sewer district. However, the graywater would be recycled by collecting, storing (optional), and then distributing it via a dedicated piping system to toilets, landscape irrigation, or any other function that does not require potable water.

Typically, for a commercial graywater system (e.g., for toilet flushing in a hotel), a means of short-term on-site storage, or, more appropriately, a surge tank, is required. Graywater can only be held for a short period of time before it naturally becomes blackwater. Often some treatment of the graywater is done, such as with a bleach solution or other means, to prolong storage time. The surge tank would be provided with an overflow to the blackwater waste system and a potable makeup line for when the end-use need exceeds stored capacity.

Distribution would be accomplished via a pressurized piping system requiring pumps and some low level of filtration. Usually, there is a requirement for the graywater system to be a supplemental system. Therefore, systems will still need to be connected to the municipal or localized well service. Plumbing codes require that a colorant be added to graywater used within buildings, such as for toilet flushing, to help distinguish it as nonpotable water.

WHEN/WHERE IT'S APPLICABLE

Careful consideration should be given before pursuing a graywater system. While a graywater system can be applied in any facility that has a nonpotable water demand and a usable waste stream, the additional piping and energy required to provide and operate such a system may outweigh any benefits. Such a system is best applied where the ratio of demand for nonpotable water to potable water is relatively high and consistent, as in restaurants, laundries, and hotels.

Some facilities have a more reliable graywater volume than others. For example, a school would have substantially less graywater in the summer months. This may not be a problem if the graywater was being used for flushing, since it can be assumed that toilet use would vary with occupancy. However, it would be detrimental if graywater were being used for landscape irrigation.

PRO

- A graywater system reduces a building's potable water use, in turn reducing demand on the municipal water supply and lowering costs associated with water.
- It reduces a building's overall wastewater generation, thus putting less tax on the existing sewage systems.

CON

- There is an added first cost associated with the additional piping, pumping, filtration, and surge tank required.
- There are additional materials and their associated embodied energy costs.
- There is negative public perception of graywater and health concerns regarding ingestion of nonpotable water.
- Costs include maintenance of the system, including the pumps, filters, and surge tank.
- Local health code authority has jurisdiction, potentially making a particular site infeasible due to that authority's definition of blackwater versus graywater.

KEY ELEMENTS OF COST

The following provides a possible breakdown of the various cost elements that might differentiate a building utilizing a graywater system from one that does not and an indication of whether the net incremental cost is likely to be lower (L), higher (H), or the same (S). This assessment is only a perception of what might be likely, but it obviously may not be correct in all situations. There is no substitute for a detailed cost analysis as part of the design process. The listings below may also provide some assistance in identifying the cost elements involved.

First Cost

- Collection systems   H
- Surge tank   H
- Water treatment   H
- Distribution system   H
- Design fees   H

Recurring Costs

- Cost of potable water   L
- Cost related to sewer discharge   L
- Maintenance of system   H
- Training of building operators   H
- Orientation of building occupants S
- Commissioning cost   H

SOURCES OF FURTHER INFORMATION

Advanced Buildings Technologies and Practices
   www.advancedbuildings.org.
American Society of Plumbing Engineers
   www.aspe.org.
Del Porto, D., and C. Steinfeld. 1999. *The Composting Toilet System Book*. New Bedford, MA: The Center for Ecological Pollution Prevention.
Ludwig, A. 1997. *Builder's Greywater Guide and Create an Oasis with Greywater*. Oasis Design.
USGBC. 2009. *LEED 2009 Green Building Design and Construction Reference Guide*. Washington, DC: U.S. Green Building Council.

ASHRAE GreenTip #14-3

Point-of-Use Domestic Hot-Water Heaters

GENERAL DESCRIPTION

As implied by the title, point-of-use domestic hot-water heaters provide small quantities of hot water at the point of use, without tie-in to a central hot water source. A cold-water line from a central source must still be connected as well as the energy source for heating the water, which could be electricity or gas.

There is some variation in types. Typically, the device may be truly instantaneous (e.g., lavatories), or it may have a small amount of storage capacity. With the instantaneous type, the heating source is sized such that it can heat a normal-use flow of water up to the desired hot-water temperature (e.g., 120°F [49°C]). When a small tank (usually 3 to 10 gal) is incorporated in the device, an electric heating coil can be built into the tank and can be sized somewhat smaller because of the small amount of stored water available. The device is usually installed under the counter of the sink or bank of sinks. Slightly larger-sized units are available that can provide instantaneous heating for a residential house without requiring a storage tank and its associated thermal losses.

A similar type of device boosts the water supply (which is cold water) up to near boiling temperature (about 190°F [88°C]). This is typically used, for example, to make a cup of coffee or tea without having to brew it separately in a coffeepot or teapot.

WHEN/WHERE IT'S APPLICABLE

These devices are applicable wherever there is a need for a hot-water supply that is low in quantity and relatively infrequently used and is excessively inconvenient or costly to run a hot-water line (with perhaps a recirculation line as well) from a central hot-water source. Typically, these are installed in lavatories or washrooms that are isolated or remote, or both. However, they can be used in any situation where there is a hot-water need but where it would be too inconvenient and costly to tie in to a central source. (There must, of course, be a source of incoming water and a source of electricity.)

PRO

- A point-of-use device is a simple and direct way to provide small amounts of domestic hot water per use.
- Long pipe runs—and, in some cases, a central hot-water heating source—can be avoided.
- Energy is saved by avoiding heat loss from hot-water pipes and, if not needed, from a central water heater.
- In most cases, where applicable, it has a lower first cost.
- It is convenient—especially as a source of 190°F to 210°F (88°C to 99°C) water supply.
- When installed in multiple locations, central equipment failure does not knock out all user locations.
- It may save floor space in the central equipment room if no central heater is required.
- Water is saved by not having to run the faucet until the water warms up.

CON

- This is a more expensive source of heating energy (though cost may be trivial if usage is low and may be exceeded by heat losses saved from a central heating method).
- Water impurities can cause caking and premature failure of electrical heating coils.
- It cannot handle changed demand for large hot-water quantities or too-frequent use.
- Maintenance is less convenient (when required), since it is not centralized.
- A temperature and pressure relief valve and floor drain may be required by some code jurisdictions.

KEY ELEMENTS OF COST

The following provides a possible breakdown of the various cost elements that might differentiate a point-of-use domestic hot-water heater from a conventional one and an indication of whether the net incremental cost for the system is likely to be lower (L), higher (H), or the same (S). This assessment is only a perception of what might be likely, but it obviously may not be correct in all situations. There is no substitute for a detailed cost analysis as part of the design process. The listings below may also provide some assistance in identifying the cost elements involved.

First Cost

- Point-of-use water-heater equipment   H
- Domestic hot-water piping to central source
  (including insulation thereof)   L
- Central water heater (if not required) and associated fuel
  and flue gas connections   L
- Electrical connection   H
- Temperature and pressure relief valve and floor drain
  (when required by code jurisdiction)   H

Recurring Costs

- Energy to heat water to appropriate temperatures   H
- Energy lost from piping not installed   L
- Maintenance/repairs, including replacement   H

SOURCES OF FURTHER INFORMATION

American Society of Plumbing Engineers
 www.aspe.org.
ASPE. 1998. *Domestic Water Heating Design Manual*. Chicago: American Society of Plumbing Engineers.
Fagan, D. 2001. A comparison of storage-type and instantaneous heaters for commercial use. *Heating/Piping/Air Conditioning Engineering*, April.

ASHRAE GreenTip #14-4

Direct-Contact Water Heaters

GENERAL DESCRIPTION

A direct-contact water heater consists of a heat exchanger in which flue gases are in direct contact with the water. It can heat large quantities of water for washing and/or industrial process purposes. Cold supply water enters the top of a heat exchanger column and flows down through stainless steel rings or other devices. Natural gas is burned in a combustion chamber, and the flue gases are directed up the heat exchanger column. As the gases move upward through the column, they transfer their sensible and latent heat to the water. A heat exchanger or water jacket on the combustion chamber captures any heat loss from the chamber. The gases exit only a few degrees warmer than the inlet water temperature. The heated water may be stored in a storage tank for on-demand use. Direct-contact water heaters can be 99% efficient when the inlet water temperature is below 59°F (15°C).

The low-temperature combustion process results in low emissions of NO_x and CO; thus, the system is, in effect, a low-NO_x burner. It is also a low-pressure process, since heat transfer occurs at atmospheric pressure.

Although there is direct contact between the flue gases and the water, there is very little contamination of the water. Direct-contact systems are suitable for all water-heating applications, including food processing and dairy applications; the water used in these systems is considered bacteriologically safe for human consumption.

WHEN/WHERE IT'S APPLICABLE

The high cost of direct-contact water heaters (due to stainless steel construction) restricts their use to where there is a large, almost continuous, demand for hot water. Appropriate applications include laundries, food processing, washing, and industrial processes. The system can also be used for closed-loop (or recirculating) applications (e.g., space heating). However, efficiency—the primary benefit of direct-contact water heating—will be reduced because of the higher inlet water temperature resulting from recirculation.

PRO

- Increases part-load and instantaneous efficiency.
- Reduces NO_x and CO emissions.
- Increases safety.
- Increases system response time.

CON

- High cost.
- Less effective in higher-pressure or closed-loop applications or where inlet water temperatures must be relatively high.
- Due to high evaporation rates, results in considerable water usage beyond that required for the process.

KEY ELEMENTS OF COST

The following provides a possible breakdown of the various cost elements that might differentiate a direct-contact water heater from a conventional one and an indication of whether the net cost for the alternative option is likely to be lower (L), higher (H), or the same (S). This assessment is only a perception of what might be likely, but it obviously may not be correct in all situations. There is no substitute for a detailed cost analysis as part of the design process. The listings below may also provide some assistance in identifying the cost elements involved.

First Cost

- Water heater   H
- Operator training (unfamiliarity)   H

Recurring Costs

- Water heating energy   L

Direct-contact boilers are two to three times the price of indirect or conventional boilers, primarily because of the stainless steel construction. In high and continuous water use applications, however, the payback period can be less than two years.

SOURCES OF FURTHER INFORMATION

NSF. 2000. *NSF/ANSI 5-2000e: Water Heaters, Hot Water Supply Boilers, and Heat Recovery Equipment.* Ann Arbor, MI: National Sanitation Foundation and American National Standards Institute.

QuikWater, High Efficiency Direct Contact Water Heaters www.quikwater.com.

ASHRAE GreenTip #14-5

Rainwater Harvesting

GENERAL DESCRIPTION

Rainwater harvesting has been around for thousands of years. Rainwater harvesting is a simple technology that can stand alone or augment other water sources. Systems can be as basic as a rain barrel under a downspout or as complex as a pumped and filtered graywater system providing landscape irrigation, cooling tower makeup, and/or building waste conveyance.

Systems are generally composed of five or fewer basic components: a catchment area, a means of conveyance from the catchment, storage (optional), water treatment (optional), and a conveyance system to the end use.

The catchment area can be any impermeable area from which water can be harvested. Typically this is the roof, but paved areas (e.g., patios, entries, and parking lots) may also be considered. Roofing materials that are metal, clay, or concrete-based are preferable for roofs planned for rainwater harvesting compared to those with potential contaminants, such as asphalt or those with lead-containing materials. Similarly, care should be given when considering a parking lot for catchment due to oils and residues that can be present.

Conveyance to the storage will be gravity-fed, like any stormwater piping system. The only difference is that now the rainwater is being diverted for useful purposes instead of literally going down the drain.

Commercial systems will require a means of storage. Cisterns can be located outside the building (e.g., above-grade or buried) or placed on the lower levels of the building. The storage tank should have an overflow device piped to the storm system and a potable water makeup if the end-use need is ever greater than the harvested volume.

Depending on the catchment source and the end use, the level of treatment will vary. For simple site irrigation, filtration can be achieved through a series of graded screens and paper filters. If the water is to be used for waste conveyance, then an additional sand filter may be appropriate. Parking lot catchments may require an oil separator. The local code authority will likely decide acceptable water standards, and, in turn, filtration and chemical polishing will be a dictated parameter, not a design choice.

Distribution can be via gravity or pump depending on the proximity of the storage tank and the end use.

WHEN/WHERE IT'S APPLICABLE

If the building design is to include a graywater system, condensate collection, or landscape irrigation—and space for storage can be found—rainwater harvesting is a simple addition to those systems.

When a desire exists to limit potable water demand and use, depending on the end-use requirement and the anticipated annual rainfall in a region, harvesting can be provided as a stand-alone system or to augment a conventional makeup water system.

Sites with significant precipitation volumes may determine that reuse of these volumes is more cost-effective than creating stormwater systems or on-site treatment facilities.

Rainwater harvesting is most attractive where municipal water supply is either nonexistent or unreliable, hence its popularity in rural regions and developing countries.

PRO

- Rainwater harvesting reduces a building's potable water use, and reduces demand on the municipal water supply, lowering costs associated with water.
- Rainwater is soft and does not cause scale buildup in piping, equipment, and appliances. It could extend the life of systems.
- It can reduce or eliminate the need for stormwater treatment or conveyance systems.

CON

- There is added first cost associated with the cisterns and the treatment system.
- There are additional materials and their associated embodied energy costs.

- The storage vessels must be accommodated. Small sites or projects with limited space allocated for utilities would be bad candidates.
- Costs include maintenance of the system (e.g., maintaining the catchments, conveyance, cisterns, and treatment systems).
- There is no U.S. guideline on rainwater harvesting. The local health code authority has jurisdiction, potentially making a particular site infeasible due to backflow prevention requirements, special separators, or additional treatment.

KEY ELEMENTS OF COST

The following provides a possible breakdown of the various cost elements that might differentiate a building utilizing rainwater harvesting from one that does not and an indication of whether the net cost is likely to be lower (L), higher (H), or the same (S). This assessment is only a perception of what might be likely, but it obviously may not be correct in all situations. There is no substitute for a detailed cost analysis as part of the design process. The listings below may also provide some assistance in identifying the cost elements involved.

First Cost

- Catchment area — S
- Conveyance systems — S
- Storage tank — H
- Water treatment — S/H
- Distribution system — S/H
- Design fees — H

Recurring Costs

- Cost of potable water — L
- Maintenance of system — H
- Training of building operators — H
- Orientation of building occupants — S
- Commissioning cost — H

SOURCES OF FURTHER INFORMATION

American Society of Plumbing Engineers
  www.aspe.org.
Irrigation Association
  www.irrigation.org.
Texas Manual on Rainwater Harvesting
  www.twdb.state.tx.us/publications/reports
  /rainwaterharvestingmanual_3rdedition.pdf.
USGBC. 2009. *LEED 2009 Green Building Design and Construction Reference Guide*. Washington, DC: U.S. Green Building Council.
Waterfall, P.H. 1998. *Harvesting Rainwater for Landscape Use.* http://ag.arizona.edu/pubs/water/az1052/.
Environmental Protection Agency, Water Sense Program
  http://epa.gov/watersense/.

ASHRAE GreenTip #14-6

Air-Handling Unit (AHU) Condensate Capture and Reuse

GENERAL DESCRIPTION

As air passes across cold cooling coils in an AHU, if the coil surface temperature is less than the dew point of the airstream passing by, then the potential exists for water condensation on the coils. Water that condenses on the coils will collect and drop to the drain pan below. In conventional building systems, this water typically drains, unused, to the sewer system or elsewhere, but in some situations it can be worthwhile to capture and reuse it.

Condensate collection can either be designed into new construction or retrofitted into existing buildings. While the former is preferable (due to lower costs and fewer complications), the latter presents the highest potential, since existing buildings comprise about 98% of the building stock (the other 2% being new construction).

The best end use for the collected water will depend on the particular circumstances of the location. In a building with its own chiller and cooling tower, the most logical choice is to route collected condensate to the cooling tower sump to reduce the need to use fresh water for makeup. In most cases, peak condensate production will occur at the same times as peak makeup water demands, creating an elegant synergy. This is also the simplest retrofit, involving reasonably inexpensive equipment and piping; water can be routed directly to the tower with no need for treatment.

Condensate collection can also be integrated with a rainwater collection system, a scheme often referred to as *rainwater plus*. This will usually involve a storage tank or cistern and can require considerably more expense and engineering than using the condensate in a cooling tower. Depending on the intended use (e.g., irrigation, ornamental fountains, or other internal uses including toilet flushing), different amounts of further treatment will be required. In all cases, local building codes must be followed.

WHEN/WHERE IT'S APPLICABLE

Factors that determine whether condensate collection should be considered include location (e.g., climate), building type (particularly relating to amount of outdoor air required), the size, number, and accessibility of air handlers that condition outdoor air, location of potential uses for the condensate, etc. Location determines both the potential to collect a significant amount of condensate and the value of the water to the local community. In periods of drought, the actual value of a unit of water to the local society and economy may be worth much more than the rate currently paid to the local utility.

A building or space that requires a lot of outdoor air on an ongoing basis (e.g., laboratories) is an ideal candidate for condensate collection. Other obvious candidates include spaces with indoor water features, natatoriums, gymnasiums, or locker rooms. DOAS units are also prime candidates for consideration, since they are typically designed for optimal latent load removal.

Using typical meteorological year data, assumptions about the air-handling system, and the following equation, it is possible to estimate the amount of condensate that can be collected annually in a particular location.

Condensate collection is required by ANSI/ASHRAE/USGBC/IES Standard 189.1 for air-conditioning units that are above a certain cooling capacity and in more humid climates.

LESSONS LEARNED

Attention must be paid to the cleanliness of the water and the system components. For example, any external condensate collection pan should be covered to prevent foreign particles from getting into the system. The potential is also there for biological growth and contamination. Also, there may be an increase in corrosion potential in the cooling tower loop, if that is where the condensate is sent.

Sweating on the outside of the condensate piping can be an issue, particularly in semiconditioned mechanical rooms, so all lines (as well as perhaps the collection basin itself) should be insulated. If the condensate line is tied into rain downspouts, you may want to consider running a smaller pipe or tube inside of the downspout for the condensate to avoid moisture buildup on the outside downspout surface.

The dimensions of the U-trap in the existing condensate drain pipe between the AHU and the floor drain should be maintained when connecting the drain pipe to the external collection pan. It is also highly recommended that a condensate flow meter be installed and that it be located to facilitate easy reading. The additional cost of the meter is worthwhile because of the good feedback on functionality and the education potential it provides. It's also a good way to verify water and cost savings.

PRO

- If condensate is routed to a cooling tower, demand for makeup water will reduce and so will the need for treatment chemicals. Blowdown frequency should decrease, and sewer costs could be reduced with appropriate metering.
- Cool condensate routed to a cooling tower will provide residual free cooling for condenser water.
- Incorporating condensate collection into a rainwater collection and storage system can reduce the cistern size requirement by providing a supplemental water source during long periods between rain events.

CON

- Complications arise when dealing with district cooling systems with satellite chillers, because it is possible to produce condensate in an AHU while the chiller and cooling tower (for that particular building) are idle. This leads to the risk that treatment chemicals in the sump will be diluted and needlessly washed away via the overflow drain.
- In general, less-efficient systems have higher condensate production potential. Enthalpy wheels, for example, will greatly improve system efficiency but will dramatically reduce condensate production. A building with 100% outdoor air supply that is overpressurized will produce more condensate but will waste energy. Energy efficiency should always take precedence over water production.

KEY ELEMENTS OF COST

The following capital cost issues list the various cost elements associated with either building condensate collection into new construction or retrofitting an existing building. This assessment is only a perception of what might be likely, but it obviously may not be correct in all situations. There is no substitute for a detailed cost analysis as part of the design process.

Pipe. Depending on the distance between the AHU and the end use, and whether a storage tank is involved, the material and labor costs of the pipe installation are likely to be the most expensive part of the system. For new construction the additional cost should be minimal, since a condensate drain pipe would need to be furnished, regardless of the end use.

Storage. If condensate is to be stored for later use, a cistern or storage tank can represent a considerable part of the system cost. Additional costs will be incurred for system design (i.e., tank sizing) and tank site selection and installation. Finally, treatment of stored water prior to end use, if necessary, will add equipment, design, and maintenance costs.

Metering. A totalizing meter is a relatively inexpensive but important component of a condensate collection system. Once in place, a condensate meter will help verify payback on the investment in the system.

SOURCES OF FURTHER INFORMATION

Guz, K. 2005. Condensate water recovery. *ASHRAE Journal* 47(6):54–56.

Lawrence, T., J. Perry, an P. Dempsey. 2010. Making every drop count: retrofitting condensate collection on HVAC air-handling units. *ASHRAE Journal* 52(1): 48–54.

Lawrence, T,M., J. Perry and T. Alsen. 2012, AHU Condensate Collection Economics. *ASHRAE Journal* 54(5):12–17.

HOW MUCH WATER WILL COLLECT AT DESIGN CONDITIONS?

For simplicity, consider the process of a unit conditioning 100% outdoor air (where the unit is a dedicated outdoor air system). The psychrometric chart represents a path of outdoor air as it passes across the cooling coil for the 0.4% cooling design condition in Athens, Georgia. Assuming a supply air condition of 55°F (12.8°C) and 85% relative humidity (wet-bulb temperature = 52.5°F [11.4°C]), the humidity ratio changes across the coil from 0.0141 to 0.0078 lb/lb$_{air}$ (kg/kg$_{air}$). The difference in absolute humidity (ω) between the incoming outdoor air and supply air leaving the unit represents the amount of condensation that occurs. Thus, for every lb (kg) of air supplied by the unit, 0.0141 − 0.0078 or 0.0063 lb (0.00286 kg) of water are condensed.

The total amount of condensate expected is determined by the equation below:

$$\text{Condensate} = \text{Airflow} \times \text{Density} \times 60\tfrac{\min}{h} \text{ (I-P) or } 3600\tfrac{s}{h} \text{ (SI)} \times \Delta\omega$$

Assuming 1000 cfm (472 L/s) of outdoor air is being conditioned, the total amount of condensate expected would be the following:

$$\text{Condensate} = 1000\frac{ft^3}{\min} \times \frac{lb}{13.133\ ft^3} \times 60\frac{\min}{h} \times (0.0141 - 0.0078)\frac{lb_{water}}{lb_{dry\ air}} \quad \text{(I-P)}$$

$$= 28.8\frac{lb}{h}$$

$$\text{Condensate} = 0.472\frac{m^3}{s} \times \frac{kg}{0.820\ m^3} \times 3600\frac{s}{h} \times (0.0141 - 0.0078)\frac{kg_{water}}{kg_{dry\ air}} \quad \text{(SI)}$$

$$= 13.05\frac{kg}{h}$$

This is approximately 3.5 gal (13.1 L)/h at the cooling design condition.

Similar calculations can be run for any locality, and the result can vary widely depending on the climate. For example, when the calculation is run for other representative cities the condensate yields are:

- Boston, Massachusetts (90.8°F [32.6°C] dry-bulb/73.3°F [22.8°C] mean coincident wet-bulb [MCWB] temperature)
= 3.2 gal (12.1 L)/h

- Sacramento, California (100.4°F [37.9°C] dry-bulb/70.7°F [21.4°C] MCWB)
  = 0.8 gal (3.1 L)/h
- Denver, Colorado (94.3°F [34.5°C] dry-bulb/60.3°F [15.6°C] MCWB)
  = no condensate collected

Interestingly, the total annual rainfall is only a partial indicator of how much condensate might be collected, as shown in Table 14-5. The two comparative eastern and western U.S. cities have similar rainfall totals, but they vary significantly in terms of the total amount of condensate collection potential.

What to do with Collected Water

The best end use for the collected water will depend on the particular circumstances of the location. In many locations the primary use of city water is for makeup water in cooling towers. Therefore, it may be the most logical choice to collected condensate to its cooling tower sump. In most cases, peak condensate production will occur at the same times as peak makeup water demands, creating an elegant feedback loop.

Table 14-5: Annual Condensate Collection Compared to Total Annual Rainfall

	Annual Condensate for Continuous Outdoor Air, gal/cfm (L/L/s)	Average Annual Rainfall, in. (m)
Athens, Georgia	12.5 (100.4)	47.8 (1.21)
Boston, Massachusetts	4.5 (36.1)	42.5 (1.08)
Sacramento, California	1.3 (10.4)	17.9 (0.45)
Denver, Colorado	0.5 (4)	15.4 (0.39)

> This is also the simplest retrofit, involving reasonably inexpensive equipment and piping; water can be routed directly to the tower with no need for treatment.
>
> Complications arise when dealing with district cooling systems with satellite chillers, because it is possible to produce condensate in an air-handling unit while the chiller and cooling tower for that particular building are idle. While it is no tragedy that condensate sent to the cooling tower will simply overflow to the sewer (where it would have gone prior to retrofit), there is the risk that treatment chemicals in the sump will be diluted and needlessly washed away. In this scenario, care should be taken to prioritize condensate retrofits in buildings with baseline chiller plants.
>
> Condensate collection can also be integrated into a rainwater collection system, a scheme often referred to as *rainwater plus*. This will usually involve a storage tank or cistern and can require considerably more expense and engineering than using the condensate in a cooling tower. Depending on the intended use (e.g., for irrigation, fountains, toilet flushing, or potable water), different amounts of further treatment will be required. In all cases, local building codes must be followed.

CHAPTER FIFTEEN

BUILDING AUTOMATION SYSTEMS

Building control systems play an important part in the operation of a building and determine whether many of the green design aspects included in the original plan actually function as intended. Controls for HVAC and related systems have evolved over the years, but in general, they can be described as either distributed (local) or centralized. Local controls are generally packaged devices that are provided with the equipment. A building automation system (BAS), on the other hand, is a form of central control capable of coordinating local control operation and controlling HVAC and other systems (e.g., life-safety, lighting, water distribution, and security from a central location).

Control systems are at the core of building performance. When they work well, the indoor environment promotes productivity with the lighting, comfort, and ventilation people need to carry out their tasks effectively and efficiently. When they break down, the results are higher utility bills, loss of productivity, and discomfort. In modern buildings, direct digital control systems operate lights, chilled- or hot-water plants, ventilation, space temperature and humidity control, plumbing systems, electrical systems, life-safety systems, and other building systems. These control systems can assist in conserving resources through the scheduling, staging, modulation, and optimization of equipment to meet the needs of the occupants and systems that they are designed to serve. The control system can assist with operation and maintenance through the accumulation of equipment runtimes, display of trend logs, use of part-load performance modeling equations, and automated alarms. Finally, the control system can interface with a central repository for building maintenance information where operation and maintenance manuals or equipment ratings, such as pump curves, are stored as electronic documents available through a hyperlink on the control system graphic for the appropriate system. This chapter presents the key issues to designing, commissioning, and maintaining control systems for optimal performance.

This chapter is divided into seven sections as follows:

- *Control System Role in Delivering Energy Efficiency.* Through scheduling, optimal loading and unloading, optimal setpoint determination, and fault detection, controls have the capability of significant reductions in building energy usage in a typical commercial building.
- *The Interaction of a Smart Building with the Coming Smart Grid.* This section describes new concepts and technologies needed for a smooth interaction between the building automation system and the coming smart grid.
- *Control System Role in Delivering Water Efficiency.* Used primarily in landscape irrigation and leak detection, controls can significantly reduce water usage compared to systems with simplistic control (such as time clock-based irrigation controllers). Building controls can also provide trending and alarming for potable and nonpotable water usage.
- *Control System Role in Delivering Indoor Environmental Quality (IEQ).* In most commercial buildings, controls play a crucial role in providing IEQ. Controls can regulate the quantities of outdoor air brought into the building based on occupancy levels, zone ventilation, zone temperature, and relative humidity, and can monitor the loading of air filters.
- *Control System Commissioning Process.* Of all the building systems, controls are the most susceptible to problems in installation. These can be addressed by a thorough process of commissioning and postcommissioning performance verification.
- *Designing for Sustained Efficiency.* Control systems help ensure continued efficient building operation by enabling measurement and verification (M&V) of building performance and serving as a repository of maintenance procedures.

CONTROL SYSTEM ROLE IN DELIVERING ENERGY EFFICIENCY

A U.S. Department of Energy report (DOE 2005) featured the following information:

- Seventeen quad/yr (17.9 EJ/yr) of energy are used in commercial buildings.
- Only 10% of commercial buildings (33% of the floor space) have a building management system
- Less than 10% of the commercial building space has automated lighting control.
- A potential for nearly 1 quad/yr of energy savings (11% of total building energy) exists through installing automation systems in all buildings and fully commissioning them.

Thus, the potential for energy savings by installing building automation systems is huge. The study indicated that commercial building energy usage could be reduced by 2% to 11% (0.34 to 1.8 quad/yr [0.36 to 1.9 EJ/yr]) with proper management and control.

Building automation can save energy through a variety of methods, including the following:

- *Reduction of Equipment Runtime.* Examples include scheduled control of lighting and air-conditioning systems inside buildings and photoelectric controls for site (exterior) lighting.
- *Efficient Unloading of Equipment.* Examples include daylight control with dimming or stepped lighting in spaces with access to natural light. This also would include resetting setpoints for chilled, condenser water supply, hot-water temperature, coil discharge air temperature, or variable-air-volume (VAV) fan static pressure. Variable-speed control of pumps, fans, and compressors is another common method employed for cooling systems. Also, consider methods for optimization among building HVAC systems, such as in static pressure reset from VAV direct digital controllers with variable-frequency drives (VFDs), which often end up saving more energy than a VFD by itself.
- *Automated Fault Detection and Diagnostic Systems.* Examples include controls that report when dampers or valves are stuck open or closed.

It may be desirable to reset the minimum outdoor airflow based on changes in occupancy or changes in zone air distribution effectiveness. This concept is known as demand-controlled ventilation (DCV). DCV is required for densely occupied spaces by ANSI/ASHRAE/IESNA Standards 90.1-2013 and ANSI/ASHRAE/USGBC/IES 189.1-2011, with the definition of a densely occupied space based on the design occupancy density of 25 people per 1000 ft^2 (100 m^2). Control of the ventilation air provided is based on occupancy or occupancy level. For example, occupancy can be estimated directly by a card reader system, provided that there are card checks as the person enters and leaves a zone. Occupancy can be estimated indirectly by measuring the CO_2 concentration in the occupied space. If CO_2 levels are used to estimate occupancy, it is important to keep in mind the interaction of the various spaces in multizone systems. This is a topic of current ASHRAE research.

THE INTERACTION OF A SMART BUILDING WITH THE COMING SMART GRID

While the application of building automation and controls systems is now mainstream in applications across the globe, there is a growing need for developing tools that will allow these new smart buildings to interact smoothly with the coming smart grid. Electric utilities have been working for some time now in developing the tools and procedures for what is needed on their end. Research is only now really beginning on topics within the built environment and particularly with commercial buildings, and there is a need for a grand unifying focus. Such a unifying focus is becoming increasingly important with the spread of distributed generation of electricity through on-site renewable energy and as equipment manufacturers begin to develop grid-ready devices. Prior research has demonstrated that information gathering and modeling can help reduce energy consumption within individual buildings and the electrical grid, but the coming smart grid will require a new and more complex integration of information gathering, decision, and control.

One concept description for how the built environment will interact with the coming smart grid is given in Figure 15-1. Technology innovations are leading to the potential for a number of items interacting with a smart grid that may or may not be controlled through the normal building automation system.

The recent release of the Open Automated Demand Response (OpenADR) Standard will help pave the way for building energy management and automation systems to communicate with the electric utility provider and implement demand response measures. The development of the OpenADR Standard during the past decade was the result of national research labs and companies, with Lawrence Berkeley National Laboratory providing a significant, early role. The OpenADR

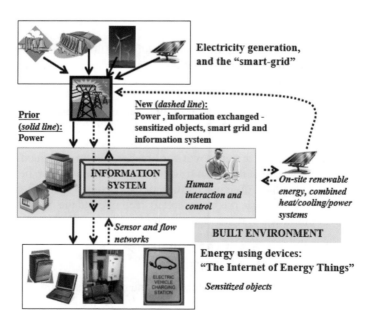

Figure 15-1 A concept for how the built environment will interact with a smart grid.

Alliance was formed to support the development and deployment of commercial demand response through adoption of the OpenADR Standard. We are just now beginning to see products enter the market that will allow demand response programs to be implemented in commercial buildings, and this area will continue to grow in importance in society and with green buildings in particular. More information on OpenADR can be found in the OpenADR primer (OpenADR Alliance).

CONTROL SYSTEM ROLE IN DELIVERING WATER EFFICIENCY

Control systems can deliver building water efficiency in two main areas: landscape irrigation and leak detection. Additionally, controls can be used to regulate and monitor on-site wastewater treatment plants where those systems are in place.

While the most obvious contribution to a sustainable design solution is to not include irrigated landscaping, when landscape irrigation is done, smart irrigation controllers using embedded sensors in the ground can make a significant contribution to reducing the annual water use of a building. (Note that this is a requirement for landscaping irrigation in ANSI/ASHRAE/USGBC/IES Standard 189.1-2011, *Standard for the Design of High-Performance Green Buildings* [ASHRAE 2011].) The level of contribution depends on the amount and type of landscaping, as well as on geographical location. The simplest irrigation controls are based on time clocks that open valves for a set duration, for a set number of periods per week. Unless the duration and frequency of watering is adjusted throughout the irrigation season, the use of time clock controllers often results in excessive water use. In order to automatically address seasonally varying irrigation requirements, some time clock controllers allow for 365 day programming.

An improvement on the time clock controller is to add moisture sensors, which can enable the system to bypass a watering period (if ground moisture levels are above a setpoint). Still more sophisticated controllers can gather data about local weather conditions—either directly via sensors, indirectly via a remote weather station, or by direct input from a remote weather station—and use that data to adjust the amount of water delivered to the landscape.

Integrating landscape irrigation controls into the building automation system (BAS) provides a number of advantages. Among these are the ability to adjust schedules and setpoints from a single location, the ability to perform remote diagnostics, and the ability to track system performance and water use. Integrating water meters into the BAS enables continuous measurement of water consumption. Water consumption data can be analyzed during unoccupied periods to determine whether leaks are present in the water distribution system. The judicious placement of submeters can allow building maintenance staff to find the system or location in which the leak or leaks are present. Continuous water meter data can also be used to identify processes and areas of high water use and guide postconstruction water conservation efforts.

Finally, some advanced green buildings feature on-site wastewater treatment plants. Such plants generally include pumps, fans, and sensors to monitor characteristics of the treated water (e.g., dissolved oxygen levels and total suspended solids). Depending on the goals and complexity of the wastewater treatment plant, an industrial supervisory control and data acquisition system may be necessary to ensure maximization of throughput, efficient energy use, and code compliance. The supervisory control and data acquisition system may be stand-alone or it may be integrated into the BAS.

CONTROL SYSTEM ROLE IN DELIVERING IEQ

Factors regulated by building control systems that impact IEQ include operative temperature, relative humidity, outdoor airflow rates, and light levels. The first two factors are addressed in the latest edition of ANSI/ASHRAE Standard 55, *Thermal Environmental Conditions for Human Occupancy* (ASHRAE 2010), the third factor is addressed in the latest edition of ANSI/ASHRAE Standard 62.1, *Ventilation for Acceptable Indoor Air Quality* (ASHRAE 2013a), and the fourth factor is addressed in Chapter 10 of the IES *Lighting Handbook* (IES).

Thermal Comfort

The operative temperature is approximately equal to the average of the room air temperature and the mean radiant temperature (MRT) for most building interior environments. Radiant heating systems raise the MRT, and thus, lower the room air temperature required to provide a comfortable operative temperature. Radiant cooling systems lower the MRT, and thus, raise the room air temperature required to provide a comfortable operative temperature. Stand-alone thermostats and BASs typically measure and control room air temperature. When designing radiant systems, it is important to keep the distinction between operative temperature and air temperature in mind, as well as the impact of building thermal mass. The current level of control technology is generally not sophisticated enough to optimize energy consumption with MRT, but this is one area for future control system development. For a discussion of these two quantities, see Chapter 54 of the 2011 *ASHRAE Handbook—HVAC Applications* (ASHRAE 2011). ASHRAE Standard 55 describes methods for determining acceptable operative temperatures that depend on occupant clothing insulation levels and metabolic rates.

While they are less common than thermostats and temperature sensors, stand-alone humidistats and humidity sensors can be used to control room humidity levels. Standard 55 specifies a maximum humidity ratio of 0.012. The standard does not specify a minimum humidity ratio for thermal comfort (although it had in earlier versions) but notes that nonthermal comfort factors (e.g., skin drying) may be used to establish such a minimum.

Air Quality and Ventilation

Procedures for determining minimum outdoor airflow rates are described in ASHRAE Standard 62.1 (ASHRAE 2013a). Building controls play a critical role in ensuring that minimum outdoor airflow rates are achieved in three areas: VAV system control, mixed-mode ventilation systems, and dynamic reset of outdoor air intake flows.

It is difficult to ensure that minimum outdoor airflow rates are met by VAV systems over the entire range of operating conditions in the absence of controls designed, installed, and maintained specifically for that purpose. One means of achieving such control is to measure the supply airflow rate and CO_2 concentrations in the occupied zones, and outdoor air intake on a continuous basis. Another potential method is by measuring the outdoor air intake flow. Control systems based on CO_2 levels, as well as those that measure outdoor air intake flow, all have their potential errors and design difficulties. In addition, like any measurement device, maintenance such as calibration must be done as recommended by the manufacturer to ensure accuracy of that information. The BAS can determine if the outdoor airflow rate is sufficient based on this information and can adjust the mixed-air dampers accordingly. The preferred method of controlling outside air intake is to control the mixing plenum pressure based on a measured, mixing plenum, static pressure sensor. This will indirectly control the needed variations in outside air flow. If airflows are also measured at each VAV box, this control routine can be further improved upon. As described in Standard 62.1, zone air distribution effectiveness can change when the temperature of the supply air changes. Therefore, it may be necessary for the control system to reset the minimum outdoor airflow after a seasonal switchover of supply air temperature.

Mixed-mode ventilation refers to the combination of mechanical ventilation and operable windows that provide natural ventilation. During some times of the year, if the windows are sized, located, and operated (by automatic control) properly, it is not necessary to provide mechanical ventilation when the windows are open. Sizing must include factors such as, for example, location of the window with respect to prevailing wind patterns and solar loads. Perhaps the most straightforward way to control mixed-mode ventilation systems is to use the output of a window switch to shut down the terminal unit (e.g., the VAV box) when the window is open. This is most applicable for single-occupant zones or areas with a relatively small number of occupants. When the occupants decide that it is preferable to shut the window, the mechanical HVAC system is brought back online. It may be necessary to install an alarm or override based on space temperature, wind annoyance, wet weather, or in order to provide freeze protection.

Lighting Levels

The BAS can maintain desired light levels by either adjusting electric light output or controlling the amount of daylight entering the building. One or more photocells may be used to measure the light level in the occupied space and the output used to brighten or dim electric lights. Alternatively, photocell output may be used to switch between multiple light levels. Photocell output may also be used to raise or lower window blinds, or adjust louvers to keep daylighting at comfortable levels and eliminate glare. Alternately, equations that calculate the sun position angle for various wall or roof orientations can be input into the BAS and used to vary the tilt angle of window blinds, in order to minimize glare and maximize daylighting. Of course, human perceptions of thermal and visual comfort vary considerably among individuals. For this reason, a good design principle to keep in mind is that building occupants should be given as much control over their thermal and visual environments as practical. Room thermostats, operable windows, dimming switches, and adjustable blinds are all means of giving people this control. Integrating these manually operable controls into the BAS can contribute significantly toward optimizing both IEQ and energy efficiency.

CONTROL SYSTEM COMMISSIONING PROCESS

The commissioning process is defined in ASHRAE Guideline 0-2005, *The Commissioning Process* (ASHRAE 2005), as follows:

> A quality focused process for enhancing the delivery of a project. The process focuses upon verifying and documenting that the facility and all of its systems and assemblies are planned, designed, installed, tested, operated, and maintained to meet the Owner's Project Requirements.

The full process from project planning through occupancy and operations is explained in ASHRAE Guideline 0 (ASHRAE 2005) and in Chapter 43 of the 2011 *ASHRAE Handbook—HVAC Applications* (ASHRAE 2007). Guidelines for applying the commissioning process to green buildings in general are found in Chapter 6 of this guide. Additional information relative to design reviews and commissioning for designers and commissioning providers can be found on the Energy Design Resources website (www.energydesignresources.com/publication/gd/).

This section will cover salient elements of applying the commissioning process to controls in new-construction green buildings and will focus on what the design engineer and commissioning provider can do during the design phase to facilitate a successful commissioning program. For commissioning of existing buildings, the reader is referred to sources of information, such as U.S. Green Building Council's (USGBC) LEED® program for Existing Buildings (see "References and Resources" section at the end of the chapter for website).

Include Commissioning Engagement in Design Fees

In estimating the fees for the design process, owners and the design team should include sufficient allowance to be fully engaged in the commissioning process during design. This includes responding to design review comments and incorporating commissioning requirements into the project specifications.

Conduct and Participate in Design Reviews

During the design reviews, the commissioning provider should see that the following control-related elements are included in the contract documents. This list of items is not comprehensive, but it provides an idea of the type of issues that should be addressed.

Provide Detailed Control Descriptions. One of the most prevalent reasons why control systems fail to perform as intended is that insufficient forethought is given to the sequence of operation prior to the contractor programming setup. The designers, and later, the control contractor's programmer often do not think the sequence out and consider how it will (or will not) function during all possible modes and scenarios of weather, loads, staging, and interactions. This issue can be mitigated by the following:

- Ensure that the designer provides flow diagrams of the major controlled systems, showing interfaces and control authorities between local and central control.
- Ensure that the designer includes detailed sequences of operation that include brief system narrative, points list, alarms, what initially starts equipment, staging, failure and standby functions, power outage response and reset requirements, interlocks to other systems, control authorities with local (packaged) controls, trending requirements, and energy efficiency strategies with given setpoints.
- Develop a graphical test simulation of the control program(s) to ensure that the mechanical equipment sequences on and off as load increases and decreases, according to the sequences of operation.

Match Control Strategies to Operator Capabilities. If the operators do not understand the features or sequences sufficiently and there is not a qualified controls technician maintaining the system, the advanced features or sequences that have problems will likely be overridden or disabled. Designers and design reviewers should make sure the complexity of control schemes matches the expected level of technical expertise of the operators. (It is a known fact that the operators will reduce the complexity of the control system to the operator's level of understanding.) It is critical that operator training be conducted in a timely manner, including follow-up sessions as needed.

Strategies Relying on Drift-Prone Sensors. Control sensor and loop recalibrations are a necessary function for maintaining high-performing systems. The Owner's Project Requirements document should define the operator training and skill sets needed to maintain the system functioning at design efficiency. Major control strategies that depend on sensors that are known to drift should be avoided or, if called for, the necessary training and recalibration programs must be institutionalized into the building maintenance culture. For example, consider the case of a chiller staging sequence utilizing supply and return temperatures and a flowmeter(s). The sensors may well drift over time, and typical accepted errors in these types of sensors will yield load calculations that may disrupt proper staging. This strategy can result in high overall efficiency, but it requires a regular calibration and maintenance check.

Requirements for System Architecture Rationale. Ensure that in the requirements for the controls submittal, the controls contractor is required to provide calculations and rationale for the number and layout of the primary (peer-to-peer) and local (application-specific) controllers in relation to the total number of points and other network traffic. Require that the contractor describe how many points can be reasonably trended without appreciably affecting point value refresh rates, and describe the impacts on network speed that alternative layouts would have.

The BAS performance requirements should be defined in the contract documents (i.e., specifications such as those set forth in ASHRAE Guideline 13-2007, *Specifying Direct Digital Control Systems* [ASHRAE 2007]). The performance requirements are defined during the predesign phase and should be contained in the Owner's Project Requirements document in accordance with ASHRAE Guideline 0 (ASHRAE 2005).

Requirements for Clear Control Sequences. Ensure that the requirements for the control drawing submittals in the specifications include statements requiring the following:

- A brief overview narrative of the system, generally describing its purpose, components, and function (i.e., the design intent for the controls)
- All interactions and interlocks with other systems
- Detailed delineation of interaction between any localized controls and the BAS, listing what points only the BAS monitors and what BAS points are control points and are adjustable
- Start-up, warm-up, cooldown, occupied and unoccupied operating modes, plus power failure recovery and alarm sequences
- Capacity control sequences and equipment staging
- Initial and recommended values for all adjustable settings, setpoints, and parameters that are typically set or adjusted by operating staff and any other control settings or fixed values, delays, etc., that will be useful during testing and operating the equipment
- Rogue zone analysis requirements to assure reset strategies are effective
- Energy mapping requirements to help operators see the building efficiency at a glance

- System override abilities and requirements
- Description of building isolation areas for off-hours operation
- Front-end graphics requirements including summary screens by system type, zone, and plant
- To facilitate review and referencing in testing procedures, all sequences written in short statements, each with a number for easy reference

Requirements for Clear Control Drawings. The specifications must ensure that the control drawing submittal requirements include at least the following:

- The control drawings must contain graphic schematic depictions of all systems showing each component (e.g., valves, dampers, actuators, coils, filters, fans, pumps, speed controllers, piping, ducting, etc.), each monitored or control point and sensor, and all interlocks to other equipment. Drawings may include fan and pump flow rates as well as horsepower.
- The schematics will include the system and component layout of any equipment that the control system monitors, enables, or controls, even if the equipment is primarily controlled by packaged controls.
- Provide a full points list, including point abbreviation key, point type, system point with which it is associated, point description, units, panel ID, and field device.
- Network architecture drawing showing all controllers, workstations, printers, and other devices in a riser format and including protocols and speeds for all trunks. Include the network buses with the bus speeds.
- Sketches of all graphics screens for review and approval.

Specify a Systems Manual

Ensure that the commissioning scope for the commissioning provider, contractor, and designer includes a systems manual that, among other things, includes narratives explaining all energy-efficiency features and strategies, a setpoint and parameter table that indicates the impacts of changing the values, a recalibration and retesting frequency table, suggested smart alerts in the control system to send alerts on malfunctioning sensors and actuators, and a list of standard trend logs to view to verify proper performance. The building operators need to be trained on the systems manual and its contents in order to properly operate and maintain the system.

DESIGNING FOR SUSTAINED EFFICIENCY

This chapter contains a number of recommendations on how building and system controls can be used to obtain a good, green design. Getting a good design up front is important but is only the beginning. Sustainability also includes continued efficient building operation over its entire lifetime. Three factors that are critical to sustaining the efficiency level of a new building are (1) a well-designed M&V process,

(2) implementing a commissioning program that will evaluate the function of all key systems and equipment on a regular basis, and (3) good operator training on the control system functions. The first two factors allow building operators to monitor performance on a regular basis and to intervene when problems are detected. As discussed in this chapter, the control system is essential in implementing a good M&V process. Good training can help ensure that all of the capabilities of the control system, including those related to M&V, are used to their full potential over the lifetime of the building.

REFERENCES AND RESOURCES

Published

ASHRAE. 2002. ASHRAE Guideline 14-2002, *Measurement of Energy and Demand Savings*. Atlanta: ASHRAE.

ASHRAE. 2005. ASHRAE Guideline 0-2005, *The Commissioning Process*. Atlanta: ASHRAE.

ASHRAE. 2007. ASHRAE Guideline 13-2007, *Specifying Direct Digital Control Systems*. Atlanta: ASHRAE.

ASHRAE. 2010. ANSI/ASHRAE Standard 55-2010, *Thermal Environmental Conditions for Human Occupancy*. Atlanta: ASHRAE.

ASHRAE. 2011. *ASHRAE Handbook—HVAC Applications*. Atlanta: ASHRAE.

ASHRAE. 2011. ANSI/ASHRAE/USGBC/IES Standard 189.1-2011, *Standard for the Design of High-Performance Green Buildings*. Atlanta: ASHRAE.

ASHRAE. 2013a. ANSI/ASHRAE Standard 62.1-2013 *Ventilation for Acceptable Indoor Air Quality*. Atlanta: ASHRAE.

ASHRAE. 2013b. ANSI/ASHRAE/IES Standard 90.1-2013, *Energy Standard for Buildings Except Low-Rise Residential Buildings*. Atlanta: ASHRAE.

CBE. 2007. *Summary report: Control strategies for mixed-mode buildings*. Center for the Built Environment. www.cbe.berkeley.edu/research/pdf_files/SR_MixedModeControls2007.pdf.

DOE. 2005. Energy impact of commercial building controls and performance diagnostics: Market characterization, energy impact of building faults and energy savings potential. U.S. Department of Energy, Washington, D.C.

IES. 2000. *Lighting Handbook*. New York: Illuminating Engineering Society of North America.

USGBC. 2009. *LEED® 2009 Green Building Operations and Maintenance*. Washington, D.C.: U.S. Green Building Council.

USGBC. 2009. *LEED® 2009 for New Construction and Major Renovations*. Washington, D.C.: U.S. Green Building Council.

Online
DDC Online
 www.ddc-online.org.
Energy Design Resources
 www.energydesignresources.com/publication/gd/.
International Performance Measurement and Verification Protocol
 www.evo-world.org.
National Building Controls Information Program
 www.buildingcontrols.org.
OpenADR Alliance
 www.openadr.org.
OpenADR Primer
 www.openadr.org/assets/docs/openadr_primer.pdf.
U.S. Green Building Council
 www.usgbc.org.

CHAPTER SIXTEEN

COMPLETING DESIGN AND DOCUMENTATION FOR CONSTRUCTION

DRAWINGS/DOCUMENTATION STAGE

Once the project has reached the working drawing/construction document stage, the green design concepts and resulting configurations should be well set, and the task of incorporating them into the documents that contractors will use to build the project should be relatively routine. However, quality control at this stage is especially important in green design projects.

Many firms have a routine procedure to review the documents for quality before they are released to contractors (to ensure that concepts are adequately depicted and described and to catch errors and omissions). This process should also include a green design concept review, preferably by one or more design team members that were in on the early stages of the project. This is particularly true if those preparing the construction documents were not part of that process. This is not the time to allow an excellent green design to become diluted or slip away.

SPECIFYING MATERIALS/EQUIPMENT

Green-Building Materials

Sources for guidance on selecting and specifying materials for a green project are as follows:

- Athena Institute, www.athenasmi.ca
- BuildingGreen, www.buildinggreen.com

Also see the DiggingDeeper sidebar titled "One Design Firm's Materials Specification Checklist."

Controlling Construction Quality

It is far easier to control construction quality in the design and specification stage of a project than during its construction. During preparation of the final construction drawings and specifications, it is critical to be diligent in spelling out the quality expected in the field to carry through the green design concepts developed throughout the early design stages. Some further thoughts on this subject, many applicable to the design phase, are covered in Chapter 17 of this guide.

COST ESTIMATING AND BUDGET RECONCILIATION

It is very important to have the cost estimator involved right from the start of the project to ensure that the project budget reflects the decisions made by the rest of the project team throughout the integrated design process.

The chapters, case studies, resources, and cost data included in the latest edition of the RS Means' publication *Green Building: Project Planning & Cost Estimating, Second Edition* (Adler et al. 2002) are good resources that include information on the following:

- Green -building approaches, materials, systems, and standards
- Evaluating the cost vs. value of green products over their life cycle
- Specifying green-building projects—complete with a list of often-specified products/materials and a sample spec
- Low-cost green strategies—and special economic incentives and funding
- Deconstruction—featured in a new chapter on this key element of sustainable building

This publication has been completely updated with the latest in green-building technologies, design concepts, standards, and costs.

BIDDING

The following is an excerpt from the Harvard University Office for Sustainability, Green Building Resource, Design Phase Guide website (www.green.harvard.edu/theresource/new-construction/design-phase/bidding-construction/):

"The expectations of general and subcontractors on Leadership in Energy and Environmental Design (LEED®) and other green projects are different than those associated with conventional projects. Practices such as recycling, erosion, and sedimentation control, indoor air quality (IAQ), and filling out LEED (and other rating system) submittals require special attention, and should be communicated to the team early in the construction process.

Include sustainability language in the Request for Proposals and Owner's Project Requirements document. Consider green-building expertise and LEED (or other green rating system) project experience as key criteria in the selection process. Ask to see evidence of experience with LEED, material tracking, construction and demolition waste management, and plans for IAQ during construction."

The contractor should be required to have a LEED coordinator if the project is going to LEED and a commissioning coordinator. Having experienced people involved in these tasks is critical to success. Also, the contractor should report on regional materials at a minimum, but reuse and/or rapidly renewable materials as well.

According to the same website, the contractor is responsible for conveying the following to subcontractors: workplace practice expectations for recycling and erosion and sedimentation control, and construction IAQ practices. The Owner's Sustainability Representative (OSR) is responsible for coordinating regular LEED meetings to track the LEED documentation process and to collect submittal and audit requirements from contractors. Also, a submittal review process should be established among the commissioning authority, the architect, and OSR.

MANAGING RISK

Green Design Documentation Issues

Reviewing green design work done by others allows engineers to see how built projects have created new opportunities and can guide efforts to sell and provide green engineering services. Performing sustainable design can yield important benefits to design engineers individually and to their firms. Improved client service, more repeat work, improved market position, enhanced public relations, and better employee satisfaction and retention are among the many benefits that numerous architectural firms and several pioneering engineering firms have derived from informing their practices with sustainable design expertise.

On the other hand, failure to address client concerns about green design can harm reputation. Disproportionate start-up costs, risk, or other perceived obstacles (e.g., the educational investment and commitment required to change engineering thinking and culture) may be perceived as being associated with undertaking sustainable design. Managing each of these issues, communicating frankly, and crafting creative solutions, as required, can reduce exposure to these potentially negative issues.

However, when documenting the project requirements for bidding by contractors and subcontractors, it is important that the engineer not guarantee that the results predicted by building simulation modeling will actually be achieved. It is important that the engineer not take more than his or her fair share of the risk in incorporating newer technologies into a building design project. Sustaining fair and practical business practices is critical to being a successful green engineer.

Contract Provisions

The following is an excerpt from the online essay, "Green Buildings and Risk," by Tim Corbett (www.greenbiz.com/blog/2007/11/26/managing-risk-green-building-projects):

> A good contract is the best line of defense when it comes to mitigating your risk. The contract is an excellent method for defining your scope of services (e.g., what will be provided, when, and what will not). Contracts are also an excellent method for qualifying clients and managing and establishing expectations. Contracts should address the following:
>
> - New and innovative products and technology may be used; they may lack proven history of successful application. Owner understands and agrees that project objectives may not be realized. A caveat is that the decision to try new, unproved products or technologies needs to be a decision by the owner and the design team together to make sure the owner is informed of the risks and opportunities.
>
> - Ordinary skill and care will be used to achieve project objectives; however there is no warranty or guarantee the project will achieve LEED certification.
>
> - Verify the level of investigation and analysis that will be performed for new material and technologies, with no expressed or implied warranty or guarantees of results.
>
> - Client agrees to measure the potential risks related to incorporating the innovation product and/or systems and accepts the risks.
>
> - Limit your exposure to consequential damages by including appropriate language in your contract.
>
> - Specify whether a shortage or unavailability of green materials must be treated as a compensable or noncompensable delay under the contract documents.
>
> - Monthly progress report should be submitted to indicate project progress along with the required documentation such as photographs, periodic waste records, and periodic material purchase records.

ONE DESIGN FIRM'S MATERIALS SPECIFICATION CHECKLIST

Choose at least one portion of materials/products that have the following characteristics:

- Local and/or indigenous, reducing the environmental impacts resulting from transportation and supporting the local economy
- Extracted, harvested, recovered, manufactured regionally within a radius of 500 miles (805 km)
- Low-embodied energy
- Reused, recycled, and/or recyclable, reducing the impacts resulting from extraction of new resources
- Salvaged building materials (e.g., lumber, millwork, plumbing fixtures, hardware, etc.)
- Postconsumer recycled content material and/or postindustrial recycled content (recovered) material
- Nonhazardous to recycle, compost, or dispose of
- Renewable and sustainably harvested (no old-growth timber), preferably with minimal associated environmental burdens (reducing the use and depletion of finite raw and long-cycle renewable materials)
- Rapidly renewable building materials, including any nonwood materials that are typically harvested within a 10 yr or shorter cycle
- Nontoxic/nonpolluting in manufacture, use, and disposal

Use finished materials, products, and furnishings that are free of known, probable, and suspected carcinogens, mutagens, teratogens, persistent toxic organic pollutants, and toxic heavy metals pursuant to the U.S. Environmental Protection Agency's (EPA) Toxicity Characteristic Leaching Procedure Test, 40 CFR Part 260. (Based on the EPA's Universal Waste Rule and Part 260, building owners and their contractors must use the toxicity characteristic leaching procedure to determine if they are generators of hazardous waste and subject to EPA and state hazardous waste regulations when disposing of mercury lamps and other waste products.)

Specify lead-free solder for copper water supply tubing; do not use plastic for supply water.

Avoid plastic foam insulation.

Specify natural, nontoxic, low-volatility organic compound–emitting, nonsolvent-based finishes, paints, stains, and adhesives.

When specifying painting, remember the following:

- Consider surfaces that don't require painting.
- Choose paints that have been independently certified (e.g., Green Seal).
- Choose latex over oil-based paint.

> - Increase direct-to-outdoors ventilation when painting.
> - Dispose of oil-based paints like hazardous waste.
> - Recycle latex paint or save for touch-ups.
>
> Avoid the use of materials containing or produced with ozone-depleting chlorofluorocarbons (CFCs) or hydrochlorofluorocarbons (HCFCs) (often embedded in foam insulation and refrigeration/cooling systems).
>
> Avoid CFC-based refrigerants in new base-building HVAC systems. When reusing existing base-building HVAC equipment, complete a comprehensive CFC phaseout conversion.
>
> Commit to working with manufacturing teams who provide performance/service contracts of integrated systems for service delivery, product longevity, adaptability, and/or recycling.
>
> Favor suppliers who will minimize packaging and will take back excess packaging (e.g., pallets, crates, cardboard, and excess building materials).
>
> Maximize the use of materials that retain a high value in future life cycles
>
> Specify products or systems that extend manufacturer responsibility through a lease or take-back program that ensures future reuse or recycling into a product of similar or higher value.
>
> To guide material selection choices during the design process, educate designers and elicit life-cycle information from manufacturers to encourage selection of building materials and furnishings with favorable life-cycle performance. Select materials from manufacturers that are committed to improving their overall environmental performance at their manufacturing facilities and their suppliers' facilities.

REFERENCES AND RESOURCES

Published

Adler, A., J.E. Armstrong, S.K. Fuller, G.B. Guy, M. Kalin, A. Karolides, M. Lelek, B.C. Lippiatt, J. Macaluso, G. Spencer, H.A. Walker, and P. Waier. 2002. *Green Building: Project Planning & Cost Estimating,* 2d ed. Kingston, MA: RS Means.

Corbett, T. *Green Buildings and Risk.* http://www.greenbiz.com/blog/2007/11/26/managing-risk-green-building-projects

EPA. Toxicity Characteristic Leaching Procedure (TCLP) Test, 40 CFR Part 260. U.S. Environmental Protection Agency, Washington, D.C.

EPA. Universal Waste Rule, 40 CFR Part 260. U.S. Environmental Protection Agency, Washington, D.C.

Online
Athena Institute
   www.athenasmi.ca.
BuildingGreen
   www.buildinggreen.com.
Harvard University Office for Sustainability, Green Building Resource, Design Phase Guide
   www.green.harvard.edu/theresource/new-construction/design-phase/bidding-construction/.

Section 3: Postdesign—Construction and Beyond

CHAPTER SEVENTEEN

CONSTRUCTION

To minimize gaps between design intent and what is actually built, the green-conscious engineer should provide a level of construction administration that is beyond the norm. That engineer (or an independent agent) may also provide mechanical/electrical system commissioning. (See Chapter 6 and Chapter 18 as well as a related section in Chapter 4 on project delivery methods and contractor selection.)

SITE PLANNING AND DEVELOPMENT

Green-building design includes the design of, and implementation of, the changes needed on the site to accommodate the building. While not generally the role of a mechanical or electrical engineer or a typical ASHRAE member, this is an important function that if not done right can have major negative consequences on the surrounding environment.

THE ENGINEER'S ROLE IN CONSTRUCTION QUALITY

In construction, the engineer's key roles encompass shop drawing review, review of equipment substitutions, handling of change order and value engineering requests, and site visits and inspections. Leadership in Energy and Environmental Design (LEED®) or other green design process documentation coordination and processing are sometimes required when the project team has decided to achieve LEED certification or another green design process recognition for the building.

Shop Drawing Review

There should be a thorough review of shop drawing submittals. The specifications should require their timely submission to allow time for proper review without delaying the project. The purpose of the review is to ensure that the contractor is correctly interpreting the specifications and drawings and is not missing any important details. Areas for emphasis include checking motor power and efficiency ratings, checking air filter details, checking for proper clearances around equipment for servicing, and checking anything else relevant on the project concerning green design elements. Thorough review of all control sequences submitted by the building automation system subcontractor is critical.

Alternative Equipment Substitutions and or Equivalent

Specifications may be written with several named manufacturers (often three) of a given product, or it may name one and say "or equivalent." In both cases, the system designed is usually based on a single manufacturer's product. If others are named, it usually means they are acceptable, as long as they meet the specified conditions and fit in the space provided on the project, including all required service and maintenance access clearances. The shop drawing reviewer must determine whether alternative equipment is truly equal to the base design, including meeting the green design aspects of the equipment and systems.

If "or equivalent" language is used to specify equipment, determining equivalence can be more difficult. It is advisable to be certain that what the contractor proposes is truly equivalent, if not superior, to the specified components or systems. Contractors may have legitimate reasons for proposing alternative products, such as better delivery times or negative experiences with the products or manufacturers specified. Such equivalent, or outright substitution, proposals from the contractor should be treated seriously and examined carefully to ascertain that they will not adversely impact the project goals. If they do not meet the requirements, the objections should first be discussed with the construction manager, owner's representative, and subcontractor. Only if no satisfactory alternative can be agreed on should they then be rejected. It is very important to clearly identify the substitution process requirements when specifying the equipment during the design phase of the project.

Request for Information (RFI)

During the project's construction phase the construction team may prepare and submit requests for information based on their need for clarification or interpretation of the construction documents. The following definition describes examples of an RFI's content:

> An RFI is used in the construction industry in cases where it is necessary to confirm the interpretation of a detail, specification, or note on the construction drawings or to secure a documented directive or clarification from the architect or client that is needed to continue work. An RFI raised by the general contractor that has been answered by the client or architect and distributed to all stakeholders is generally accepted as a change to the scope of work unless further approval is required for costs associated with the change. It is common and accepted practice for a subcontractor or supplier to use an RFI to state

his/her concern related to the omission or misapplication of a product and seek further clarification of the building owner's intended use or the building official acceptance of the specified product. It is also acceptable for the subcontractor to use an RFI to call attention to an inferior product that may not meet the building owner's needs, and use his/her expertise to recommend the better/correct product. (Construction Management Solutions.)

Normally, an RFI is prepared by the construction team and submitted to the design team for purposes of obtaining additional information and/or details concerning a construction or building system specification and/or details that may have been omitted or inadequately conveyed in the bid/construction documents. RFIs have increasingly become a form of notification to the design team and owner that a change order request will be submitted once the RFI has been responded to by the design team. Thus, an RFI can create great concern and cost to the design team in their review and design activity needed to provide a clear and comprehensive response to the RFI.

Change Order Requests/Value Engineering and Value Management

Many change orders are legitimate, such as an owner changing the scope, expansion or reduction of project scope, or the construction manager and/or subcontractors encountering unforeseen conditions. However, any change order that only cheapens the project or lowers the project's green design standards without counteracting benefits should be regarded with skepticism.

Likewise, any value engineering (VE) or value management (VM) suggestions should be carefully studied for their impact on the project's green design goals. (VE/VM is often offered under the assumption that first-cost savings are paramount to the owner and project team.) The need for careful study remains even in the case of genuine VE/VM done by trained professionals who perform real trade-off analyses to arrive at the best value for a project. Before beginning such a VE/VM exercise, VE/VM facilitators should clearly understand the green design objectives to ensure that their suggestions and recommendations are consistent with project goals, as well as the priority those goals have with respect to first cost, life-cycle cost, or other building operational considerations. Each VE/VM item must include the first-cost impact to properly evaluate the suggestion. The impact on life-cycle costs and predicted building operation and performance must also be clearly understood.

Site Visits/Observations

Site visits should be planned for key times and should involve the engineer's best personnel. HVAC work should be viewed before it is covered up by ceiling, floor, and wall installations. Check that equipment nameplates are correct. For example, if high-efficiency motors were specified, check in the field to be certain that is what is being installed. Look up the manufacturer's data on motor efficiency and see that it matches the specification. Check that absorptive construction materials (e.g., insulation) are being stored in accord with the IAQ Construction Management Plan and that no contamination has occurred. Check for air filters in any operating air-handling units. Ensure that air-handling equipment has been stored properly, cleaned (internals), and that filters were installed before start-up. If air-handling units were used for construction activities, check that return air paths have proper temporary filtration at the inlets to the ductwork.

Work with the commissioning authority to review their site observations and address any design-based questions that arise during the final commissioning process. More information on this is provided in the following chapter.

Final punch list preparation, follow-up, and the final sign-off observation are particularly important. If the earlier site visits and observations were done thoroughly and at appropriate intervals, and if the construction manager and subcontractors have been part of the team and in accord with the green design goals, the final punch list should be minimal and the final observation should go smoothly.

CONSTRUCTION PRACTICES AND METHODS

In the design phase, the specifications should prescribe that certain construction methods and procedures must be followed to ensure a fully realized green project. This would include topics such as reduced site disturbance, handling of construction waste, control of rainwater runoff, and indoor air quality (IAQ) management during construction. This section includes a brief discussion of several items that are key to the construction and delivery of a green building.

Indoor Air Quality During Construction

It is recommended that the engineer and architect work with the construction manager or general contractor and associated subcontractors to develop an IAQ construction management plan that the contractor would be required to follow. The Sheet Metal and Air Conditioning National Association's *IAQ Guideline for Occupied Buildings Under Construction* (SMACNA) is cited by U.S. Green Building Council (USGBC) as a source document. The contractor should particularly protect installed or stored absorptive materials (e.g., insulation or sheetrock) from water damage or other contamination. Water damage is especially insidious if materials get wet, are installed wet, and are then covered up. If air-handling units (AHUs) run during construction, the construction manager and associated subcontractor should be required to protect AHU components and the duct distribution system from dirt and debris, clean any components and ductwork that are damaged, and replace filters before occupancy. (ANSI/ASHRAE Standard 52.2-2007, *Method of Testing General Ventilation Air-Cleaning Devices for Removal Efficiency by Particle Size* [ASHRAE 2007] deals with this subject.)

Even if your project is not going to pursue certification in a green-building rating program (such as LEED) or is not subject to a green-building code (such as ANSI/ASHRAE/USGBC/IES Standard 189.1 or the *International Green Construction Code*), it is a good idea to also conduct a building flush out or indoor air quality test after construction is complete and before occupancy begins. Established criteria for indoor air quality construction management practices can be found in Section 10 of Standard 189.1.

The construction team should also consider the issue of whether permanent HVAC systems should be operated during construction. This practice is not allowed in high-performance, green-building situations (such as a project needing to comply with Standard 189.1), and the use of permanent HVAC is not recommended for any green-building project if at all possible.

Construction Waste

One of the more positive side benefits from an increased adoption of green building practices has been a better awareness of the need to minimize waste sent to a landfill from new construction or renovation projects. Specific criteria again are contained in the LEED program descriptions and in Section 10 of Standard 189.1. Many waste haulers now will gladly cooperate with a waste minimization program that includes sorting of waste for reuse or recycling, as this can reduce the tipping fees that they will have to pay and can provide another revenue stream from the sale of recyclable materials, such as metals.

The ability to readily recycle or reuse materials from demolition and construction can depend on the location of the building project; some areas have better established recycling programs than others. As an engineer, contractor, or other professional on the project, you can help encourage the development of a waste minimization and recycling program as part of the building project.

Other Important Areas

In addition to the above, a green building should have and implement a good erosion and sedimentation control program. Many local jurisdictions have requirements in this area. However, the requirements are often very rudimentary and will provide minimal erosion and sedimentation control if not properly implemented. A green-building construction project should take this aspect seriously.

Worker health and safety measures, while not specifically considered green topics, are important as well and particular concern can be given to items such as respiratory protective equipment.

In addition, there is a concern about the potential for vehicular pollution from idling construction vehicles causing IAQ problems in occupied buildings. A well-thought-out construction vehicle staging and idling program plan should be part of any high-performance green-building construction planning.

COMMISSIONING DURING CONSTRUCTION

Once construction starts, the commissioning process enters a new phase. Depending on the commissioning plan and the activities that were identified in the plan, the commissioning team will be active during construction with tasks such as verification that equipment is being installed properly. It is during construction that the building envelope commissioning is done, and there is a growing recognition of the importance and value of commissioning the building envelope. The chapter on commissioning (Chapter 6) contains a thorough discussion on the commissioning process.

MOVING INTO OCCUPANCY AND OPERATION

Plans for Operation

An increased focus is being placed on the need to have a good operation and maintenance program in place in order for the benefits from energy and water efficiency and other green features of a new building project to continue long after occupancy takes place. Standard 189.1, Section 10 requires that plans for operation be developed as part of the construction documents. Although responsibility for final approval and implementation of these plans obviously is with the owner and the building operations staff, the project design team may be asked to help prepare these plans.

A more detailed discussion on the concept of a plan for operation is contained in Chapter 18.

REFERENCES AND RESOURCES

Published

ASHRAE. 2007. ANSI/ASHRAE Standard 52.2-2007, *Method of Testing General Ventilation Air-Cleaning Devices for Removal Efficiency by Particle Size.* Atlanta: ASHRAE.

ASHRAE. 2011. ANSI/ASHRAE/USGBC/IES Standard 189.1-2011, *Standard for the Design of High-Performance Green Buildings.* Atlanta: ASHRAE.

EPA. Storm water management for construction activities. EPA document No. EPA-8320R-92-005. U.S. Environmental Protection Agency, Washington, D.C.

SMACNA. 2007. *IAQ Guidelines for Occupied Buildings Under Construction.* Chantilly, VA: Sheet Metal and Air Conditioning Contractors' National Association, Inc.

Online

Construction Management Solutions
    www.contractorform.net/Request-for-Information-Form-Template.html.

CONSTRUCTION FACTORS TO CONSIDER IN A GREEN DESIGN

- Determine locations for construction vehicle parking, temporary piling of topsoil, and building material storage in order to minimize soil compaction and other site impacts.
- Control erosion to reduce negative impacts on water and air quality.
- Design a site sediment and erosion control plan that conforms to best management practices specified in the U.S. Environmental Protection Agency's *Storm Water Management for Construction Activities* (EPA) or local erosion and sedimentation control standards and codes, whichever is more stringent.
- Conserve existing natural areas and restore damaged areas to provide habitat and promote biodiversity.
- Schedule construction carefully to minimize impacts.
- Avoid leaving disturbed soil exposed for extended periods.
- Fill trenches quickly to minimize damage to severed tree roots.
- Avoid building when the ground is saturated and easily damaged.
- Estimate the amount of material needed to avoid excess.
- Design to accommodate standard lumber and drywall sizes.
- Assess the construction site waste stream to determine which materials can be reduced, reused, and recycled.
- Conduct a waste audit, quantifying material diversion by weight.
- Recycle and/or salvage construction and demolition debris.
- Specify materials that minimize waste and reduce shipping impacts through bulk packaging, dry-mix shipping, reused bulk packaging, recycled-content packaging, or elimination of packaging.
- Develop and implement a waste management plan, quantifying material diversion by weight.
- Research markets in area for salvaged materials.
- Establish on-site construction material recycling areas and recycle and/or salvage construction, demolition, and land clearing waste.
- Contract with licensed haulers and processors of recyclables.
- Require subcontractors to be responsible for their waste (including lunch wastes); create incentives for minimizing waste.
- Educate employees and subcontractors.
- Monitor and evaluate waste/recycling program.
- Review the IAQ Construction Management Plan to ensure that all subcontractors understand the process and goals desired.

CHAPTER EIGHTEEN

OPERATION, MAINTENANCE, AND PERFORMANCE EVALUATION

PLANS FOR OPERATION

The best-designed, constructed, and commissioned building will, over time, degrade in performance if it is not properly operated and maintained. The concept of preparing a plan for operation of a high-performance, green building was initiated with release of ANSI/ASHRAE/USGBC/IES Standard 189.1-2011. Section 10 of this standard includes a description of the minimum amount of information that should be contained in such a plan.

The plans for operation as specified in Standard 189.1 are focused mostly on maintaining equipment and systems that are included in the building design to comply with specific high-performance requirements set by the standard. These plans can, however, be used as guidelines for outlining how a green building should be operated in general. For example, modern control and automation systems provide an opportunity for monitoring the performance of systems as they relate to energy and water consumption, indoor air quality (ventilation rates and perhaps CO_2 levels), and thermal comfort. A well-thought-out and written plan for operation would include the planned response to take when temperature excursions outside the design or setpoint range occur or if CO_2 levels exceed a predetermined value in a zone. It is also recommended that the building operations team include an evaluation of energy consumption in a more detailed fashion than just the aggregated monthly utility bill. This evaluation should at least include monitoring of the total energy consumption for key systems such as the HVAC or lighting, and preferably would be done on a finer level of detail to include the energy consumption of key equipment, such as a chiller.

COMMISSIONING

Commissioning of existing buildings focuses on identifying current facility requirements needed by owners and occupants to efficiently and effectively deliver their daily mission. Building performance generally declines after two to five years. As a result, commissioning of existing buildings can return the building to its peak performance. It is important that designers have a deeper understanding of why performance wanes and how design decisions affect building operation efficiencies.

Commissioning can and should be an ongoing process. Opportunities to improve performance through optimizing HVAC system operation and control, lighting modifications, and identification and control of phantom vampire loads are brought to the owner's attention, along with cost-benefit analysis, to assist the owner with selection of an improvement strategy. Commissioning of existing buildings has proven to be very cost effective and has the greatest promise for substantially reducing energy consumption in the country and reducing dependency on foreign oil. Most existing buildings could immediately reduce energy consumption by 10% or more and could realize a return on investment in as little as six months to two years, depending on the types of projects undertaken. Commissioning of existing buildings can also identify improvements for further reductions in energy usage and carbon emissions so that owners can include these improvements in their capital improvement plans.

Commissioning of existing buildings can take different forms, and some confusion can occur in the definitions, depending on whether the building had been previously commissioned and the mechanisms for conducting that commissioning. Repeating a commissioning process for a building that had already been commissioned is generally known as *recommissioning*, while a commissioning process on a building that had not been commissioned is called *retrocommissioning*. Recently, a variation on the commissioning process has appeared that uses equipment that tracks the operation and performance of building systems, called *monitoring-based commissioning*, which has been discussed previously in this guide. Use of monitoring-based commissioning provides operators and owners a benchmark for continually assessing building performance under dynamic conditions. It also notifies operators and owners when building performance has drifted from peak performance, signaling that action is need. Monitoring-based commissioning is beyond the scope of most new construction projects but is becoming a common service owners are employing to manage costs, maintain occupant satisfaction, and minimize the environmental impact of existing buildings.

Educating operators to understand building operation can reduce operation errors but does not eliminate them. Operators typically respond to complaints and often override systems to eliminate complaints. Often, the operators forget or are busy "fighting fires" and never determine the root cause of the complaint so that it can be permanently resolved. The resolution of these complaints frequently results in changes that negatively impact facility performance. A properly designed and run commissioning process can help alleviate this problem.

ENERGY EFFICIENCY IN EXISTING BUILDINGS

New construction each year adds roughly 1% to the U.S. building stock. Considering that an average commercial building in the United States has an estimated life of 40 years, even governmental regulations requiring that all new buildings be constructed to green specifications would not transform the commercial building stock to cleaner and greener until well past 2050. However, the creation and adoption of building energy-efficiency standards over the past several decades is slowly leading to a more efficient building stock overall. Figure 18-1 shows the potential that these new codes and standards offer. For example, from the 2003 Commercial Building Energy Consumption Survey (CBECS) data, the overall weighted Energy Utilization Index (EUI) of 90 kBtu/ft^2·yr (1020 MJ/m^2·yr). If all these buildings were rebuilt according to the 2010 version of ANSI/ASHRAE/IES Standard 90.1, the overall EUI could be reduced by 40%, to 52.2 kBtu/ft^2·yr (592 MJ/m^2·yr).

Because energy and operations account for approximately 75% of a building's costs over its lifetime, whereas design and construction costs are 11% and financing is approximately 14%, it is clear that existing buildings are perfect candidates for green retrofit (see von Paumgartten).

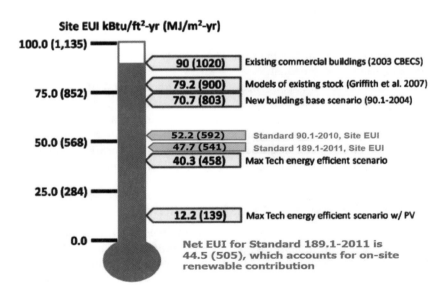

Figure 18-1 Trends in EUI with the improved energy efficiency standards.

Engineers can help owners and managers of existing buildings understand the economic benefits of improving systems and operations. "Energy Efficiency in Existing Buildings, Our Greatest Opportunity for a Sustainable Future" was ASHRAE President Gordon Holness' theme during his 2009–2010 term. In his Presidential Address to ASHRAE, Mr. Holness emphasized that the path forward for ASHRAE was an evaluation and updating of the following key guidance produced by the society:

- The Advanced Energy Design Guide series (AEDG)
- Standard 90.1, *Energy Standard for Buildings Except Low-Rise Residential Buildings*
- Standard 189.1, *Standard for the Design of High-Perfor-mance, Green Buildings Except Low-Rise Residential Buildings*
- Standard 100, *Energy Conservation in Existing Buildings*
- Commissioning and retrocommissioning guidance
- Operations and maintenance guidance

A summary of his address can be viewed at www.ashrae.org/File%20Library/docLib/PastPresidentsGallery/Holness.pdf.

The guidelines for making existing buildings more energy efficient include the following:

- Benchmarking the current energy utilization index (EUI) (kBtu/ft^2· yr [MJ/m^2· yr]).
- Establishing a target energy utilization index and an initial budget estimate for achieving this goal.
- Conducting an internal energy audit or having the facility retrocommissioned by a certified retrocommissioning firm. This activity may result in a modification to the estimated budget amount.
- Identifying energy efficiency measures with attractive rates of return on energy retrofit or renovation investments.
- Implementing the recommended energy conservation measures that will get the facility to the desired goal within the stipulated budget.
- Commissioning the energy conservation measures by a certified commissioning firm. This process should include training of facility personnel on properly operating and maintaining equipment and systems.

RETROFIT STRATEGIES FOR EXISTING BUILDINGS

Buildings that are more than five years old are often good candidates for retrofit projects that make them more energy and water efficient and improve their indoor environmental quality (IEQ). Many retrofit projects can be justified by the attendant cost savings. Common retrofit strategies that offer good savings to investment ratios include HVAC and control system retrocommissioning, energy recovery, air- and water-side free cooling, variable speed control of fans and pumps, lighting system improvements, and plumbing fixture upgrades.

HVAC and control system retrocommissioning is addressed in detail in the next section of this chapter. Energy recovery includes such diverse strategies as glycol runaround loops and entropy wheels to transfer energy between exhaust and outdoor airstreams, preheaters to transfer energy from boiler stack gases to combustion air, and heat pipes to reheat air using energy from upstream air entering the cooling coil. Air- and water-side free cooling refers to the use of outdoor air to cool interior spaces, either via air-handling unit economizer controls or a plate-and-frame heat exchanger piped between the chilled-water system and a cooling tower. And installing variable-speed drives (together with ancillary equipment and controls) makes it possible to match air and water flows to heating and cooling demands more effectively than was the case in the past.

Moving beyond the realm of HVAC, lighting systems can be upgraded by applying relatively new technologies (e.g., electronic ballasts, T5 and T8 lamps, compact fluorescent lamps, light-emitting diode (LED) signage, and occupancy sensors). In addition to saving energy, lighting retrofits often improve light quality significantly, by improving color rendering and eliminating flicker. Finally, new low-flow plumbing fixtures, including water closets, urinals, lavatory faucets, and shower heads, can reduce both water and energy use without compromising performance or comfort.

When major mechanical equipment (e.g., chillers and boilers) reach the end of their service lives, they should of course be replaced with more efficient equipment. Such retrofits generally cannot be justified in terms of energy cost savings alone. Replacing a standard efficiency motor with a premium efficiency motor is another retrofit strategy that usually cannot be cost-justified unless the motor is close to failure and needs to be replaced or rewound.

Renewable energy technologies should not be overlooked during a building retrofit project. In areas with high fossil fuel costs, solar service water and pool heating are likely to have the most attractive savings to investment ratios of these technologies.

MEASUREMENT AND VERIFICATION (M&V)

Several guidelines have been published on M&V energy savings. Each of the guidelines listed in this section are unique, albeit similar, and are intended for use in different instances. All of these documents provide standard M&V methods that are proven and accepted strategies.

INTERNATIONAL PERFORMANCE MEASUREMENT AND VERIFICATION PROTOCOL

The International Performance Measurement and Verification Protocol (IPMVP) is a document that discusses procedures that, when implemented, allow building owners, energy service companies, and financiers of building energy efficiency projects to quantify energy conservation measure performance and energy savings. The IPMVP provides an overview of current best-practice techniques available for verifying savings from both traditional and third-party-financed energy- and water-efficiency projects.

The IPMVP is now in three volumes, available at the Efficiency Valuation Organization's website (www.evo-world.org):

- Volume I—Concepts and Options for Determining Savings
- Volume II—Concepts and Practices for Improving Indoor Environmental Quality
- Volume III—Applications Cover Renewable Energy, New Construction, Emissions Trading, and Demand Reduction Baseline Calculation Methods)

FEDERAL ENERGY MANAGEMENT PROGRAM (FEMP) GUIDELINES VERSION 2.2

The FEMP Guidelines provide procedures and guidelines for quantifying the savings resulting from the installation of energy conservation measures. The FEMP Guidelines are fully compatible and consistent with the IPMVP. Intended for use in Energy Savings Performance Contracting and utility program projects, the guidelines provides the methodology for establishing energy cost savings called for in the Energy Savings Performance Contracting rule. These guidelines are available at www1.eere.energy.gov/femp/pdfs/mv_guidelines.pdf.

FEMP M&V OPTION A DETAILED GUIDELINES

FEMP has developed the *M&V Option A Detailed Guidelines* (FEMP 2008), which provides recommended practices for using the Option A methods described in FEMP's *M&V Guidelines for Federal Energy Projects, Version 3.0*. http://mnv.lbl.gov/keyMnVDocs/femp.

ASHRAE GUIDELINES AND STANDARDS

ASHRAE Guideline 14-2002, *Measurement of Energy and Demand Savings* (ASHRAE 2002) was developed by ASHRAE to provide guidance on the minimum acceptable level of performance in the measurement of energy and demand savings for the purpose of a commercial transaction based on that measurement. It only deals with the measurement of energy and demand savings.

The best-designed building and systems will only keep their high efficiency if they are properly maintained and operated. ASHRAE has recognized this, and part of the response was the creation of ASHRAE Standard 180-2008, *Standard Practice for Inspection and Maintenance of Commercial Building HVAC Systems*.

BUILDING LABELING

Information on a building's energy use is the critical first step in making the necessary changes and choices that reduce energy use and costs. The ASHRAE Building Energy Quotient (bEQ) program provides an easily understood scale to convey a building's energy use in comparison to similar buildings, occupancy types, and climate zone, while also providing building owners with building-specific information that can be used to improve building energy performance.

Based on predicted and actual energy usage, bEQ was introduced by ASHRAE in 2009, and a pilot program was initiated in early 2010. It includes both "As Designed" (asset) and "In Operation" (as operated) ratings for all building types, except residential. A sample label is shown in Figure 18-2. It also provides a detailed certificate with data on actual energy use, energy demand profiles, indoor air quality and other information that will enable building owners to evaluate and reduce their building's energy use.

This program is similar to a program initiated in 2002 by European member countries called the Energy Performance of Buildings Directive (EPBD). Its BUILD UP web portal (http://www.buildup.eu) has been created to help facilitate the exchange of information on building energy efficiency.

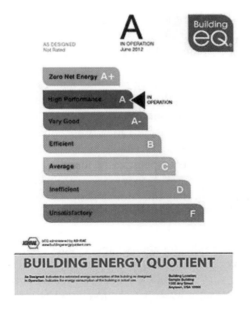

Figure 18-2 ASHRAE bEQ rating scale.

OCCUPANT SURVEYS

Occupant surveys can alert owners to specific problems with IEQ, which could potentially have implications for occupant health and productivity. Occupants are an important and often underutilized source of information about IEQ and its effect on comfort, satisfaction, and productivity. Poor thermal comfort is a common occupant complaint in buildings. Relying on complaint logs only provides an indication of local, personal, or sporadic dissatisfaction. Surveys are much more effective, as they provide a systematic mechanism for occupants to provide feedback about a broader range of aspects of the indoor environment. They are an important tool for identifying the specific nature and location of any problems, guiding postoccupancy retrocommissioning and corrective actions, and helping owners decide how to prioritize their investments in building improvements. Everyone in the building process benefits from learning how a building actually performs in practice. Occupant surveys are an informative and critical link in closing the feedback loop and allowing building owners, facility managers, and the design team to understand more completely how the building design and operation is affecting occupant productivity and well-being.

A survey should be designed so that participation is voluntary, occupants' responses remain anonymous, and results are reported only in aggregate. The survey should ask occupants for general location information only, so one can identify if problems are occurring in particular zones of the building. After asking about basic demographics and workstation characteristics, a survey should then ideally address a wide variety of IEQ features, including thermal comfort, air quality, lighting, acoustics, office layout and furnishings, and building cleanliness and maintenance. A common form of satisfaction question asks occupants to respond on a seven-point satisfaction scale ranging from very dissatisfied, to neutral, to very satisfied. Ideally, occupants who are dissatisfied with a particular aspect of their environment are presented with follow-up questions that allow them to more specifically identify the nature and potential source of their dissatisfaction. This is important for providing diagnostic information, and helping the building operators become more informed about how to respond.

Surveys can be administered in a variety of ways, but once the method is selected it should be consistently applied and available for all normal occupants of the building. Surveys can be administered directly, either by phone or in person, although this is very time-intensive and raises potential issues about privacy and accuracy of results. Web-based surveys are becoming a more common alternative to the traditional paper-based surveys and offer many advantages. They can be far less expensive to administer to a large number of people or to multiple buildings. The cost and potential errors of manually entering data from a paper survey are not present. They allow for more interactive branching features that provide diagnostic information, while keeping the survey to a reasonable length. Lastly, they offer the potential for automated reporting so that building owners and professionals can get quick access to the survey results.

Sources of sample surveys include, but are not limited to (see "References and Resources" section at the end of the chapter for more information):

- Center for the Built Environment (CBE) Occupant IEQ Survey
- Usable Buildings Trust

ONE DESIGN FIRM'S OPERATIONS, MAINTENANCE, AND PERFORMANCE EVALUATION CHECKLIST

Provide for the ongoing accountability and optimization of building energy and water consumption performance over time.

Design and specify equipment to be installed in base building systems to allow for comparison, management, and optimization of actual vs. estimated energy and water performance.

Provide for the ongoing accountability and optimization of building energy and water consumption performance over time.

Comply with the installed equipment requirements for continuous metering as stated in *Option B: Methods by Technology* of the U.S. Department of Energy's (DOE) IPMVP (www.energyautomation.com/pdfs/ipmvp-vol2.pdf) for the following:

- Lighting systems and controls
- Constant and variable motor loads
- Variable-frequency drive operation
- Chiller efficiency at variable loads (kW/ton)
- Cooling load
- Air and water economizer and heat recovery cycles
- Air distribution static pressures and ventilation air volumes
- Boiler efficiencies
- Building-specific process energy efficiency systems and equipment
- Indoor water risers and outdoor irrigation systems.

Allocate an appropriate percentage of building funds for ongoing monitoring of environmental performance, product purchasing, maintenance, and improvements.

Digging Deeper

Provide for the ongoing accountability of waste streams, including hazardous pollutants.

Use environmentally safe cleaning materials.

Educate operation and maintenance workers.

Facilitate the reduction of waste generated by building occupants.

Provide an easily accessible dedicated area for the collection and storage of materials for recycling of paper, glass, plastics, metals, and hazardous substances.

After six months, evaluate existing ecosystems to determine if they have remained undisturbed.

- Assess building energy use to ensure it is at predicted levels.

Determine if indoor air quality levels are at predicted levels (particularly CO_2 and airborne particulates).

Measure water consumption and evaluate against target usage in original plan.

Monitor water levels and determine if recycling and reuse of materials meet expectations.

Monitor and evaluate additional sustainability goals for project including the following:

- Building specific process, energyefficient systems, and equipment.
- Indoor water risers and outdoor irrigation systems.

REFERENCES AND RESOURCES

Published

ASHRAE. 2002. ASHRAE Guideline 14-2002, *Measurement of Energy and Demand Savings.* Atlanta: ASHRAE.

ASHRAE. 2005. ASHRAE Guideline 0-2005, *The Commissioning Process.* Atlanta: ASHRAE.

ASHRAE. 2008. ASHRAE Standard 180-2008, *Standard Practice for Inspection and Maintenance of Commercial Building HVAC Systems.* Atlanta: ASHRAE.

ASHRAE. 2011. ANSI/ASHRAE/USGBC/IES Standard 189.1-2011, *Standard for the Design of High-Performance Green Buildings.* Atlanta: ASHRAE.

FEMP. 2008. *M&V Option A Detailed Guidelines.* Washington, DC: U.S. Department of Energy, Federal Energy Management Program.

Leaman, A., B. and Bordass. 1993. Building design, complexity and manageability. *Facilities*, Vol. 11.

Zagreus, L., C. Huizenga, E. Arens, and D. Lehrer. 2004. Listening to the occupants: A Web-based indoor environmental quality survey. *Indoor Air 2004* 14(suppl 8):65–74.

Online

ASHRAE Building Energy Quotient
  www.buildingeq.com.
Center for the Built Environment, Occupant Indoor Environmental Quality Survey
  www.cbe.berkeley.edu/research/survey.htm.
European Commission, BUILD UP program
  www.buildup.eu/.
von Paumgartten, P. *Existing Buildings Hold the Key.* Alliance for Sustainable Built Environments.
  www.awarenessintoaction.com/whitepapers/how-existing-buildings-high-performing-green-leed-certified.html (28 Sep. 2010).
The Federal Energy Management Program Continuous Commissioning (SM) Guidebook
  www1.eere.energy.gov/femp/program/om_guidebook.html.
International Performance Measurement and Verification Protocol
  www.evo-world.org.
M&V Guidelines for Federal Energy Projects, Version 3.0
  http://mnv.lbl.gov/keyMnVDocs/femp.
Option B: Methods by Technology
  www.energyautomation.com/pdfs/ipmvp-vol2.pdf.
Usable Buildings Trust
  www.usablebuildings.co.uk.
U.S. Department of Energy, Federal Energy Management Program Guidelines
  www1.eere.energy.gov/femp/pdfs/mv_guidelines.pdf.

REFERENCES AND RESOURCES

These listings do not necessarily imply endorsement or agreement by ASHRAE or the authors with the information contained in the documents.

BOOKS AND SOFTWARE PROGRAMS

ACGIH. 1999. *Bioaerosols: Assessment and Control.* J. Nacher, ed. Cincinnati, Ohio: American Conference of Governmental Industrial Hygienists. www.acgih.org.

AGCC. 1994. *Applications Engineering Manual for Direct-Fired Absorption.* Washington, DC: American Gas Cooling Center.

Adler, A., J.E.Armstrong, S.K. Fuller, G.B. Guy, M. Kalin, A. Karolides, M. Lelek, B.C. Lippiatt, J. Macaluso, G. Spencer, H.A. Walker, and P. Waier. 2002. *Green Building: Project Planning & Cost Estimating,* 2d ed. Kingston, MA: RS Means.

AIA. 1996. *Environmental Resource Guide.* J. Demkin, ed. American Institute of Architects. New York: John Wiley & Sons, Inc.

AIA. 2006. *Guidelines for Design and Construction of Health Care Facilities.* Washington, DC: American Institute of Architects.

Alvarez, S., E. Dascalaki, G. Guarracino, E. Maldonado, S. Sciuto, and L. Vandaele. 1998. *Natural Ventilation of Buildings—A Design Handbook.* F. Allard, ed. London: James & James Science Publishers Ltd.

Argiriou, A., A. Dimoudi, C.A. Balaras, D. Mantas, E. Dascalaki, and I. Tselepidaki. 1996. Passive Cooling of Buildings. M. Santamouris and A. Asimakopoulos, eds. London: James & James Science Publishers Ltd.

ARI. 1997. *ARI STANDARD 275-97, Standard for Application of Sound Rating Levels of Outdoor Unitary Equipment.* Arlington, VA: Air-Conditioning and Refrigeration Institute.

Arthus-Bertrand, Y. 2001. *Earth from Above: 365 Days.* New York: Harry N. Abrams.

Arthus-Bertrand, Y. 2002. *Earth From Above.* New York: Harry N. Abrams.

ASHRAE. 1975. ASHRAE Standard 90-75, *Energy Conservation in New Building Design*. Atlanta: ASHRAE.
ASHRAE. 1988. *Active Solar Heating Systems Design Manual*. Atlanta: ASHRAE.
ASHRAE. 1995. *Commercial/Institutional Ground-Source Heat Pump Engineering Manual*. Atlanta: ASHRAE.
ASHRAE. 1996. ASHRAE Guideline 1-1996, *The HVAC Commissioning Process*. Atlanta: ASHRAE.
ASHRAE. 1998. *Fundamentals of Water System Design*. Atlanta: ASHRAE.
ASHRAE. 1999. ANSI/ASHRAE/IESNA Standard 90.1-1999, *Energy Standard for Buildings Except Low-Rise Residential Buildings*. Atlanta: ASHRAE.
ASHRAE. 2000. ANSI/ASHRAE Standard 52.2-1999, *Method of Testing General Ventilation Air-Cleaning Devices for Removal Efficiency by Particle Size*. Atlanta: ASHRAE.
ASHRAE. 2000. Interpretation IC-62-1999-30 (August) of *ANSI/ASHRAE Standard 62-1999, Ventilation for Acceptable Indoor Air Quality*. Atlanta: ASHRAE.
ASHRAE. 2002. ASHRAE Guideline 14-2002, *Measurement of Energy and Demand Savings*. Atlanta: ASHRAE.
ASHRAE. 2003. *HVAC Design Manual for Hospitals and Clinics*. Atlanta: ASHRAE.
ASHRAE. 2004. *Advanced Energy Design Guide for Small Office Buildings*. Atlanta: ASHRAE.
ASHRAE. 2004. ANSI/ASHRAE Standard 55-2004, *Thermal Environmental Conditions for Human Occupancy*. Atlanta: ASHRAE.
ASHRAE. 2004. ANSI/ASHRAE/IESNA Standard 90.1-2004, *Energy Standard for Buildings Except Low-Rise Residential Buildings*. Atlanta: ASHRAE.
ASHRAE. 2005. *ASHRAE Handbook—Fundamentals*. Atlanta: ASHRAE.
ASHRAE. 2005. ASHRAE Guideline 0-2005, *The Commissioning Process*. Atlanta: ASHRAE.
ASHRAE. 2006. ANSI/ASHRAE/IESNA Standard 100-2006, *Energy Conservation in Existing Buildings*. Atlanta: ASHRAE.
ASHRAE. 2006. *Advanced Energy Design Guide for Small Retail Buildings*. Atlanta: ASHRAE.
ASHRAE. 2007. *ASHRAE Handbook—HVAC Applications*. Atlanta: ASHRAE.
ASHRAE. 2007. ASHRAE Guideline 1.1-2007, *HVAC&R Technical Requirements for the Commissioning Process*. Atlanta: ASHRAE.
ASHRAE. 2007. ASHRAE Guideline 13-2007, *Specifying Direct Digital Control Systems*. Atlanta: ASHRAE.
ASHRAE. 2007. ANSI/ASHRAE Standard 62.1-2007, *Ventilation for Acceptable Indoor Air Quality*. Atlanta: ASHRAE.
ASHRAE. 2007. ANSI/ASHRAE/IESNA Standard 90.1-2007, *Energy Standard for Buildings Except Low-Rise Residential Buildings*. Atlanta: ASHRAE.

ASHRAE. 2007. ANSI/ASHRAE Standard 90.2-2007, *Energy Efficient Design of Low-Rise Residential Buildings*. Atlanta: ASHRAE.

ASHRAE. 2008. *ASHRAE Handbook—HVAC Systems and Equipment*. Atlanta: ASHRAE.

ASHRAE. 2009. *ASHRAE Handbook—Fundamentals*. Atlanta: ASHRAE.

ASHRAE. 2009. *ANSI/ASHRAE/USGBC/IES Standard 189.1-2009, Standard for the Design of High-Performance Green Buildings*. Atlanta: ASHRAE.

ASHRAE. 2009. *The ASHRAE Guide for Buildings in Hot and Humid Climates*. Atlanta: ASHRAE.

ASHRAE. 2009. *Indoor Air Quality Guide: Best Practices for Design, Construction and Commissioning*. Atlanta: ASHRAE.

ASHRAE. 2010. *ASHRAE Handbook—Refrigeration*. Atlanta: ASHRAE.

ASHRAE. 2013. ANSI/ASHRAE Standard 62.1-2010, *Ventilation for Acceptable Indoor Air Quality*. Atlanta: ASHRAE.

ASHRAE. 2013. ANSI/ASHRAE Standard 62.2-2010, *Ventilation and Acceptable Indoor Air Quality in Low-Rise Residential Buildings*. Atlanta: ASHRAE.

ASPE. 1998. *Domestic Water Heating Design Manual*. Chicago: American Society of Plumbing Engineers.

ASTM. 1998. *ASTM D 6245-1998: Standard Guide for Using Indoor Carbon Dioxide Concentrations to Evaluate Indoor Air Quality and Ventilation*. Philadelphia: American Society for Testing and Materials.

Baker, N. 1995. *Light and Shade: Optimising Daylight Design*. European Directory of sustainable and energy efficient building. London: James & James Science Publishers Ltd.

Balcomb, J.D. 1984. *Passive Solar Heating Analysis: A Design Manual*. Atlanta: ASHRAE.

Bauman, F.S., and A. Daly. 2003. *Underfloor Air Distribution Design Guide*. Atlanta: ASHRAE.

Beckman, W., S.A. Klein, and J.A. Duffie. 1977. *Solar Heating Design*. New York: John Wiley & Sons, Inc.

BRESCU, BRE. 1999. *Natural Ventilation for Offices Guide and CD-ROM*. ÓBRE on behalf of the NatVent Consortium, Garston, Watford, UK, March.

CEC. 2001. *Nonresidential Alternative Calculations Methods Manual*. Sacramento, CA: California Energy Commission.

CEC. 2002. *Part II: Measure Analysis and Life-Cycle Cost 2005, California Building Energy Standards*, P400-02-012. Sacramento, CA: California Energy Commission.

Crosbie, M.J., ed. 1998. *The Passive Solar Design and Construction Handbook*. Steven Winter Associates. New York: John Wiley & Sons, Inc.

Del Porto, D., and C. Steinfeld. 1999. *The Composting Toilet System Book*. New Bedford, MA: The Center for Ecological Pollution Prevention.

Dorgan, C., and J.S. Elleson. 1993. *Design Guide for Cool Thermal Storage*. Atlanta: ASHRAE.

Elleson, J.S. 1996. *Successful Cool Storage Projects: From Planning to Operation*. Atlanta: ASHRAE.

EPA. 2010. *Guide to Purchasing Green Power*. Washington, DC: U.S. Environmental Protection Agency. www.epa.gov/grnpower/buygreenpower/guide.htm.

Esty, D.C., and A.S. Winston. 2006. *Green to Gold: How Smart Companies Use Environmental Strategy to Innovate, Create Value, and Build Competitive Advantage*. New Haven, CT: Yale University Press.

Evans, B. 1997. Daylighting design. In *Time-Saver Standards for Architectural Design Data*. New York: McGraw-Hill.

FEMP. 2008. *M&V Option A Detailed Guidelines*. Washington, DC: U.S. Department of Energy, Federal Energy Management Program.

Givoni, B. 1994. *Passive and Low Energy Cooling of Buildings*. New York: Van Nostrand Reinhold.

Gottfried, D. (Continuous updating). *Sustainable Building Technical Manual, Green Building Design, Construction, and Operations*. Washington, DC: U.S. Green Building Council. www.usgbc.org.

Hawkin, P., A. Lovins, and L.H. Lovins. 1999. *Natural Capitalism*. Little Brown.

Henning, H-M., ed. 2004. *Solar Assisted Air-Conditioning in Buildings—A Handbook For Planners*. Wien: Springer-Verlag.

Heubach, J.G., J.C. Montgomery, W.C. Weimer, and J.H. Heerwagen. 1995. Pacific Northwest Laboratory. *Assessing the Human and Organizational Impacts of Green Buildings*.

Houghton, J.T., G.J. Jenkins, and J.J. Ephraums, eds. 1990. *Climate Change: The IPCC Scientific Assessment*. Intergovernmental Panel on Climate Change. New York: Cambridge UP.

Howell, J., R. Bannerot, and G. Vliet. 1982. *Solar-Thermal Energy Systems: Analysis and Design*. New York: McGraw-Hill.

IECC. 2006. *International Energy Conservation Code*. Available from www.icc safe.org.

IES. 2000. *Lighting Handbook*. New York: Illuminating Engineering Society of North America.

IES. 1998. *RP-33-99, Lighting for Exterior Environments; RP-20-98, Lighting for Parking Facilities*. New York: Illuminating Engineering Society of North America. www.iesna.org.

IESNA. 2002. Proposed revisions to ASHRAE/IESNA Standard 90.1-2001. Illuminating Engineering Society of North America, Energy Management Committee.

Kavanaugh, S.P., and K. Rafferty. 1997. *Ground-Source Heat Pumps: Design of Geothermal Systems for Commercial and Institutional Buildings*. Atlanta: ASHRAE.

Kinsley, M. 1997. *Economic Renewal Guide*. Colorado: Rocky Mountain Institute.

Klein, S., and W. Beckman. 2001. *F-Chart Software*. The F-Chart Software Company., Madison, WI. www.fchart.com.

Kowalski, W.J. 2006. *Aerobiological Engineering Handbook: Airborne Disease and Control Technologies*. New York: McGraw-Hill.

Kreith, F., and J. Kreider. 1989. *Principles of Solar Engineering*, 2d ed. Wasington, DC: Hemisphere Pub.

Ludwig, A. 1997. *Builder's Greywater Guide and Create an Oasis with Greywater*. Oasis Design.

Macher, J., ed. 1999. *Bioaerosols: Assessment and Control*. Cincinatti, Ohio: American Conference of Governmental Industrial Hygenists.

McIntosh, I.B.D., C.B. Dorgan, and C.E. Dorgan. 2002. *ASHRAE Laboratory Design Guide*. Atlanta: ASHRAE.

Mendler, S.F., and W. Odell. 2000. *HOK Guidebook to Sustainable Design*. New York: John Wiley & Sons, Inc.

NADCA. *General Specifications for the Cleaning of Commercial Heating, Ventilating and Air Conditioning Systems*. Washington, DC: National Air Duct Cleaners Association.

Nattrass, B., and M. Altomare. *The Natural Step for Business; Wealth, Ecology and the Evolutionary Corporation*. Gabriola Island, British Columbia: New Society Publishers.

NBI. 1998. *Gas Engine Driven Chillers Guideline*. Fair Oaks, CA: New Buildings Institute and Southern California Gas Company. www.newbuildings.org /downloads/guidelines/GasEngine.pdf.

NBI. 2001. *Advanced Lighting Guidelines*. Fair Oaks, CA: New Buildings Institute. www.newbuildings.org/lighting.htm.

NFPA. 2004. *NFPA 45, Standard on Fire Protection for Laboratories using Chemicals*. Quincy, MA: National Fire Protection Association.

NFPA. 2005. *NFPA 99, Standard for Health Care Facilities*. Quincy, MA: National Fire Protection Agency.

NRC. *Photovoltaic Systems Design Manual*. Ottawa, Ontario, Canada: Natural Resources Canada, Office of Coordination and Technical Information.

NRC. RETScreen (Renewable Energy Analysis Software). Natural Resources Canada, Energy Diversification Research Laboratory, Ottawa, Ontario, Canada. www.retscreen.net.

NSF. 2000. *NSF/ANSI 5-2000e: Water Heaters, Hot Water Supply Boilers, and Heat Recovery Equipment*. Ann Arbor, MI: National Sanitation Foundation.

PG&E. 2002. *Codes and Standards Enhancement Report, Code Change Proposal for Cooling Towers*. California: Pacific Gas & Electric.

Savitz, A.W., and K. Weber. 2006. *The Triple Bottom Line*. San Fransisco: John Wiley & Sons, Inc.

Schaffer, M. 2005. *Practical Guide to Noise and Vibration Control for HVAC Systems, Second Edition*. Atlanta: ASHRAE.

SMACNA. 2005. *HVAC Duct Construction Standards*, 3d ed., Chapter 1. Chantilly, VA: Sheet Metal and Air Conditioning Contractors National Association, Inc.

SMACNA. 2007. *IAQ Guidelines for Occupied Buildings Under Construction*. Chantilly, VA: Sheet Metal and Air Conditioning National Association, Inc.

Spitler, J. 2009. *Load Calculation Applications Manual*. Atlanta: ASHRAE.

Taylor, S., P. Dupont, M. Hydeman, B. Jones, and T. Hartman. 1999. *The CoolTools Chilled Water Plant Design and Performance Specification Guide*. San Francisco, CA: PG&E Pacific Energy Center.

Trane. 1994. *Water-Source Heat Pump System Design*. Trane Applications Engineering Manual, SYS-AM-7. Lacrosse, WI: Trane Company.

Trane Co. 1999. *Trane Applications Engineering Manual, Absorption Chiller System Design*, SYS-AM-13. Lacrosse, WI: Trane Company.

Trane. 2000. *Multiple-Chiller-System Design and Control Manual*. Lacrosse, WI: Trane Company.

Trane. 2005. *Trane Installation, Operation, and Maintenance Manual: Series $R^{®}$ Air-Cooled Rotary Liquid Chillers*, RTAA-SVX01A-EN. Lacrosse, WI: Trane Company. www.trane.com/Commercial/Uploads/Pdf/1060/RTAA-SVX01A-EN_09012005.pdf.

UG. 2010. *Wise Energy Guide*. Chatham, ON: Union Gas. www.uniongas.com/residential/energyconservation/education/wiseEnergyGuide/WEG_Booklet_Web_Final.pdf/.

USGBC. 2001. *LEED Reference Guide, Version 2.0*. Washington, DC: U.S. Green Building Council.

USGBC. 2005. *LEED Reference Guide, Version 2.2*. Washington, DC: U.S. Green Building Council.

USGBC. 2009. *LEED 2009 for New Construction and Major Renovations*. Washington, D.C.: U.S. Green Building Council.

USGBC. 2009. *LEED 2009 Green Building Operations and Maintenance*. Washington, DC: U.S. Green Building Council.

Watson, D., and G. Buchanan. 1993. *Designing Healthy Buildings*. Washington, DC: American Institute of Architects.

Watson, R.D., and K.S. Chapman. 2002. *Radiant Heating and Cooling Handbook*. New York: McGraw-Hill.

Watsun. 1999. *WATSUN-PV—A computer program for simulation of solar photovoltaic systems, User's Manual and Program Documentation*, Version 6.1. Watsun Simulation Laboratory, University of Waterloo, Ontario, Canada.

Wolverton, B.C. 1996. *How to Grow Fresh Air: 50 Plants That Purify Your Home of Office*. Baltimore, MD: Penguin Books.

PERIODICALS AND REPORTS

Adamson, R. 2002. Mariah Heat Plus Power™ Packaged CHP, applications and economics. Second Annual DOE/CETC/CANDRA Workshop on Microturbine Applications, January 17–18, University of Maryland, College Park, MD.

Afify, R. 2008. Designing VRF systems. *ASHRAE Journal* 50(6):52–55.

Alliance for Sustainable Built Environments. Existing buildings hold the key. www.awarenessintoaction.com/whitepapers/how-existing-buildings-high-performing-green-leed-certified.html.

Andersson, L.O., K.G. Bernander, E. Isfält, and A.H. Rosenfeld. 1979. Storage of heat and coolth in hollow-core concrete slabs. Swedish experience and application to large, American style buildings. Second International Conference on Energy Use and Management, Lawrence Berkeley National Laboratory, LBL-8913.

ASHE. 2002. Green healthcare design guidance statement. American Society of Healthcare Engineering, Chicago.

Aynur, T., Y. Hwang, and R. Radermacher. 2008. Experimental evaluation of the ventilation effect on the performance of a VRV system in cooling mode—Part I: Experimental evaluation. *HVAC&R Research* 14(4):615–30.

Aynur, T., Y. Hwang, and R. Radermacher. 2008. Simulation evaluation of the ventilation effect on the performance of a VRV system in cooling mode—Part II: Simulation evaluation. *HVAC&R Research* 14(5):783–95.

Bahnfleth, W.P., and W.S. Joyce. 1994. Energy use in a district cooling system with stratified chilled water storage. *ASHRAE Transactions* 100(1):1767–78.

Balaras, C.A. 1995. The role of thermal mass on the cooling load of buildings. An overview of computational methods. *Energy and Buildings* 24(1):1–10.

Balaras, C.A., E. Dascalaki, P. Tsekouras, and A. Aidonis. 2010. High solar combi systems in Europe. *ASHRAE Transactions* 116(1).

Balaras, C.A., H.-M. Henning, E. Wiemken, G. Grossman, E. Podesser, and C.A. Infante Ferreira. 2006. Solar cooling—An overview of european applications and design guidelines. *ASHRAE Journal* 48(6):14–22.

Bauman, F., and T. Webster. 2001. Outlook for underfloor air distribution. *ASHRAE Journal* 43(6):18–25.

Beebe, J.M. 1959. Stability of disseminated aerosols of *Pastuerella tularensis* subjected to simulated solar radiations at various humidities. *Journal of Bacteriology* 78:18–24.

Bisbee, D. 2003. Pulse-power water treatment systems for cooling towers. Energy Efficiency & Customer Research & Development, Sacramento Municipal Utility District.

Braun, J.E. 1990. Reducing energy costs and peak electrical demand through optimal control of building thermal storage. *ASHRAE Transactions* 96(2):876–88.

CBE. 2007. Summary report: Control strategies for mixed-mode buildings. Center for Built Environment, Berkeley, California. www.cbe.berkeley.edu/research/pdf_files/SR_MixedModeControls2007.pdf.

CEC. 1996. Source energy and environmental impacts of thermal energy storage. Completed by Tabors, Caramanis & Assoc. for the California Energy Commission.

CEC. 2004. Monitoring the energy-use effects of cool roofs on California commercial buildings, CEC PIER study report. California Energy Commission.

CEC. 2005. Title 24 2005 Reports and Proceedings. California Energy Commission. www.energy.ca.gov.
Cendón, S.F. 2009. *New and Cool: Variable Refrigerant Flow Systems*. AIArchitect.
CETC. 1996a. Technical report on Bentall Corporation Crestwood 8 C-2000 Building. CANMET Energy Technology Centre, Natural Resources Canada, Ottawa, Ontario, Canada.
CETC. 1996b. Technical report on Green on the Grand C-2000 Building. CANMET Energy Technology Centre, Natural Resources Canada, Ottawa, Ontario, Canada.
CFR. 1995. Energy conservation voluntary performance standards for commercial and multi-family high rise residential buildings; mandatory for new federal buildings. Code of Federal Regulations 10 CFR 435, Office of the Federal Register, National Archives and Records Administration, Washington, DC.
Chapman, K.S. ASHRAE RP-907, Final report, Development of design factors for the combination of radiant and convective in-space heating and cooling systems. Atlanta: ASHRAE.
Chapman, K.S. 2002. Development of radiation transfer equation that encompasses shading and obstacles. *ASHRAE Transactions* 108(2):997–1004.
Chapman, K.S., J.M. DeGreef, and R.D. Watson. 1997. Thermal comfort analysis using BCAP for retrofitting a radiantly heated residence. *ASHRAE Transactions* 103(1):959–65.
Chapman, K.S., J.E. Howell, and R.D. Watson. 2001. Radiant panel surface temperature over a range of ambient temperatures. *ASHRAE Transactions* 107(1):383–89.
Chapman, K.S., J. Rutler, and R.D. Watson. 2000. Impact of heating systems and wall surface temperatures on room operative temperature fields. *ASHRAE Transactions* 106(1):506–14.
Cho, Y.I., S. Lee, and W. Kim. 2003. Physical water treatment for the mitigation of mineral fouling in cooling-tower water applications. *ASHRAE Transactions* 109(1):346–57.
Coad, W.J. 1999. Conditioning ventilation air for improved performance and air quality. *Heating/Piping/Air Conditioning*, September.
Corbett, T. *Green Buildings and Risk*. www.aepronet.org/ge/no43.html.
Darlington, A., M. Chan, D. Malloch, C. Pilger, and M.A. Dixon. 2000. The biofiltration of indoor air: Implications for air quality. *Indoor Air 2000* 10(1):39–46.
Darlington, A.B., J.F. Dat, and M.A. Dixon. 2001. The biofiltration of indoor air: Air flux and temperature influences the removal of toluene, ethylbenzene, and xylene. *Environ Sci Technol* 35(1):240–46.
Darlington, A., M.A. Dixon, and C. Pilger. 1998. The use of biofilters to improve indoor air quality: The removal of toluene, TCE, and formaldehyde. *Life Support Biosph Sci* 5(1):63–69.

DeGreef, J.M., and K.S. Chapman. 1998. Simplified thermal comfort evaluation of MRT gradients and power consumption predicted with the BCAP methodology. *ASHRAE Transactions* 104(1):1090–97.

Deru, M., and P. Torcellini. 2007. *Source Energy and Emissions Factors for Energy Use in Buidlings*. Technical Report NREL/TP_550-38617, National Renewable Energy Laboratory, Golden, CO.

DOE. *The International Performance Measurement and Performance Protocol* (IPMPP). U.S. Department of Energy, Office of Energy Efficiency and Renewable Energy, Washington, DC.

DOE. 2002. *Energy Tip Sheet #1*, May. U.S. Department of Energy, Office of Industrial Technologies, Energy Efficiency, and Renewable Energy, Washington, DC.

DOE. 2005. Energy impact of commercial building controls and performance diagnostics: Market characterization, energy impact of building faults and energy savings potential. U.S. Department of Energy, Washington, DC.

DOE. 2006. Solar buildings. Transpired air collectors ventilation preheating. DOE/GO-102001-1288, U.S. Department of Energy, Washington, DC.

Duffy, G. 1992. Thermal storage shifts to saving energy. *Eng. Sys.* July/August:32–38.

EDR. 2009. *Chilled Water Plant Design Guide*. Energy Design Resources. www.energydesignresources.com/Resources/Publications/DesignGuidelines/tabid/73/articleType/ArticleView/articleId/422/Default.aspx.

Ehrt, D., M. Carl, T. Kittel, M. Muller, and W. Seeber. 1994. High performance glass for the deep ultraviolet range. *Journal of Non-Crystalline Solids* 177:405–19.

EIA. Table 10.8, Photovoltaic cell and module shipments by type, trade, and prices, 1982–2008. U.S. Energy Information Administration. www.eia.doe.gov/aer/txt/stb1008.xls.

El-Adhami, W., S. Daly, and P.R. Stewart. 1994. Biochemical studies on the lethal effects of solar and artificial ultraviolet radiation on *Staphylococcus aureus*. *Arch Microbiol* 161:82–87.

Ellis, M.W., and M.B. Gunes. 2002. Status of fuel cell systems for combined heat and power applications in buildings. *ASHRAE Transactions* 108(1):1032–44.

EPA. Storm water management for construction activities. EPA document No. EPA-8320R-92-005. U.S. Environmental Protection Agency, Washington, DC.

EPA. Toxicity Characteristic Leaching Procedure (TCLP) Test, 40 CFR Part 260. U.S. Environmental Protection Agency, Washington, DC.

EPA. Universal Waste Rule, 40 CFR Part 260. U.S. Environmental Protection Agency, Washington, DC.

EPA. 2010. *How to Conserve Water and Use It Effectively*. U.S. Environmental Protection Agency, Washington, DC. www.epa.gov/owow/nps/chap3.html.

EPIA. 2010. 2014 global market outlook for photovoltaics until 2014. European Photovoltaic Industry Association. www.epia.org/fileadmin/EPIA_docs/public/Global_Market_Outlook_for_Photovoltaics_until_2014.pdf.

Epoch Times. 2009. Water efficiency a priority for green buildings. www.theepochtimes.com/n2/content/view/17751.

Erickson, D.C., and M. Rane. 1994. Advanced absorption cycle: Vapor exchange GAX. ASME International Absorption Heat Pump Conference, January 19–21, New Orleans, LA.

ESTIF. 2010 Solar thermal markets in Europe—Trends and market statistics 2009. www.estif.org/fileadmin/estif/content/market_data/downloads/2009%20solar_thermal_markets.pdf.

EU. 2008. Climate change: Commission welcomes final adoption of Europe's climate and energy package. Europa press release IP/08/1998, European Union, Brussels.

European Comission. 2002. Energy Performance of Buildings, Directive 2002/91/EC of the European Parliament and of the Council. Official Journal of the European Communities, Brussels.

Fagan, D. 2001. A comparison of storage-type and instantaneous heaters for commercial use. *Heating/Piping/Air Conditioning Engineering*, April.

Fernandez, R.O. 1996. Lethal effect induced in *Pseudomonas aeruginosa* exposed to ultraviolet-A radiation. *Photochem & Photobiol* 64(2):334–39.

Fiorino, D.P. 1994. Energy conservation with stratified chilled water storage. *ASHRAE Transactions* 100(1):1754–66.

Fiorino, D.P. 2000. Six conservation and efficiency measures reducing steam costs. *ASHRAE Journal* 42(2):31–39.

Galuska, E.J. 1994. Thermal storage system reduces costs of manufacturing facility (Technology Award case study). *ASHRAE Journal*, March.

Gansler, R.A., D.T. Teindil, and T.B. Jekel. 2001. Simulation of source energy utilization and emissions for HVAC systems. *ASHRAE Transactions* 107(1):39–51.

Goetzler, W. 2007. Variable refrigerant flow systems. *ASHRAE Journal* 46(4): 24–31.

Goss, J.O., L. Hyman, and J. Corbett. 1996. Integrated heating, cooling and thermal energy storage with heat pump provides economic and environmental solutions at California State University, Fullerton. *Proceedings of the EPRI International Conference on Sustainable Thermal Energy Storage, Minneapolis, MN*, pp. 163–67.

Grondzik, W.T. 2001. The (mechanical) engineer's role in sustainable design: Indoor environmental quality issues in sustainable design. HTML presentation available at www.polaris.net/~gzik/ieq/ieq.htm.

Guz, K. 2005. Condensate water recovery. *ASHRAE Journal* 47(6):54–56.

GWEC. 2010. Global wind 2009 report. Global Wind Energy Council, Brussels, Belgium. www.gwec.net/index.php?id=167.

Hayter, S., P. Torcellini, and R. Judkoff. 1999. Optimizing building and HVAC systems. *ASHRAE Journal* 41(12):46–49.
Heerwagen, J. 2001. Do green buildings enhance the well being of workers? www.edcmag.com.
Hilten, R.N. 2005. An analysis of the energetics and stormwater mediation potential of greenroofs. Master's thesis, University of Georgia, Athens, GA.
HPAC. 2004. Innovative grocery store seeks LEED certification. *HPAC Engineering* 27:31.
ICC. 2010. International green construction code™, Public version 1.0. International Code Council, Washington, DC.
IUVA. 2005. General guideline for UVGI air and surface disinfection systems. IUVA-G01A-2005 International Ultraviolet Association, Ayr, Ontario, Canada, www.iuva.org.
Jarnagin, R.E., R.M. Colker, D. Nail, and H. Davies. 2009. ASHRAE Building eQ. *ASHRAE Journal* 51(12):18–21.
Jones, B., and K.S. Chapman. ASHRAE RP-657, Final report, Simplified method to factor mean radiant temperature (MRT) into building and HVAC system design. Atlanta: ASHRAE.
Kainlauri, E.O., and M.P. Vilmain. 1993. Atrium design criteria resulting from comparative studies of atriums with different orientation and complex interfacing of environmental systems. *ASHRAE Transactions* 99(1):1061–69.
Keeney, K.R., and J.E. Braun. 1997. Application of building precooling to reduce peak cooling requirements. *ASHRAE Transactions* 103(1):463–69.
Kintner-Meyer, M., and A.F. Emery. 1995. Optimal control of an HVAC system using cold storage and building thermal capacitance. *Energy and Buildings* 23:19–31.
Kosik, W.J. 2001. Design strategies for hybrid ventilation. *ASHRAE Journal* 43(10):18–19, 22–24.
Larsson, N., ed. 1996. *C-2000 Program Requirements*. Natural Resources Canada, Ottawa, Ontario, Canada.
Lawrence, T.M. 2004. Demand-controlled ventilation and sustainability. *ASHRAE Journal* 46(12):117–21.
Lawrence, T. 2008. Selecting CO_2 criteria for space monitoring. *ASHRAE Journal* 50(12):18–27.
Lawrence, T., J. Perry, and P. Dempsey. 2010. Capturing condensate by retrofitting AHUs. *ASHRAE Journal* 52(1):48–54.
Lawson, 1988. Computer facility keeps cool with ice storage. *HPAC*, August.
Leaman, A., and B. Bordass. 1993. Building design, complexity and manageability. *Facilities*, Vol. 11.
LeMar, P. 2002. Integrated energy systems (IES) for buildings: A market assessment. Final report by Resource Dynamics Corporation for Oak Ridge National Laboratory, Contract No. DE-AC05-00OR22725.

Mathaudhu, S.S. 1999. Energy conservation showcase. *ASHRAE Journal* 41(4):44–46.

Mauthner, F. and W. Weiss. 2013. Solar heat worldwide: Markets and contribution to the energy supply 2011. Edition 2013. IEA Solar Heating and Cooling Programme, May 2013. International Energy Agency, Paris. http://www.iea-shc.org/solar-heat-worldwide.

McDonough, W. 1992. The Hannover principles: Design for sustainability. Presentation, Earth Summit, Brazil.

McKurdy, G., S.J. Harrison, and R. Cooke. Preliminary evaluation of cylindrical skylights. 23rd Annual Conference of the Solar Energy Society of Canada Inc., Vancouver, British Columbia.

Morris, W. 2003. The ABCs of DOAS. *ASHRAE Journal* 45(5).

Mumma, S.A. 2001. Designing dedicated outdoor air systems. *ASHRAE Journal* 43(5):28–31.

NREL. 2010. NREL highlights utility green power leaders. NREL press release NR-1910, National Renewable Energy Laboratory, Golden, CO.

O'Neal, E.J. 1996. Thermal storage system achieves operating and first-cost savings (Technology Award case study). *ASHRAE Journal* 38(4).

Palmer, J.M., and K.S. Chapman. 2000. Direct calculation of mean radiant temperature using radiant intensities. *ASHRAE Transactions* 106(1):477–86.

Patnaik, V. 2004. Experimental verification of an absorption chiller for BCHP applications. *ASHRAE Transactions* 110(1):503–507.

Pearson, F. 2003. ICR 06 Plenary Washington. *ASHRAE Journal* 45(10).

Rautiala, S., S. Haatainen, H. Kallunki, L. Kujanpaa, S. Laitinen, A. Miihkinen, M. Reiman, and M. Seuri. 1999. Do plants in office have any effect on indoor air microorganisms? *Indoor Air 99: Proceedings of the 8th International Conference on Indoor Air Quality and Climate, Edinburgh, Scotland*, pp. 704–709.

Reindl, D.T., R.A. Gansler, and T.B. Jekel. 2001. Simulation of source energy utilization and emissions for HVAC systems. *ASHRAE Transactions* 107(1):39–51.

Reindl, D.T., D.E. Knebel, and R.A. Gansler. 1995. Characterising the marginal basis source energy and emissions associated with comfort cooling systems. *ASHRAE Transactions* 101(1):1353–63.

Rosenbaum, M. 2002. A green building on campus. *ASHRAE Journal* 44(1):41–44.

Rosfjord, T., T. Wagner, and B. Knight. 2004. UTC microturbine CHP product development and launch. Presented at the 2004 DOE/CETC Annual Workshop on Microturbine Applications, January 20–22, Marina del Rey, CA.

Ruud, M.D., J.W. Mitchell, and S.A. Klein. 1990. Use of building thermal mass to offset cooling loads. *ASHRAE Transactions* 96(2):820–29.

Ryan, W.A. 2003. CHP: The concept. Midwest CHP Application Center, Energy Resources Center, University of Illinois, Chicago. www.chpcentermw.org /presentations/WI-Focus-on-Energy-Presentation-05212003.pdf

Scheatzle, David. ASHRAE RP-1140 Final report, Establishing a base-line data set for the evaluation of hybrid HVAC systems. Atlanta: ASHRAE.

Schell, M., and D. Int-Hout. 2001. Demand control ventilation using CO_2. *ASHRAE Journal* 43(2):18–24.

Stack, A., J. Goulding, and J.O. Lewis. Shading systems: Solar shading for the European climates. DG TREN, Brussels, Belgium. http://erg.ucd.ie /down_thermie.html.

Svensson C., and S.A. Aggerholm. 1998. Design tool for natural ventilation. ASHRAE IAQ 1998 Conference, October 24–27, New Orleans, LA.

Taylor, S. 2002. Primary-only vs. primary-secondary variable flow chilled water systems. *ASHRAE Journal* 44(2):25–29.

Taylor, S.T. 2005. LEED and Standard 62.1. *ASHRAE Journal Sustainability Supplement*, September.

Taylor, S., and J. Stein. 2002. Balancing variable flow hydronic systems. *ASHRAE Journal* 44(10):17–24.

Torcellini, P.A., N. Long, and R. Judkoff. 2004. Consumptive water use for U.S. power production. *ASHRAE Transactions* 110(1):96–100.

Torcellini, P., S. Pless, M. Deru, B. Griffith, N. Long, and R. Judkoff. 2006. Lessons learned from case studies of six high-performance buildings. Technical report, NREL/TP-550-37542. National Renewable Energy Laboratory, Golden, CO.

Torcellini, P., N. Long, S. Pless, and R. Judkoff. 2005. Evaluation of the low-energy design and energy performance of the Zion National Park Visitor Center, NREL Report No. TP-550-34607. National Renewable Energy Laboratory, Golden, CO. www.eere.energy.gov/buildings/highperformance /research_reports.html.

Trane. 2000. Energy conscious design ideas—Air-to-air energy recovery. *Engineers Newsletter* 29(5).

Trane. *A Guide to Understanding ASHRAE Standard 62-2001*. Publication IAQ-TS-1, July. Trane Company, Lacrosse, WI. http://trane.com/commercial/issues /iaq/ashrae2001.asp.

USGBC. LEED-NC v 2.2 Rating System. U.S. Green Building Council, Washington, DC.

USGBC. 2009. LEED Version 3. U.S. Green Building Council, Washington, DC.

von Kempski, D. 2003. Air and well being—A way to more profitability. *Proceedings of Healthy Buildings 2003, Singapore*, pp. 248–354.

von Kempski, D. 2009. Going beyond green—green perception A paradigm change in IAQ for the well-being and health of the occupants. *Proceedings of Healthy Buildings 2009, Syracuse, New York,* paper 138.

von Paumgartten, P. *Existing Buildings Hold the Key.* Alliance for Sustainable Built Environments.www.awarenessintoaction.com/whitepapers/how-existing-buildings-high-performing-green-leed-certified.html (28 Sep. 2010).

Waterfall, P.H. 1998. *Harvesting Rainwater for Landscape Use.* http://ag.arizona.edu/pubs/water/az1052/.

Watson, R.D., K.S. Chapman, and L. Wiggington. 2001. Impact of dual utility selection on 305 m^2 (1000 ft^2) residences. *ASHRAE Transactions* 107(1):365–70.

Watson, R.D., K.S. Chapman, and J.M. DeGreef. 1998. Case study: Seven-system analysis of thermal comfort and energy use for a fast-acting radiant heating system. *ASHRAE Transactions* 104(1).

Waye, K.P., J. Bengtsson, R. Rylander, F. Hucklebridge, P. Evans, and A. Clow. 2002. Low frequency noise enhances cortisol among noise sensitive subjects during work performance. *Life Science* 70(7):745–58.

Webber, M.E. 2008. *Energy versus Water: Solving Both Crises Together.* Scientific American Special Editions. www.sehn.org/tccenergyversuswater.html.

Zagreus, L., C. Huizenga, E. Arens, and D. Lehrer. 2004. Listening to the occupants: A Web-based indoor environmental quality survey. *Indoor Air 2004* 14(suppl 8):65–74.

WEBSITES

Advanced Buildings Technologies and Practices
    www.advancedbuildings.org.

AGORES—A Global Overview of Renewable Energy Sources (the official European Commission renewable energy information centre and knowledge gateway, with a global overview of RES)
    www.agores.org.

Air-Conditioning, Heating and Refrigeration Institute
    www.ahrinet.org/.

Alliance for Sustainable Built Environments
www.awarenessintoaction.com/whitepapers/how-existing-buildings-high-performing-green-leed-certified.html.

Alliance to Save Energy
    http://ase.org .

American Council for an Energy-Efficient Economy
    www.aceee.org.

American Gas Association
    www.aga.org.

The American Institute of Architects
    www.aia.org.

American Society of Healthcare Engineering
    www.ashe.org.

American Society of Healthcare Engineering, *Green Guide for Health Care*
    www.gghc.org.
American Society of Plumbing Engineers
    http://aspe.org.
American Solar Energy Society
    www.ases.org.
American Wind Energy Association
    www.awea.org.
Architecture 2030 Challenge
    www.architecture2030.org/.
Architectural Energy Corporation
    www.archenergy.com.
Armstrong Intelligent Systems Solutions
    www.armstrong-intl.com.
ASHRAE
    www.ashrae.org.
ASHRAE Building eQ
    www.buildingeq.com.
ASHRAE Engineering for Sustainability
    www.engineeringforsustainability.org/.
ASHRAE Standard Project Committee 191
    http://spc191.ashraepcs.org/.
ASHRAE's Sustainability Roadmap
    http://images.ashrae.biz/renovation/documents/sust_roadmap.pdf.
ASHRAE TC 8.3, recently sponsored programs and presentations
    http://tc83.ashraetcs.org/programs.html.
Athena Institute
    www.athenasmi.ca.
BetterBricks
    www.betterbricks.com.
Building Energy Efficiency Research Project at the Department of Architecture, The University of Hong Kong
    www.arch.hku.hk/research/beer/.
Building for Environmental and Economic Sustainability (BEES)
    www.bfrl.nist.gov/oae/software/bees.html.
BuildingGreen
    www.greenbuildingadvisor.com.
Building Research Establishment Environmental Assessment Method (BREEAM®) rating program
    www.breeam.org.
California Debt & Investment Advisory Commission document
    www.treasurer.ca.gov/cdiac/publications/p3.pdf.

California Energy Commission
   www.energy.ca.gov.
California Energy Commission, Renewable Energy Program
   www.energy.ca.gov/renewables/.
The Canadian Renewable Energy Network
   www.canren.gc.ca/.
The Center for Health Design
   www.healthdesign.org.
Center for the Built Environment, Occupant Indoor Environmental Quality Survey
   www.cbesurvey.org.
Center of Excellence for Sustainable Development
   www.sustainable.doe.gov.
Center of Excellence for Sustainable Development, Smart Communities Network
   www.smartcommunities.ncat.org.
Centre for Analysis and Dissemination of Demonstrated Energy Technologies
   www.caddet-re.org.
Collaborative for High Performance Schools
   www.chps.net.
Cool Roof Rating Council
   www.coolroofs.org.
DDC Online
   www.ddc-online.org.
Dena
   www.dena.de/en/.
Earth Energy Society of Canada
   www.earthenergy.ca/.
ecoENERGY for Renewable Power
   www.ecoaction.gc.ca/ecoenergy-ecoenergie/power-electricite/index-eng.cfm.
EN4M Energy in Commercial Buildings (software tool)
   www.eere.energy.gov/buildings/tools_directory/software.cfm/ID=299/.
Energy Design Resources
   www.energydesignresources.com/publication/gd/.
ENERGY STAR®
   www.energystar.gov.
Energy Trust of Oregon
   www.energytrust.org.
Europe's Energy Portal
   www.energy.eu/#renewable.
European Biomass Association
   www.aebiom.org.
European Biomass Industry Association
   www.eubia.org/about_biomass.0.html.

Euporean Commission, BUILD UP program
   www.buildup.eu/.
European Commission, Concerted Action Energy Performance of Buildings Directive
   www.epbd-ca.org.
European Commission (DG TREN), SARA Project
   www.sara-project.net.
European Photovoltaic Industry Association
   www.epia.org.
European Renewable Energy Centres Agency
   www.eurec.be.
European Solar Thermal Industry Federation
   www.estif.org.
European Wind Energy Association
   www.ewea.org.
EUROSOLAR—The European Association for Renewable Energies e.V.
   www.eurosolar.org.
F-Chart Software
   www.fchart.com.
The Federal Energy Management Program Continuous Commissioning (SM) Guidebook
   www1.eere.energy.gov/femp/program/om_guidebook.html.
Geoexchange
   www.geoexchange.org.
Global Wind Energy Council
   www.gwec.net.
Green Building Advisor
   www.greenbuildingadvisor.com.
Green Energy In Europe
   www.ecofys.com/.
Green Globes
   www.greenglobes.com.
Green Roof Industry Information Clearinghouse and Database
   www.greenroofs.com.
GreenSpec® Product Guide
   www.buildinggreen.com/menus/index.cfm.
Green Venture
   www.greenventure.on.ca.
Hannover Principles
   www.mindfully.org/Sustainability/Hannover-Principles.htm.
Harvard University Office for Sustainability, Green Building Resource, Design Phase Guide

www.green.harvard.edu/theresource/new-construction/design-phase/bidding
-construction/www.aepronet.org/ge/no43.html.

Heat Pump Centre
www.heatpumpcentre.org.

Heschong Mahone Group
www.h-m-g.com/projects/daylighting/projects-PIER.htm.

Illuminating Engineering Society of North America
www.iesna.org.

Intergovernmental Panel on Climate Change
www.ipcc.ch.

International Code Council
www.iccsafe.org.

International Energy Agency, Bioenergy
www.ieabioenergy.com.

International Energy Agency, Photovoltaic Power Systems Programme
www.iea-pvps.org.

International Energy Agency, Solar Heating And Cooling Programme
www.iea-shc.org.

International Energy Agency, Task 38
www.iea-shc.org/task38.

International Hydropower Association
www.hydropower.org.

International Hydropower Association Fact Sheet
www.hydropower.org/downloads/F1_the_contribution_of_hydropower.pdf.

International Living Building Institute, The Living Building Challenge V2.0
http://ilbi.org/the-standard/version-2-0.

International Organization for Standardization family of 14000 standards,
www.iso.org/iso/en/iso9000-14000/index.html.

International Performance Measurement and Verification Protocol
www.evo-world.org.

International Solar Energy Society
www.ises.org.

Irrigation Association
www.irrigation.org.

Labs 21, Environmental Performance Criteria
www.labs21century.gov.

Lawrence Berkeley National Laboratories
www.lbl.gov/.

Lawrence Berkeley National Laboratories,
Environmental Energy Technologies Division
http://eetd.lbl.gov/.

Lawrence Berkeley National Laboratory, Radiance
   http://radsite.lbl.gov/radiance/HOME.html.
Massachusetts State Building Code
   www.mass.gov/bbrs/code.htm.
Minnesota Sustainable Design Guide
   www.sustainabledesignguide.umn.edu.
Mondaq
   www.mondaq.com/canada/article.asp?articleid=23737.
M&V Guidelines for Federal Energy Projects, Version 3.0.
   http://mnv.lbl.gov/keyMnVDocs/femp.
National Building Controls Information Program
   www.buildingcontrols.org.
National Hydropower Association
   www.hydro.org.
National Oceanic and Atmospheric Administration
   www.noaa.gov.
National Renewable Energy Laboratory
   www.nrel.gov.
National Renewable Energy Laboratory, Building-Integrated PV
   www.nrel.gov/pv/building_integrated_pv.html.
National Renewable Energy Laboratory, Buildings Research
   www.nrel.gov/buildings/.
National Renewable Energy Laboratory, Solar Energy Research Facility
   www.nrel.gov/pv/facilities_serf.html.
Natural Resources Canada, Commercial Buildings Incentive Program
   http://nrcan.gc.ca/evaluation/reprap/2001/cbip-pebc-eng.php.
Natural Resources Canada, EE4 Commercial Buildings Incentive Program
   http://canmetenergy-canmetenergie.nrcan-rncan.gc.ca/eng/software_tools/ee4.html.
Natural Resources Canada, Office of Energy Efficiency
   http://oee.nrcan.gc.ca/commercial/newbuildings.cfm?attr=20.
Natural Resources Canada, RETScreen
   (software for renewable energy analysis)
   www.retscreen.net.
The Natural Step
   www.naturalstep.org.
New Buildings Institute
   www.newbuildings.org.
New York's Battery Park City Authority
   www.batteryparkcity.org/page/index_battery.html.
New York City Department of Design and Construction, Sustainable Design
   www.nyc.gov/html/ddc/html/design/sustainable_home.shtml.

National Oceanic and Atmospheric Administration
   www.noaa.gov.
North Carolina Cooperative Extension Service, Water Quality and Waste Management
   www.bae.ncsu.edu/programs/extension/publicat/wqwm.
North Carolina Solar Center, DSIRE
   www.dsireusa.org/.
Oak Ridge National Laboratory, Bioenergy Feedstock Information Network
   http://bioenergy.ornl.gov/.
Oikos: Green Building Source
   www.oikos.com.
Option B: Methods by Technology
   www.energyautomation.com/pdfs/ipmvp-vol2.pdf.
Photovoltaic Resource Site
   www.pvpower.com.
QuikWater, High Efficiency Direct Contact Water Heaters
   www.quikwater.com.
RealWinWin, Inc.
   www.realwinwin.com.
Regional Climate Data
   www.wrcc.dri.edu/rcc.html.
Renewable Energy Deployment Initiative
   www.nrcan.gc.ca/redi.
Rising Sun Enterprises
   www.rselight.com.
Rocky Mountain Institute
   www.rmi.org.
Savings by Design
   www.savingsbydesign.com.
School of Photovoltaic and Renewable Energy Engineering, University of New South Wales
   www.pv.unsw.edu.au.
Sheet Metal and Air Conditioning Contractors' National Association
   www.smacna.org.
SkyCalc
   www.energydesignresources.com/resource/129/.
Solar Energy Industries Association
   www.seia.org.
Spirx Sarco Design of Fluid Systems, Steam Learning Module
   www.spiraxsarco.com/learn/modules.asp.
Sustainable Building Challenge
   www.iisbe.org/.

Sustainable Building Guidelines
   www.ciwmb.ca.gov/GreenBuilding/Design/Guidelines.htm.
Sustainable Buildings Industry Council
   www.sbicouncil.org.
Sustainable Communities Network
   www.sustainable.org.
Sustainable Sources
   www.greenbuilder.com.
Sustainable Sources Bookstore
   bookstore.greenbuilder.com/index.books.
Texas Manual on Rainwater Harvesting
   www.twdb.state.tx.us/publications/reports
   /rainwaterharvestingmanual_3rdedition.pdf.
"Tips for Daylighting with Windows"
   http://windows.lbl.gov/pub/designguide/designguide.html.
Tool for the Reduction and Assessment of Chemical and Other Environmental Impacts (TRACI)
   www.epa.gov/nrmrl/std/sab/traci/.
Trane Company
   www.trane.com.
Tri-State Generation and Transmission Association, Inc.
   http://www.tristategt.org/.
Usable Buildings Trust
   www.usablebuildings.co.uk.
U.S. Department of Energy, Energy Efficiency and Renewable Energy
   www.eere.energy.gov/.
U.S. Department of Energy, Energy Efficiency and Renewable Energy, Biomass program
   www1.eere.energy.gov/biomass.
U.S. Department of Energy, Energy Efficiency and Renewable Energy, Biomass program, Resources for Consumers Web page
   www1.eere.energy.gov/biomass/for_consumers.html.
U.S. Department of Energy, Energy Efficiency and Renewable Energy, Energy Savers
   www.energysavers.gov/.
U.S. Department of Energy, Energy Efficiency and Renewable Energy, Zero Energy Buildings
   http://zeb.buildinggreen.com.
U.S. Department of Energy, Federal Energy Management Program
   www.eere.energy.gov/femp.
U.S. Department of Energy, Federal Energy Management Program Guidelines
   www1.eere.energy.gov/femp/pdfs/mv_guidelines.pdf.

U.S. Department of Energy, High Performance Buildings Institute
www1.eere.energy.gov/buildings/commercial_initiative/.

U.S. Energy Information Administration,
Commercial Building Energy Consumption Survey
www.eia.doe.gov/emeu/cbecs/.

U.S. Environmental Protection Agency
www.epa.gov.

U.S. Environmental Protection Agency, Building Energy Analysis Tool, eQuest
www.energydesignresources.com/resource/130/.

U.S. Environmental Protection Agency, Green Chemistry
www.epa.gov/greenchemistry/.

U.S. Environmental Protection Agency, Water Sense Program
http://epa.gov/watersense/.

U.S. Green Building Council
www.usgbc.org.

U.S. Green Building Council, Leadership in Energy and Environmental Design, Gold building, Hewlett Foundation
www.usgbc.org/Docs/Certified_Projects/Cert_Reg67.pdf.

U.S. Green Building Council, Leadership in Energy and Environmental Design, Green Building Rating System™
www.usgbc.org/DisplayPage.aspx?CategoryID=19.

U.S. Green Building Council, Regional Priority Credit Listing
www.usgbc.org/DisplayPage.aspx?CMSPageID=1984.

The Whole Building Design Guide
www.wbdg.org.

World Meteorological Organization, World Radiation Data Centre
http://wrdc-mgo.nrel.gov/.

The World's Water
www.worldwater.org.

Terms, Definitions, and Acronyms

Δ	=	change or change in
ΔT	=	change or change in temperature
AC	=	air conditioning
ACGIH	=	American Conference of Governmental Industrial Hygienists
AHU	=	air-handling unit
AIA	=	American Institute of Architects
ANSI	=	American National Standards Institute
BAS	=	building automation system
BEES	=	Building for Environmental and Economic Sustainability
BF	=	ballast factor
BIM	=	building information modeling
bio-climatic chart	=	a psychrometric chart with a plot of the conditions for primarily passive strategies
BOMA	=	Building Owners and Managers Association
BREEAM®	=	Building Research Establishment Environmental Assessment Method
brownfield	=	real estate property that is, or potentially is, contaminated
Btu	=	British thermal unit
C	=	centigrade (temperature scale)
CAN	=	Canada
CAV	=	constant air volume
CBE	=	Center for the Built Environment
CCHP	=	combined cooling, heating, and power
CDT	=	cold deck temperature
CFC	=	chlorofluorocarbon
cfm	=	cubic feet per minute
CFR	=	current facility requirements
charette	=	collaborative and interdisciplinary effort to solve a design problem within a limited time
CHP	=	combined heating and power
CHW	=	chilled water
condenser	=	device to dissipate (get rid of) excess energy in A/C systems
COP	=	coefficient of performance
CRAC	=	computer room air conditioners
CxA	=	commissioning authority
daylighting	=	lighting (of a building) using daylight directly or indirectly from the sun

D/B	=	design/build
dB(A)	=	A-weighting
D/B/B	=	design/bid/build
DDC	=	direct digital control
DE	=	district energy
DG	=	distributed generation
DHW	=	domestic hot water
DOAS	=	dedicated outdoor air system
DOE	=	U.S. Department of Energy
DSF	=	double skin facade
DX	=	direct expansion
E	=	ventilation effectiveness
EA	=	energy and atmosphere
EDC	=	environmental design consultant
EDG	=	engine-driven generator
ENERGY STAR®	=	a government-backed program/rating system that helps consumers achieve superior energy efficiency
energy source	=	on-site energy in the form in which it arrives at or occurs on a site (e.g., electricty, gas, oil, or coal)
energy resource	=	raw energy that (1) is extracted from the earth (wellhead or mine mouth), (2) is used in the generation of the energy source delivered to a building site (e.g., coal used to generate electricty), or (3) occurs naturally and is available at a site (e.g., solar, wind, or geothermal energy)
enthalpy	=	the thermodynamic property of a system resulting from the combination of observable properties (per unit mass) thereof: namely, the sum of internal energy and flow work; flow work is the product of volume and specific mass (i.e., energy transmitted into or out of a system or transmitted across a system boundary)
entropy	=	a measure of the molecular disorder of a system, such that the more mixed a system is, the greater its entropy, and the more orderly or unmixed a system is, the lower its entropy
E&O	=	errors and omissions
EPA	=	U.S. Environmental Protection Agency
EPBD	=	Directive on the Energy Performance of Buildings
EPC	=	energy performance certificate
EU	=	European Union
F	=	fahrenheit (temperature scale)
F-chart	=	method of calculating solar fraction
fenestration	=	window treatment
GHG	=	greenhouse gas
gpf	=	gallons per fixture
gpm	=	gallons per minute
GSHP	=	ground-source heat pump
guideline	=	within ASHRAE, a document similar to a standard, but with less strict rules on developing a consensus and not written in code language
HCFC	=	hydrochlorofluorocarbons
HEPA	=	high-efficiency particulate air
HID	=	high-intensity discharge
HVAC&R	=	heating, ventilating, air-conditioning, and refrigerating
hybrid ventilation	=	combination of natural and mechanical outdoor air ventilation

hydronic	=	pertaining to liquid flow
IAQ	=	indoor air quality
IDP	=	integrated design process
IEA	=	International Energy Agency
IEC	=	indirect evaporative cooling
IEQ	=	indoor environmental quality
IES	=	Illuminating Engineering Society of North America
insolation	=	entry into a building of solar energy
IPCC	=	Intergovernmental Panel on Climate Change
IPMVP	=	International Performance Measurement and Verification Protocol
K	=	Kelvin or absolute (temperature scale)
kW	=	kilowatt
kWh	=	kilowatt-hour
kWR	=	refrigeration cooling capacity in kW
latent load	=	thermal load due strictly to effects of moisture
LCA	=	life-cycle assessment
LCCA	=	life-cycle cost analysis
LCEA	=	life-cycle environmental assessment
LD	=	liquid desiccant
LED	=	light-emitting diode
LEED	=	Leadership in Energy and Environmental Design
leeward	=	the downwind side—or side the wind blows away from
L/f	=	liters per fixture
low-E	=	low emissivity
LPD	=	lighting power density
LP	=	liquefied petroleum
L/s	=	liters per second (airflow and water flow)
media	=	energy forms distributed within a building, usually air, water, or electricity
MERV	=	minimum efficiency reporting value
mhp	=	motor horsepower
MRT	=	mean radiant temperature
M&V	=	measurement and verification
NADCA	=	National Air Duct Cleaning Association
NC	=	noise criteria
NO_x	=	oxides of nitrogen
NOAA	=	National Oceanic Atmospheric Administration
nonrenewables	=	energy resources that have definite, although sometimes unknown, quantity limitations
NR	=	natural refrigeration
NREL	=	National Renewable Energy Laboratory
NZEB	=	net zero energy building
OC	=	on center
O&M	=	operations and maintenance
OPR	=	Owner's Project Requirements
OSHA	=	Occupational Safety and Health Administration
P3	=	public-private partnership
parametric analysis	=	in situations where multiple parameters affect an outcome, an analysis that determines the magnitude of one or more parameter's impact alone on that outcome
plug loads	=	loads (electrical or thermal) from equipment plugged into electrical outlets

POTW	=	publicly owned treatment works
precooling	=	cooling done prior to the time major cooling loads are anticipated
PV	=	photovoltaic
R (as in R-19)	=	thermal heat transfer resistance
RC	=	room criteria
renewables	=	resources that can generally be freely used without net depletion or that have the potential to renew in a reasonable period of time
RES	=	renewable energy source
RFP	=	request for proposal
ROI	=	return on investment
sensible load	=	thermal load due to temperature but not moisture effects
sg	=	specific gravity
SHW	=	space hot water
skin	=	building envelope
SMACNA	=	Sheet Metal and Air Conditioning Contractors National Association
SR	=	synthetic refrigeration
SS	=	sustainable sites
standard	=	within ASHRAE, a document that defines properties, processes, dimensions, materials, relationships, concepts, nomenclature, or test methods for rating purposes or code enforcement
sustainability	=	providing for the needs of the present without detracting from the ability to fulfill the needs of the future
TBZ	=	thermal buffer zone
TC	=	technical committee (an ASHRAE group with a common interest in a particular technical subject)
TCLP	=	toxicity characteristic leaching procedure
TES	=	thermal energy storage
Title 24	=	shortened form for California's Building Energy Efficiency Standards (Title 24, Part 6 of the California State Building Code)
ton	=	cooling capacity, equal to 12,000 Btu/h
TRACI	=	Tool for the Reduction and Assessment of Chemical and Other Environmental Impacts
USGBC	=	U.S. Green Building Council
VAV	=	variable air volume
VE	=	value engineering
VFD	=	variable-frequency drive
VOC	=	volatile organic compound
VRF	=	variable refrigeration flow
WE	=	water efficiency
windward	=	the upwind side, or side the wind blows toward

INDEX

A

absorption chiller 321
absorption 274, 281, 284, 285, 288, 289, 295, 296, 297, 298, 299, 320, 321
acoustical quality 56
acoustics 4, 90, 98, 181, 182, 230, 231, 324, 478
air
  ambient 24, 25
  building 189
  circulated 168
  combustion 300, 302
  conditioned 243, 260
  cool 168, 212
  dehumidified 326
  dry 263, 327, 330
  exhaust 101, 151, 208, 219, 326, 330, 331
  fresh 138, 206
  humid 184, 196, 327
  incoming 344, 409
  indoor 4, 42, 65, 130, 144, 160, 166, 349, 444, 454, 477, 480
  mixed 263, 264
  moist 327
  outdoor 43, 91, 98, 102, 104, 107, 109, 111, 113, 117, 148, 152, 153, 168, 169, 184, 185, 187, 188, 189, 196, 198, 199, 201, 206, 209, 210, 211, 212, 213, 214, 217, 218, 219, 221, 222, 223, 257, 263, 269, 270, 277, 307, 326, 330, 331, 349, 356, 402, 408, 409, 433, 434, 436, 437, 440, 445
  process 327
  reactivation 327
  return 249, 260, 326, 466
  room 201, 203, 207, 261, 311, 312, 444, 505
  space 170, 358
  supply 104, 119, 189, 200, 212, 214, 219, 248, 260, 261, 326, 330, 331, 436, 445
  variable-flow 238, 249
  ventilation 160, 168, 169, 208, 209, 210, 211, 212, 222, 263, 266, 326, 408, 441, 479
  wetted 330
  zone 211, 212, 260, 441, 445
air conditioning 9, 80, 145, 186, 198, 209, 235, 236, 249, 297, 346, 466, 508
air distribution 65, 91, 92, 101, 109, 161, 165, 166, 167, 211, 212, 219, 247, 260, 261, 316, 441, 479
air distribution effectiveness 260, 441, 445
air distribution system 92, 247, 260, 261, 316

air filter 213, 214, 440, 464, 466
air movement 84, 225, 367
air pollution 18, 188, 286
air quality 4, 56, 59, 98, 102, 109, 111, 114, 130, 144, 145, 160, 166, 181, 182, 183, 185, 186, 188, 189, 193, 197, 198, 202, 203, 205, 206, 207, 208, 210, 212, 213, 215–218, 220, 222, 223, 226, 231, 232, 247–9, 260, 263, 264, 266, 301, 343, 349, 360, 436, 444, 445, 450, 454, 466, 467, 469, 471, 477, 478, 480, 506
air-coolant system 358
airflow 82, 94, 102, 105, 110, 112, 150, 152, 162, 186, 187, 196, 197, 199, 200, 208, 209, 210, 211, 212, 213, 217, 219, 221, 222, 223, 249, 261, 264, 271, 280, 406, 409, 441, 444, 445, 507
air-handling equipment 94, 107, 466
air-handling system 101, 199, 210, 432, 433
air-handling unit (AHU) 90, 91, 150, 199, 213, 237, 238, 249, 345, 438, 466, 475, 505
ANSI/ASHRAE Standard 55, *Thermal Environmental Conditions for Human Occupancy* 15, 20, 114, 116, 154, 197, 224, 225, 232, 444, 450
ANSI/ASHRAE Standard 62.1, *Ventilation for Acceptable Indoor Air Quality* 114, 116, 144, 152, 154, 160, 179, 186, 188, 189, 197, 205, 208, 209, 210, 211, 212, 213, 221, 222, 223, 232, 260, 262, 264, 265, 266, 268, 444, 445, 450
architectural design impacts 168
arrangement of spaces 79
ASHRAE Guideline 0, *The Commissioning Process* 78, 123, 126, 138, 144, 154, 232, 446, 446, 448, 450, 480

B

background 14, 128, 182, 219, 236
BacNET 75
ballast 113, 228, 371, 372, 374, 375, 377, 378, 380, 383, 387, 389, 392, 475, 505
    dimming 379, 384, 389
    electronic 371, 375, 376
    high-efficiency electronic 370, 378
    high-efficiency lamp 114, 384
    instant-start 378
    low-ballast-factors 371, 378
    metal halide 379
    overdrive 383
    pulse-start type 379, 380
ballast factor 505
bEQ 45, 46, 88, 145, 477
bio-climatic chart 80
biomimicry 87, 88
blackwater 419, 420
BREEAM 45, 140, 505
brownfield 43, 191, 505
building automation system xii, 132, 439, 440, 441, 442, 443, 444, 445, 446, 448, 464, 505
building energy quotient 88, 145, 146, 154, 477, 481
building envelope 51, 58, 65, 72, 76, 79, 96, 97, 128, 130, 135, 136, 161, 162, 177, 182, 189, 194, 195, 196, 197, 198, 207, 230, 231, 363, 365, 468, 508
building form/geometry 58, 65, 81, 349
building information modeling (BIM) 72, 73, 74, 87, 505

C

CALGreen 147, 148
capital cost 55, 56, 62, 65, 68, 91, 92, 119, 304, 316, 334, 341, 363, 384,

396, 411, 414, 435
change order request 465
charrette 43, 55, 70, 77, 505
checklist
   commissioning 123, 127, 137
   construction 70, 124, 131
   green building design process 56, 77
   materials specification 48, 453, 457
   operations, maintenance, and performance evaluation 479
Chilled Water Plant Design Guide 179, 275, 290
chilled-water
   panels 243
   plant 239, 273, 275, 400
   pumps 237, 247, 273
   system 43, 237, 243, 246, 251, 252, 254, 320, 323, 475
   temperature 237, 238, 273, 275, 320
   temperature difference 238
   treatment chemicals 282
chiller 274, 281, 284, 285, 288, 289, 295, 296, 297, 298, 299, 320, 321
   absorption 289, 290, 299, 320, 322
   chlorofluorocarbon 8, 458, 505
   controls 246, 275
   electric 274, 281, 289, 295, 297, 298, 321, 323, 324
   heat-driven 346
clerestory 385
coefficient of performance (COP) 237, 288, 296, 307, 320, 323, 505
combined cooling, heating, and power (CCHP) 281, 288, 289, 295, 505
combined heating and power (CHP) 281, 288, 289, 295, 296, 297, 298, 299, 315, 505
comfort
   general 56
   occupant 91, 98, 102, 105, 107, 109, 111, 114, 149, 349

   thermal 15, 50, 51, 56, 59, 166, 171, 183, 186, 198, 216, 217, 218, 219, 220, 223, 224, 261, 263, 264, 365, 384, 444, 471, 478
   visual 377, 446
commissioning 6, 40, 41, 43, 45, 46, 49, 50, 53, 54, 57, 61, 65, 70, 123, 124, 125, 126, 127, 128, 129, 130, 131, 132, 133, 134, 135, 136, 137, 138, 144, 148, 152, 154, 166, 171, 172, 175, 183, 184, 206, 211, 219, 232, 254, 262, 265, 268, 272, 317, 335, 415, 421, 430, 439, 440, 446, 447, 449, 450, 455, 463, 466, 468, 471, 472, 474, 480, 481
   activities 123, 126, 130, 134
   ASHRAE Guideline 0, *The Commissioning Process* 78, 123, 126, 138, 144, 154, 232, 446, 446, 448, 450, 480
   authority 40, 54, 58, 59, 66, 78, 123, 126, 128, 129, 130, 132, 136, 84, 455, 466
   checklist 127, 137
   general contractor's responsibility 136
   Owner's Project Requirements 54, 123, 446, 447, 448
   phases
      construction 125, 126, 127, 129, 131
      design 35, 43, 62, 123, 125, 127, 128, 129, 137, 158, 163, 231, 316, 446, 454, 459, 464, 466
      measurement and verification 50, 133, 155, 440, 451, 475, 481
      predesign 6, 39, 40, 55, 77, 123, 125, 126, 127, 184, 448
   warranty 124, 131

project design professional's responsibility 14, 27, 61, 86, 468
provider team 128
recommissioning 50, 137, 472
system selection 96, 129, 185, 239
computer room air conditioner 505
condenser 62, 118, 189, 238, 239, 240, 243, 245, 247, 251, 252, 253, 254, 273, 274, 275, 276, 282, 283, 291, 311, 312, 321, 324, 331, 332, 402, 434, 441
constant air volume (CAV) 212, 505
construction 4, 5, 6, 14, 16, 19, 40, 41, 42, 44, 45, 49, 54, 55—63, 66, 69, 70, 73, 76, 77, 79, 81, 90, 94, 96, 99, 123–6, 127—32, 134, 136, 137, 140–4, 148, 149, 152, 153, 154, 158, 160, 166, 170, 185, 186, 188, 191, 193, 204, 206, 209, 213, 230, 233, 234, 259, 262, 283, 357, 378, 387, 421, 425, 426, 431, 432, 435, 443, 446, 450, 453, 454, 455, 461, 463–9, 472, 473, 476
    checklist 124, 129, 131
    cost 6, 49, 54, 58, 62, 131, 387, 473
    documents 49, 60, 61, 77, 128, 129, 137, 453, 464, 465, 468
    manager, 57, 62, 63, 90, 130, 464, 465, 466
    materials, 4, 79, 127, 158, 160, 170, 204, 466, 469
    practices and methods 466, 463
    process 40, 59, 79, 124, 125, 131, 141, 188, 454
    project insurance 45
    quality 453, 463
    site 193, 469
    team 40, 63, 96, 124, 127, 132, 136, 137
    technologies 14
    waste 148, 152, 466, 467

contractor 40, 57, 58, 59, 60, 62, 63, 64, 66, 77, 78, 124, 125, 129, 130, 131, 134, 136, 137, 138, 145, 186, 209, 249, 259, 289, 344, 360, 447, 448, 449, 453, 455, 457, 463, 464, 466, 467, 468
    selection 62, 463
control
    building automation system (BAS) 132, 439, 440, 441, 442, 443, 464
    controls 50, 62, 65, 76, 101, 102, 104, 109, 113, 114, 118, 132, 136, 151, 164, 165, 167, 171, 175, 178, 211, 216, 219, 222, 230, 237, 246, 251, 254, 255, 256, 257, 262, 265, 267, 268, 269, 271, 272, 275, 298, 303, 304, 309, 311, 313, 316, 317, 322, 324, 328, 331, 338, 360, 377, 384, 386, 387, 388, 389, 390, 391, 392, 393, 408, 439, 440, 441, 443, 445, 446, 447, 448, 449, 450, 451, 475, 479
    daylight 441
    daylighting 136
    demand-based pressure reset 251, 253
    direct digital 251, 439, 441, 448, 450
    general contractor's responsibility 136
    lighting 62, 114, 136, 230, 381, 387, 388, 389, 390, 392, 393, 440
    National Building Controls Information 451
    steam 240, 241
    system 43, 171, 221, 237, 273, 322, 324, 445, 449
    user 65, 229
    variable-speed 43, 252, 441
controlling construction quality 453
CoolTools 179, 237, 238, 250, 259
cost 4, 5, 6, 8, 9, 16, 18, 25, 32, 35, 36, 39, 41–44, 46–50, 53–66, 68, 69, 70,

74, 75, 76, 77, 79, 85, 86, 91–4, 96, 98, 99, 101, 103, 105, 108, 110–113, 115, 118, 119, 124–31, 133, 134, 135, 136, 146, 151, 152, 157, 162, 165, 166, 167, 168, 171–75, 213, 216, 218, 219, 222, 223, 231, 235–41, 243–9, 251–4, 257, 258, 261, 262, 264–75, 280, 286, 287, 288, 292, 293, 295, 297, 298, 299, 301–5, 307, 309, 310, 312, 313, 315–24, 328, 329, 331, 332, 334, 335, 339, 341, 344, 346, 347, 349, 351, 356, 357, 359, 360–365, 367, 371, 377, 378, 379, 384, 387, 391, 392, 393, 396, 399, 408, 411, 414, 415, 420–26, 429, 430, 432, 434, 435, 454, 455, 458, 464, 465, 472–7, 479

    capital 55, 56, 62, 65, 68, 91, 92, 119, 304, 316, 334, 341, 363, 384, 411, 414, 435

    first 5, 8, 32, 35, 54, 55, 59, 61, 63, 68, 79, 98, 105, 108, 110, 111, 165, 171, 174, 175, 237, 238, 239, 246, 248, 251, 254, 257, 262, 264, 265, 267, 268, 271, 272, 286, 295, 298, 301, 303, 305, 310, 313, 315, 317, 321, 324, 328, 332, 335, 346, 363, 371, 392, 414, 420, 421, 423, 424, 426, 429, 430, 465

    recurring 8, 32, 35, 165, 172, 175, 254, 258, 262, 265, 268, 272, 299, 301, 303, 305, 310, 313, 318, 322, 324, 329, 332, 335, 364, 415, 421, 424, 426, 430

    cost effectiveness 25, 76, 349

D

daylight harvesting control 50
daylight/daylighting 42, 46, 48, 50, 51, 57, 58, 62, 65, 66, 67, 68, 69, 72, 73, 74, 76, 77, 82, 85, 98, 102, 104, 105, 109, 111, 114, 115, 116, 129, 136, 160, 162, 163, 166, 179, 183, 225, 226, 227, 230, 231, 338, 365, 368, 369, 370, 376, 377, 384, 385, 386, 387, 388, 389, 392, 393, 395, 441, 446, 505

    consideration 129, 368

    harvesting xii, 50, 51, 73, 74, 384, 385, 386, 387, 392

    penetration 42, 377

dedicated outdoor air system (DOAS) 98, 109, 111, 198, 212, 217, 218, 263, 265, 409, 436, 506

design

    basis of 5, 61, 69, 124, 127, 137, 379

    building envelope 51

    checklist 127

    considerations 42, 120, 219, 230, 275

    cost 53

    element 128, 183, 385, 464

    firm 6, 453, 457, 479

    goals 39, 54, 55, 60, 76, 126, 273, 465, 466

    good 5, 54, 139, 175, 346, 359, 446, 449

    green 3, 5, 6, 8, 14, 15, 19, 23, 39, 40, 41, 42, 45, 49, 50, 51, 54, 55, 56, 57, 58, 59, 60, 61, 62, 67, 68, 69, 70, 71, 74, 77, 125, 132, 136, 140, 169, 183, 235, 248, 285, 338, 399, 400, 402, 439, 449, 453, 454, 455, 463, 464, 465, 466, 469

    HVAC 99, 166, 216, 221, 276, 284, 346, 369

    integrated 4, 15, 23, 46, 59, 65, 66, 68, 88, 94, 97, 123, 125, 127,

166, 183, 185, 270, 288, 349, 454, 507
interior 14, 77, 377
iterative 66, 67
lighting 26, 57, 58, 115, 163, 369, 370, 371, 377, 385, 387
mechanical/electrical building system 3
phase 62, 126, 128, 137, 158, 163, 316, 446, 454, 464
process x, xxi, 3, 15, 37, 39, 46, 53, 56, 58, 60, 62, 63, 65, 66, 67, 68, 70, 73, 74, 77, 157, 158, 166, 171, 253, 257, 293, 298, 301, 303, 305, 310, 313, 317, 321, 324, 331, 332, 363, 414, 421, 424, 426, 430, 435, 447, 454, 458, 463, 507
professional 6, 58, 62, 63, 135, 136, 183, 240, 246, 275, 369
schematic 125, 129
sustainable xvii, xix, 4, 5, 8, 16, 20, 21, 23, 40, 41, 42, 43, 46, 50, 54, 72, 73, 83, 87, 88, 94, 120, 141, 147, 154, 155, 443, 455
team 5, 15, 16, 19, 27, 43, 45, 54, 56, 57, 58, 60, 61, 63, 66, 69, 70, 76, 77, 78, 79, 93, 94, 96, 124, 125, 126, 127, 128, 129, 158, 160, 170, 183, 269, 270, 320, 323, 338, 407, 447, 453, 456, 465, 468, 478
tool 14, 106, 110, 112, 121, 340, 359
design/bid/build 62, 506
design/bid-build 506
direct digital control (DDC) 251, 439, 441, 448, 450, 451, 506
directive on the Energy Performance of Buildings 8, 506
distributed generation 284, 289, 297, 442, 506
district cooling (DC) 281, 318, 434, 438
district energy (DE) xi, 281, 283, 506
district heating (DH) 281, 340
domestic water heating 284, 304, 344, 360, 407, 424
double skin façade 84, 506

E

electric lighting 68, 163, 228, 369, 384, 387, 396
energy
 distribution 247, 248
 efficiency 7, 8, 9, 19, 42, 44, 45, 49, 51, 52, 68, 76, 84, 113, 119, 120, 121, 124, 135, 139, 146, 148, 150, 151, 163, 164, 165, 166, 173, 175, 177, 183, 213, 225, 231, 234, 282, 294, 300, 301, 303, 304, 306, 307, 338, 352, 353, 370, 371, 434, 440, 446, 447, 473, 474, 475, 477, 479
 green 337
 minimum energy performance 9, 44
 nonrenewable xvii, 4, 8, 337, 338, 362
 renewable 4, 18, 20, 21, 42, 44, 46, 47, 48, 52, 55, 58, 76, 114, 115, 121, 135, 151, 153, 154, 155, 163, 303, 306, 310, 337, 338, 347, 348, 350, 351, 352, 353, 356, 359, 364, 442, 476, 507
 solar 72, 82, 88, 175, 176, 338, 339, 340, 342, 344, 347, 348, 352, 353, 355, 356, 357, 358, 359, 361, 364, 397, 507
 thermal 65, 91, 169, 170, 274, 276, 281, 284, 285, 286, 302, 318, 320, 321, 339, 340, 342, 355, 358, 360, 506, 508
 use 18, 20, 24, 44, 46, 47, 48, 55, 58, 76, 79, 83, 84, 85, 86, 87, 93, 96,

97, 113, 114, 146, 150, 153, 154, 157, 158, 159, 160, 161, 163, 169, 170, 172, 178, 181, 213, 221, 225, 226, 227, 236, 237, 239, 241, 244, 261, 264, 267, 272, 281, 308, 313, 318, 356, 359, 369, 377, 388, 391, 392, 399, 414, 444, 475, 477, 480
wind 82, 85, 349, 352, 353
energy and atmosphere 143, 506
energy cost budget (ECB) 75
energy performance certificate 9, 10, 11, 12, 506
Energy Policy Act of 1992 411, 412, 414
energy resource xvii, xviii, 4, 55, 86, 290, 315, 316, 347, 506, 507
energy source xii, 8, 44, 48, 55, 85, 86, 88, 151, 157, 158, 337, 338, 339, 350, 356, 359, 422, 506, 508
ENERGY STAR 88, 165, 167, 177, 179, 506
engine-driven generator (EDG) 65, 286, 506
enthalpy 46, 92, 103, 105, 108, 218, 288, 434, 506
entropy 475, 506
environmental design consultant 57, 58, 506

F

F-Chart method 344
F-Chart Software 353, 359, 360
fenestration 85, 87, 384, 506

G

graywater 62, 407, 419, 420, 428, 429
green
    building xvii, xix, xxi, 3–7, 9, 13, 14, 15, 20, 22, 24, 26, 27, 44, 46, 49, 51, 53, 56, 58, 60, 61, 65, 83, 94,

115, 120, 124, 125, 132, 136, 137, 139, 140–3, 147–9, 153–6, 164, 181, 182, 213, 224, 225, 227, 230–4, 262, 273, 291, 370, 393, 399, 409, 415, 421, 431, 443, 444, 446, 450, 451, 454, 456, 458, 459, 466, 467, 468, 471, 474, 480
energy 337, 352
engineering 3, 45, 455
design xviii, xx, xxi, 3, 5, 6, 8, 14, 15, 19, 23, 39, 40, 41, 42, 45, 49, 50, 51, 54, 55, 56, 57, 58, 59, 60, 61, 62, 67, 68, 69, 70, 71, 74, 77, 132, 136, 140, 169, 183, 235, 248, 285, 338, 399, 400, 402, 439, 449, 453, 454, 455, 463, 464, 465, 466, 469
power 337, 351, 352
green roof xx, 28, 29, 34, 35, 36, 83, 116, 198, 199
greenfield 77

H

hydronic xi, 102, 243, 235, 237, 239, 243, 246, 248, 251, 252, 254, 263, 275, 296, 304, 311, 312, 338, 342, 507

I

Illuminating Engineering Society (IES) 75, 145, 147, 370, 371, 383, 444, 450, 507
indoor air quality (IAQ) 4, 42, 65, 114, 116, 130, 144, 145, 154, 160, 166, 179, 181, 182, 183, 185, 186, 189, 197, 202, 203, 205, 206, 207, 208, 210, 212, 213, 216, 217, 222, 223, 231, 232, 233, 247, 262, 265, 266, 268, 349, 444, 450, 454, 466, 467, 471, 477, 480, 507

ANSI/ASHRAE Standard 62.1, *Ventilation for Acceptable Indoor Air Quality*, 114, 116, 144, 152, 154, 160, 179, 186, 188, 189, 197, 205, 208, 209, 210, 211, 212, 213, 221, 222, 223, 232, 260, 262, 264, 265, 266, 268, 444, 445, 450
indoor environmental quality 20, 41, 58, 136, 139, 143, 152, 154, 181, 230, 266, 440, 474, 476, 481, 507
integrated design process 23, 46, 66, 183, 185, 454, 507
International Performance Measurement and Verification Protocol (IPMVP) 475

L

latent load 218, 264, 433, 507
Leadership in Energy and Environmental Design (LEED) 9, 22, 56, 125, 130, 132, 134, 136, 140, 149, 152, 153, 156, 234, 291, 446, 450, 454, 456, 463, 467, 507
    certification, 24, 56, 127, 130, 143, 147, 294, 456, 463
    lighting 23, 26, 42, 44, 46, 48, 51, 57, 62, 65–8, 72, 73, 76, 77, 82, 85, 86, 91, 96, 97, 98, 104, 109, 113, 114, 129, 135, 136, 137, 158, 160–7, 177, 178, 183, 226–31, 234, 235, 338, 361, 362, 365, 369–72, 374–81, 383–9, 391–6, 439, 441, 444, 446, 450, 471, 472, 475, 478, 505, 507
    Advanced Lighting Guidelines, 393
    control 62, 114, 230, 381, 387, 388, 389, 390, 392, 393, 440
    design 26, 57, 58, 115, 163, 369, 370, 371, 377, 385, 387

natural 61, 62, 384, 388, 395, 396, 441
system 114, 163, 183, 368, 369, 370, 371, 374, 377, 378, 379, 380, 381, 383, 384, 386, 388, 389, 392, 393, 396, 474, 475, 479
lighting power density 165, 167, 369, 379, 507
liquid desiccant 236, 316, 317, 326, 327, 507

M

materials specification checklist 453, 457
measurement and verification (M&V) 50, 133, 138, 155, 440, 449, 450, 451, 475, 476, 480, 481, 507
    International Performance Measurement and Verification Protocol 155, 451, 475, 481
mean radiant temperature (MRT) 444, 507

N

natural refrigerant 236
night precooling 168, 169, 171
nonrenewable energy xvii, 4, 8, 337, 338, 362

O

Open Automated Demand Response (OpenADR) 442, 443, 451
Owner's Project Requirements (OPR) 39, 40, 41, 54, 55, 58, 69, 70, 77, 78, 123, 124, 126, 127, 128, 129, 130, 132, 133, 135, 137, 446, 447, 448, 454, 507

P

parametric analysis 76, 161, 507
photovoltaic 48, 82, 85, 86, 114, 287, 338, 347, 352, 353, 364, 508
plug loads 48, 72, 76, 164, 177, 178, 389, 507
point-of-use domestic hot-water heaters 422
potable water 46, 62, 305, 358, 360, 408, 414, 415, 419, 420, 421, 428, 429, 430, 438, 440
precooling 165, 168, 167, 169, 170, 171, 172, 330, 508
predesign 6, 39, 40, 55, 77, 123, 125, 126, 127, 184, 448
publicly owned treatment works 276, 279, 403, 405, 508

R

rain garden xx, 28, 30, 31, 32, 33
rainwater harvesting 28, 35, 408, 428, 429, 430, 431
recommissioning 50, 137
renewable energy xii, 4, 18, 20, 21, 42, 44, 46, 47, 48, 52, 55, 58, 76, 114, 115, 121, 135, 151, 153, 154, 155, 163, 303, 306, 310, 337, 338, 339, 347, 348, 350, 351, 352, 353, 356, 359, 362, 364, 442, 475, 476, 507, 508
renewable energy sources 8, 44, 55, 337, 338, 356, 359, 508
request-for-proposal (RFP) 63, 454, 508
retrocommissioning 472, 474, 475, 478
room air temperature 444

S

sensible load 243, 264, 508
skin 84, 87, 88, 161, 162, 195, 197, 204, 386, 387, 444, 506, 508
space
  heating 97, 98, 158, 256, 284, 288, 297, 301, 304, 305, 307, 339, 340, 341, 344, 355, 359, 425
  humidity 264
  temperature 212, 255, 439, 445
space/water heating 309
stormwater 4, 23, 27, 28, 29, 30, 32, 33, 35, 49, 62, 114, 141, 142, 148, 428, 429
supervisory control and data acquisition 444
sustainability 4, 5, 13, 14, 17, 19, 21, 23, 24, 28, 33, 53, 61, 74, 77, 78, 79, 87, 88, 93, 96, 99, 101, 108, 125, 130, 139, 141, 147, 154, 155, 268, 350, 353, 402, 449, 454, 455, 459, 480, 505, 508
synthetic refrigerant 236

T

thermal
  ANSI/ASHRAE Standard 55-2010, *Thermal Environmental Conditions for Human Occupancy* 15, 20, 114, 116, 154, 197, 224, 225, 232, 444, 450
  comfort 15, 50, 51, 56, 59, 166, 171, 183, 186, 198, 216, 217, 219, 223, 224, 261, 263, 264, 365, 384, 444, 471, 478
  efficiency 235, 288, 295, 303
  energy 65, 169, 170, 274, 276, 281, 284, 285, 286, 302, 318, 320, 321, 339, 340, 342, 355, 358, 360, 508

energy storage (TES) 65, 91, 169, 170, 274, 281, 285, 314, 315, 316, 317, 318, 508
envelope 9, 46, 161
load 284, 285, 288, 289, 297, 298, 355, 358, 507, 508
radiation 25, 217
regime 4
resistance 160
thermal buffer zone 84, 508
thermal energy storage (TES) 65, 91, 169, 170, 274, 281, 285, 314, 315, 316, 317, 318, 508
toxicity characteristic leaching procedure (TCLP) 380, 457, 458, 508

U

urban heat island 23, 24, 25

V

value engineering 68, 246, 391, 463, 465, 508
variable air volume (VAV) 98, 102, 104, 211, 212, 217, 221, 248, 249, 250, 258, 441, 445, 508
variable-frequency drive (VFD) 118, 238, 255, 280, 441, 479, 508
ventilation 15, 43, 46, 48, 65, 80, 82, 83, 84, 91, 99–112, 114, 116, 135, 138, 151–4, 160, 163, 168–72, 179, 181, 183, 185, 186, 188, 189, 198, 201, 203, 205–13, 216–23, 225, 226, 231, 232, 233, 236, 258, 260–272, 326, 338, 349, 351, 365, 366, 367, 408, 439, 440, 441, 444, 445, 450, 458, 466, 468, 471, 479, 506
 demand control or demand based 268
 displacement 212, 216, 260, 261
 effectiveness 219, 260, 261, 506
 hybrid 269, 270, 271, 272, 506
 mechanical 105, 110, 112, 171, 172, 186, 222, 269, 270, 272, 445
 mixed-mode 221, 226, 269, 445
 natural 15, 46, 48, 65, 80, 82, 100, 104, 105, 106, 108, 109, 110, 111, 112, 185, 221, 222, 225, 231, 338, 349, 351, 365, 366, 367, 445
 night 168, 169, 170
 outdoor air 43, 201, 206, 210, 221, 269, 270, 506
 passive 186, 269, 271
volatile organic compound (VOC) 186, 191, 203, 204, 215, 216, 326, 508

W

wastewater 64, 411, 419, 420, 443, 444
water 4, 14, 46, 49, 50, 54, 55, 58, 62, 64, 66, 69, 70, 128, 134, 138, 141, 143, 149, 153, 166, 179, 235, 236, 237, 240, 241, 242, 243, 244, 246, 247, 251, 252, 253, 254, 255, 273, 274, 275, 277, 278, 279, 280, 282, 288, 289, 291, 292, 293, 296, 297, 304
 consumption 28, 69, 134, 138, 283, 284, 399, 400, 403, 404, 411, 414, 416, 417, 418, 443, 471, 479, 480
 direct-contact water heaters 425, 426
 domestic water heating 284, 304, 344, 360, 407, 424
 drinking 407
 Energy Policy Act of 1992 411, 412
 fresh 351, 400, 407, 432
 point-of-use domestic hot-water heater 422, 423

potable 27, 28, 46, 62, 95, 174, 201, 305, 358, 360, 405, 407, 408, 414, 415, 418, 419, 420, 421, 428, 429, 430, 438, 440

pump
    chilled-water 104, 237, 247, 273
    condenser-water 273
solar water heating 338
whole building commissioning 135